VOLUME SIXTY SEVEN

ADVANCES IN MICROBIAL PHYSIOLOGY

Recent Advances in Microbial Oxygen-Binding Proteins

VOLUME SIXTY SEVEN

ADVANCES IN MICROBIAL PHYSIOLOGY

Recent Advances in Microbial Oxygen-Binding Proteins

Edited by

ROBERT K. POOLE
West Riding Professor of Microbiology
Department of Molecular Biology and Biotechnology
The University of Sheffield
Firth Court, Western Bank
Sheffield, UK

ELSEVIER

AMSTERDAM • BOSTON • HEIDELBERG • LONDON
NEW YORK • OXFORD • PARIS • SAN DIEGO
SAN FRANCISCO • SINGAPORE • SYDNEY • TOKYO
Academic Press is an imprint of Elsevier

AP

Academic Press is an imprint of Elsevier
125 London Wall, London, EC2Y 5AS, UK
The Boulevard, Langford Lane, Kidlington, Oxford, OX5 1GB, UK
525 B Street, Suite 1800, San Diego, CA 92101-4495, USA
225 Wyman Street, Waltham, MA 02451, USA

First edition 2015

Notices
Knowledge and best practice in this field are constantly changing. As new research and experience broaden our understanding, changes in research methods, professional practices, or medical treatment may become necessary.

Practitioners and researchers must always rely on their own experience and knowledge in evaluating and using any information, methods, compounds, or experiments described herein. In using such information or methods they should be mindful of their own safety and the safety of others, including parties for whom they have a professional responsibility.

To the fullest extent of the law, neither the Publisher nor the authors, contributors, or editors, assume any liability for any injury and/or damage to persons or property as a matter of products liability, negligence or otherwise, or from any use or operation of any methods, products, instructions, or ideas contained in the material herein.

ISBN: 978-0-12-803298-5
ISSN: 0065-2911

For information on all Academic Press publications visit our website at http://store.elsevier.com

CONTENTS

CONTRIBUTORS

Colin R. Andrew
Department of Chemistry and Biochemistry, Eastern Oregon University, La Grande, Oregon, USA

Paolo Ascenzi
Laboratorio Interdipartimentale di Microscopia Elettronica, Università Roma Tre, Roma, and Istituto di Bioscienze e BioRisorse, Consiglio Nazionale delle Ricerche, Napoli, Italy

Leonardo Boechi
Instituto de Cálculo, Facultad de Ciencias Exactas y Naturales, Universidad de Buenos Aires, Ciudad Universitaria, Buenos Aires, Argentina

Alberto Boffi
Dipartimento di Scienze Biochimiche, Università "Sapienza", Rome, Italy

Sylvia Dewilde
Protein Chemistry, Proteomics and Epigenetic Signalling (PPES), Department of Biomedical Sciences, University of Antwerp, Antwerp, Belgium

Alessandra di Masi
Dipartimento di Scienze, Università Roma Tre, Roma, Italy

Joke Donné
Protein Chemistry, Proteomics and Epigenetic Signalling (PPES), Department of Biomedical Sciences, University of Antwerp, Antwerp, Belgium

Dario E. Estrin
Departamento de Química Inorgánica, Analítica y Química Física and Inquimae-Conicet, Facultad de Ciencias Exactas y Naturales, Universidad de Buenos Aires, Buenos Aires, Argentina

Emanuela Frangipani
Dipartimento di Scienze, Università Roma Tre, Roma, Italy

Michael A. Hough
School of Biological Sciences, University of Essex, Colchester, United Kingdom

Barry D. Howes
Dipartimento di Chimica "Ugo Schiff", Università di Firenze, Sesto Fiorentino, Italy

Eric A. Johnson
T.C. Jenkins Department of Biophysics, Johns Hopkins University, Baltimore, Maryland, USA

Loris Leboffe
Dipartimento di Scienze, Università Roma Tre, Roma, Italy

Juliette T.J. Lecomte
T.C. Jenkins Department of Biophysics, Johns Hopkins University, Baltimore, Maryland, USA

David Lloyd
School of Biosciences, Cardiff University, Cathays Park, Cardiff, Wales, United Kingdom

Marco Nardini
Dipartimento di Bioscienze, Università di Milano, Milano, Italy

Giulietta Smulevich
Dipartimento di Chimica "Ugo Schiff", Università di Firenze, Sesto Fiorentino, Italy

Cinzia Verde
Istituto di Bioscienze e BioRisorse, Consiglio Nazionale delle Ricerche, Napoli, and Dipartimento di Scienze, Università Roma Tre, Roma, Italy

Paolo Visca
Dipartimento di Scienze, Università Roma Tre, Roma, Italy

Catrin F. Williams
School of Biosciences and School of Engineering, Cardiff University, Cardiff, Wales, United Kingdom

PREFACE

Oxygen-binding proteins are arguably the most intensively studied of all proteins. The field has been revolutionised in recent years by paradigm shifts in our understanding of these proteins. Among these developments are (i) the discovery of globins in numerous animal species, plants and microbes, with important implications for globin evolution, (ii) the recognition that the ligand binding and catalytic capabilities of 'oxygen-binding proteins' actually embrace physiologically critical ligands such as nitric oxide and hydrogen sulphide, (iii) the sheer number of globins and haem proteins encoded by 'simple' genomes and (iv) the multiple roles of oxygen-binding proteins in natural environments and disease. The subject has considerable potential for contributing to knowledge that can be applied to the development of agriculture and medicine and to understanding the impacts of human activity on living organisms and ecosystems.

Conferences on O_2-binding proteins have long been held every two years across Europe. The meeting held in Parma (Italy, 2012) generated a volume of *Advances in Microbial Physiology* (Volume 63) in which the focus was on microbial globins and their diverse functions, where these are known. The following meeting was held in Sheffield (UK, 2014) and it is that highly successful conference that has generated this volume. Again, microbial globins are well represented, but the scope is here broader, encompassing *c*-type cytochromes, the impacts of oxygen levels in biology and biofilms.

These meetings and volumes demonstrate a remarkably multidisciplinary approach to the area, with contributions extending from molecular biophysics, to biochemistry and physiology and whole organism functions. Arguably, some of these contributions appear not to be 'physiology' in a classical sense but all provide critical new information on O_2-binding proteins that will ultimately provide information on how these proteins and the organisms *work*—the real goal of physiology.

ACKNOWLEDGEMENT

With the publication of Volume 67, we would like to thank and congratulate Dr. Robert Poole for 20 years at the helm of *Advances in Microbial Physiology*. Dr. Poole has been a great asset to Academic Press and Elsevier, providing guidance for authors and commissioning important work for the research community. Thanks to his knowledge and commitment, *AMP* continues to be a highly renowned and impactful publication from year to year.

Many thanks from the whole team at Elsevier!

CHAPTER ONE

Cytochromes c': Structure, Reactivity and Relevance to Haem-Based Gas Sensing

Michael A. Hough*, Colin R. Andrew†,1
*School of Biological Sciences, University of Essex, Colchester, United Kingdom
†Department of Chemistry and Biochemistry, Eastern Oregon University, La Grande, Oregon, USA
[1]Corresponding author: e-mail address: candrew@eou.edu

Contents

Abstract

Cytochromes c' are a group of class IIa cytochromes with pentacoordinate haem centres and are found in photosynthetic, denitrifying and methanotrophic bacteria. Their function remains unclear, although roles in nitric oxide (NO) trafficking during denitrification or in cellular defence against nitrosoative stress have been proposed. Cytochromes c' are typically dimeric with each c type haem-containing monomer folding as a four-α-helix bundle. Their hydrophobic and crowded distal sites impose severe restrictions on the binding of distal ligands, including diatomic gases. By contrast, NO binds to the proximal haem face in a similar manner to that of the eukaryotic NO sensor, soluble guanylate cyclase and bacterial analogues. In this review, we focus on how structural features of cytochromes c' influence haem spectroscopy and reactivity with NO, CO and O_2. We also discuss the relevance of cytochrome c' to understanding the mechanisms of gas binding to haem-based sensor proteins.

Advances in Microbial Physiology, Volume 67
ISSN 0065-2911
http://dx.doi.org/10.1016/bs.ampbs.2015.08.001

ABBREVIATIONS

CYTOCHROMES *C′* FROM THE FOLLOWING ORGANISMS ARE ABBREVIATED AS:

ADCP *Alcaligenes denitrificans*
AXCP *Alcaligenes xylosoxidans*
CVCP *Chromatium vinosum*
NMCP *Neisseria meningitidis*
PDCP *Paracoccus denitrificans*
RCCP *Rhodobacter capsulatus*
RGCP *Rubrivivax gelatinosus*
RMCP *Rhodospirillum molischianum*
RPCP *Rhodopseudomonas palustris*
RRCP *Rhodospirillum rubrum*
RSCP *Rhodobacter sphaeroides*
SFCP *Shewanella frigidimarina*
TTCP *Thermochromatium tepidum*

1. INTRODUCTION

Bacterial cytochromes *c′* represent a biochemical enigma: first isolated in the 1950s (and coined 'pseudohaemoglobin', and subsequently '*Rhodospirillum* haem protein (RHP)', 'cytochromoid *c*' and 'cytochrome *cc′*' (Meyer & Kamen, 1982; Vernon & Kamen, 1954)), they have attracted attention due to their unusual spectroscopic properties and novel interactions with diatomic gases. Although the physiological role(s) of cytochromes *c′* are not well defined, their pentacoordinate (5c) haem centres are indicative of a ligand-binding (rather than a purely electron-transfer) function.

The majority of native cytochromes *c′* are homodimeric (~14 kDa per subunit and polypeptide length of ~130 residues), although *Rhodopseudomonas palustris* cytochrome *c′* (RPCP) is monomeric in solution (Cusanovich, 1971; Shibata et al., 1998), while ligand-induced dimer dissociation has been detected in *Chromatium vinosum* cytochrome *c′* (CVCP) (Doyle, Gill, & Cusanovich, 1986; Evers, van Dongen, Meijer, & Merkx, 2007; Motie, Kassner, Meyer, & Cusanovich, 1991; Ren, Meyer, & McRee, 1993) and *Rhodobacter capsulatus* cytochrome *c′* (RCCP) (Tsan, Hus, Caffrey, Marion, & Blackledge, 2000). The cytochromes *c′* have a number of conserved structural features including a solvent-exposed proximal pocket containing the proximal His ligand and a conserved positively charged residue (Arg or Lys) four residues beyond the C—X—X—C—H haem-binding motif. By contrast, the distal pockets of all cytochromes *c′* are buried within the protein interior and packed with hydrophobic residues, including a non-coordinating Leu or Phe (or occasionally Met or Tyr) which creates steric restrictions to the coordination of exogenous ligands.

Within the superfamily of *c*-type cytochromes (proteins with haem covalently attached via thioether linkages), cytochromes *c′* are designated as class II on account of their haem location close to the C-terminus of a four-α-helix bundle (Moore & Pettigrew, 1990). This structural motif is conserved despite sequence homologies as low as (~28%) (Ambler, 1991; Ambler et al., 1981). The chemical structure of a *c*-type haem is shown in Fig. 1.

Cytochromes *c′* are further classified as class IIa on account of their 5c mono-His ligated haem Fe, which exhibits unusual and distinctive spectroscopic properties. Accordingly, cytochromes *c′* differ significantly from class 1 mitochondrial cytochromes *c* which have a different protein architecture as well as hexacoordinate (6c) heme Fe with His/Met axial ligation.

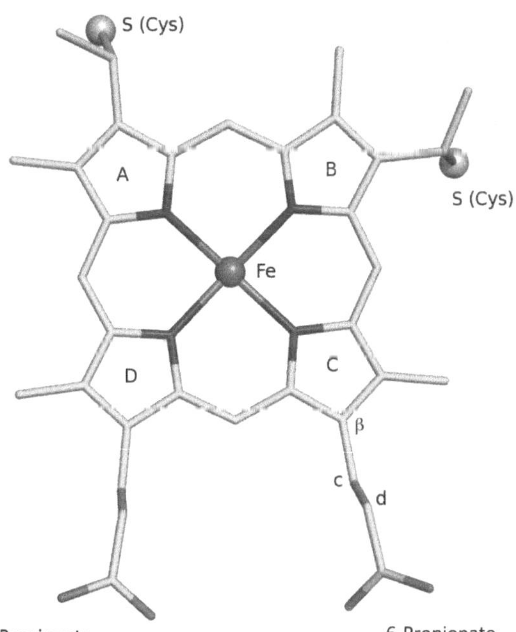

Figure 1 Haem *c* structure and nomenclature. Carbon atoms are coloured green (light grey in the print version), nitrogens are blue (dark grey in the print version), oxygens are red (grey in the print version), and sulphurs are yellow (grey in the print version).

Cytochromes *c′* are structurally related to the class IIb cytochromes *c*-556 which have spectroscopic properties consistent with hexacoordinate haem (His/Met axial ligation) (Ambler et al., 1981). Although Met is not a haem ligand in native cytochromes *c′*, a non-coordinating Met residue is located near distal Fe sites of certain cytochromes *c′*. Although the spectroscopic properties and reactivity of cytochrome *c′* from *Methylococcus capsulatus* (Bath) resemble other cytochromes *c′* (Zahn, Arciero, Hooper, & Dispirito, 1996), recent studies suggest that this protein has a predominantly β-sheet structure (rather than α-helical), with the suggestion that it be designated as *c′*-beta (Elmore, Bergmann, Klotz, & Hooper, 2007). This protein also has a much lower redox potential than other cytochromes *c′* (Table 1). Cytochrome *c″* from the obligate aerobe *Methylophilus methylotrophus* is a soluble monohaem protein (15 kDa) that exhibits bis-His coordination in the ferric state and mono-His coordination in the ferrous state (Berry, George, Thomson, Santos, & Turner, 1990; Quintas, Catarino, Todorovic, & Turner, 2011), the latter exhibiting spectroscopic features and reactivity similar to cytochromes *c′* (Quintas et al., 2011). There are also other examples of 5c *c*-type cytochromes with gas-binding reactivity. *Rhodobacter sphaeroides* haem protein (SHP) is a *c*-type cytochrome that is hexacoordinate in the ferric state (His/Asn as axial ligands), but with mono-His ligation in the ferrous state (Li et al., 2008). However, SHP differs from cytochromes *c′* in forming a stable complex with O_2.

The redox potentials of cytochromes *c′* are rather variable and have been described in detail in a previous review (Weiss, Gold, & Terner, 2006). For all but the unusual, largely β-sheet *M. capsulatus* Bath cytochrome *c′* ($E_{m\ 7.0} = -205$ mV) midpoint potentials are positive and range between +3 mV for *Rhodospirillum rubrum* cytochrome *c′* (RRCP) and +202 mV for *Paracoccus denitrificans* cytochrome *c′* (PDCP; Table 1). We note that *Alcaligenes xylosoxidans* cytochrome *c′* (AXCP) has the highest E_o (+132 to +90 mV) for any cytochrome *c′* for which a crystal structure is available.

Initial studies of cytochromes *c′* focused on their unusual spectroscopic properties, particularly the 5cFe(III) state which is one of very few biological examples of an intermediate-spin/high-spin quantum mechanical admixture. In 2000, crystal structures of the NO and CO ferrous haem complexes of AXCP revealed the first biochemical example of gas binding to opposite haem faces: CO binds at the distal site to form a 6c carbonyl complex (6cCO), whereas the NO ligand of the 5c ferrous-nitrosyl complex

Table 1 Crystal Structures of Cytochromes c'

Organism	Oligomer	Interface Type	Group	Distal Residue	Proximal Residue	Resolution (Å)	PDB Code	E_o (+mV)	References
Alcaligenes denitrificans	D	1	2	Leu	Arg	1.80	1cgo	n.d.	Dobbs, Anderson, Faber, and Baker (1996)
Alcaligenes xylosoxidans	D	1	2	Leu	Arg	1.30	4cda	190–132	Kekilli et al. (2014)
Chromatium vinosum	D	1	1	Tyr	Arg	1.80	1bbh	18	Ren et al. (1993)
Rhodobacter capsulatus	M/D	2	1	Phe	Arg	1.72	1cpq	51	Tahirov et al. (1996b)
Rhodobacter sphaeroides	M/D	Intermediate	1	Phe	Arg	1.80	1gqa	30	Ramirez et al. (2003)
Rhodopseudomonas palustris	M	2	2	Leu	Arg	2.30	1a7v	94–105	Shibata et al. (1998)
Rhodospirillum molischianum	D	1	2	Met	Lys	1.67	2ccy	14	Finzel, Weber, Hardman, and Salemme (1985)
Rhodospirillum rubrum	D	2	2	Leu	Lys	2.80	Not deposited	3	Yasui et al. (1992)

Continued

Table 1 Crystal Structures of Cytochromes *c′*—cont'd

Organism	Oligomer	Interface Type	Group	Distal Residue	Proximal Residue	Resolution (Å)	PDB Code	E_o (+mV)	References
Rubrivivax gelatinosus	D	1	1	Phe	Lys	1.50	2j9b	60	Benini, Rypniewski, Wilson, and Ciurli (2008)
Shewanella frigidimarina	D	1	2	Phe	Lys	1.29	4ulv	n.d.	Manole et al. (2015)
Thermochromatium tepidum	D	1	1	Tyr	Arg	1.00	3vrc	n.d.	Hirano, Kimura, Suzuki, Miki, and Wang (2012)

The highest resolution structure available for the as-isolated or ferric state of cytochrome *c'* from each organism is listed. Interface type refers to the classification of Shibata and co-workers based on the composition of the interface (Shibata et al., 1998). 'Group' refers to the criteria of Tahirov et al. (1996b) based on the identity of the occluding residue in the distal pocket and the presence of a solvent channel from the protein surface to the distal pocket. Proteins are indicated to be dimers if their crystal structures are of dimers unless there is evidence to the contrary. Note that structures are generally determined from protein which was ferric, but the redox state in the crystals following X-ray exposure is not known. For AXCP, single-crystal spectroscopy has been used to determine validated structures of ferric and ferrous native protein. n.d. indicates that no redox potential data are available for a particular protein. E_o values taken from Weiss et al. (2006).

(5cNO) occupies an unprecedented location on the *proximal* haem face in place of the endogenous His ligand (Antonyuk et al., 2011; Hough et al., 2011; Lawson, Stevenson, Andrew, & Eady, 2000; Lawson, Stevenson, Andrew, George, & Eady, 2003). The discovery of proximal haem-NO coordination prompted renewed interest in the gas-binding properties of cytochromes *c*′, with subsequent spectroscopic and kinetic characterization of AXCP leading to a proposed mechanism in which the proximal 5cNO product forms via a series of distal haem-NO precursors, including a 6c ferrous nitrosyl (6cNO) and a transient dinitrosyl complex (Andrew, George, Lawson, & Eady, 2002; George, Andrew, Lawson, Thorneley, & Eady, 2001; Pixton et al., 2009). Interesting parallels with 5cNO formation in other haem proteins, including the eukaryotic haem-NO receptor, soluble guanylate cyclase (sGC), provided the first glimpse that proximal NO binding might be relevant to haem-based NO sensing and signal transduction (Zhao, Brandish, Ballou, & Marletta, 1999). A similarity in the NO-binding mechanism of cytochrome *c*′ and sGC has been further emphasized in recent studies (Martin, Berka, Sharina, & Tsai, 2012). Mechanistic studies have also proposed proximal NO coordination in other 5cNO haem proteins, including various members of the haem nitric oxide/oxygen-binding(H-NOX) gas sensor family, the pro-apoptotic cytochrome *c*/cardiolipin complex and cystathione β-synthase. Very recently, a crystal structure has confirmed proximal 5cNO coordination in the H-NOX protein from *Shewanella oneidensis* (Herzik, Jonnalagadda, Kuriyan, & Marletta, 2014).

A major advancement in the study of cytochromes *c*′ has been the identification of *cycP* genes in a range of bacteria, combined with the development of heterologous overexpression systems incorporating the cytochrome *c* maturation genes in the pec86 plasmid (Arslan, Schilz, Zufferey, Künzler, & Thöny-Meyer, 1998; Harris et al., 2010) This has enabled large quantities of recombinant cytochromes *c*′ to be synthesized and structure–reactivity relationships to be investigated in detail via point mutations of residues proposed to be important in ligand binding, thermostability and dimerization. The most extensively studied cytochrome *c*′ is AXCP (previously known as *Pseudomonas denitrificans* cytochrome *c*″ and *Achromobacter xylosoxidans* cytochrome *c*′ (Cusanovich, Tedro, & Kamen, 1970; Iwasaki, Yoshimura, Suzuki, & Shidara, 1991)), with structural, spectroscopic and kinetic data available for native and variant proteins in different redox and ligation states (Andrew, Green, Lawson, & Eady, 2001;

Andrew, Rodgers, & Eady, 2003; Andrew et al., 2002, 2005; Antonyuk et al., 2011; Dobbs et al., 1996; Garton et al., 2012; Hough et al., 2011; Iwasaki et al., 1991; Lawson et al., 2000, 2003; Pixton et al., 2009; Russell et al., 2013; Servid et al., 2015; Yoo, Lamarre, Martin, Andrew, & Negrerie, 2013; Yoshimura, Suzuki, Kohzuma, Iwasaki, & Shidara, 1990). In addition to AXCP, recombinant cytochromes *c*′ have also been characterized for CVCP (de Groot, Evers, Merkx, & Koper, 2007; Evers & Merkx, 2005; Mayburd & Kassner, 2002), *Neisseria meningitidis* cytochrome *c*′ (NMCP) (Cole, Huston, & Moir, 2008), RCCP (Cole et al., 2008; Huston et al., 2006), RPCP (McGuirl et al., 2003; Pletneva et al., 2007), RSCP (Lee et al., 2009; Usov, Choi, Shapleigh, & Scholes, 2006), *Shewanella frigidimarina* cytochrome *c*′ (SFCP) (Manole et al., 2015) and *Thermochromatium tepidum* cytochrome *c*′ (TTCP) (Kimura et al., 2015). The production of heterologously expressed cytochromes *c*′ has built upon earlier work characterizing the native proteins from their original organisms. The properties of cytochromes *c*′ have been summarized previously in several reviews (Ascenzi, Santucci, Coletta, & Polticelli, 2010; Kassner, 1991; Meyer & Kamen, 1982; Moore & Pettigrew, 1990; Romão & Archer, 2001; Tsai, Berka, Martin, & Olson, 2012; Tsai, Martin, Berka, & Olson, 2012; Weiss et al., 2006). Utilizing structural, spectroscopic and reactivity data (including from recent mutagenesis studies), this review focuses on structure–reactivity relationships, proximal vs. distal haem coordination, selective responses to diatomic gases and the utility of cytochrome *c*′ as a model system for other haem proteins.

2. OCCURRENCE

The *cycP* gene encoding cytochrome *c*′ is found primarily in denitrifying, methanotrophic, sulphur-oxidizing and photosynthetic bacteria and also in some nitrogen-fixing species (Ambler, 1991; Ambler et al., 1981; Watmough et al., 1999). Cytochromes *c*′ have also been found in a number of human pathogens including *Bordatella* strains, *Neisseria meningitidis* (Huston, Lowe, Butler, & Moir, 2005) and *Neisseria gonorrhoeae* (Turner et al., 2005). However, it should be noted that *Neisseria* and certain *Bordatella* strains are capable of denitrification. The conservation of the *cycP* gene across a range of bacteria in differing environments suggests a number of possible functional roles.

3. PROPOSED FUNCTIONAL ROLES OF CYTOCHROMES C′

Despite extensive studies on the properties of cytochromes *c*′, their biological function has remained controversial. An early suggestion was that cytochromes *c*′ played a role in electron transfer in a similar manner to other *c*-type cytochromes. However, the 5c nature of the haem suggests that a gas-binding role is more likely. In the denitrification pathway of the global nitrogen cycle, nitrate is sequentially reduced to dinitrogen via nitrite, NO and N_2O in a series of metalloenzyme-catalysed reactions (Eady & Hasnain, 2003). Of these nitrogen oxides, NO is a potentially dangerous free radical which can cause cellular damage during its lifetime between production by the (periplasmic) haem or copper nitrite reductase (NiR) and its reduction by nitric oxide reductase (NOR), located at the inner membrane. Work from Yoshimura and co-workers indicated that cytochromes *c*′ from denitrifying (Yoshimura et al., 1988; Yoshimura, Shidara, Ozaki, & Kamada, 1993) and photosynthetic (Yoshimura et al., 1996) bacteria were able to bind NO that had been generated during denitrification, leading to the suggestion that its role could be to sequester and buffer free NO from the periplasm in order to limit cellular damage. Consistent with this hypothesis, binding of NO to cytochrome *c*′ *in vivo* has been shown to be reversible in *P. denitrificans* under physiological conditions (Watmough et al., 1999). Further work on *R. capsulatus* showed that inactivation of the *cycP* gene decreased the removal of exogenous NO from cultures and increased the sensitivity to added NO and to *S*-nitrosoglutathione (Cross, Aish, Paston, Poole, & Moir, 2000). This result was consistent with a later and similar result for the cytochrome *c*′ from *N. meningitidis*.

Cross and co-workers further studied the role of cytochrome *c*′ in *R. capsulatus* (Cross, Lloyd, Poole, & Moir, 2001). In this work, the bacterium was able to rapidly remove exogenous NO from a suspension under aerobic and anaerobic conditions and this ability was diminished when the *cycP* gene was knocked out. From Western blot analysis, it was suggested that the increased expression of *cycP* in the presence of NO was insufficient to explain the rate of loss of NO. From this data, together with the measurement of an N_2O product, a catalytic function was proposed, namely that cytochrome *c*′ itself acts as an NOR and also that the main role of *cycP* is to respond to exogenous NO, rather than NO produced during denitrification. For non-denitrifying organisms, detoxification of NO may be required in order to respond to NO generated by other endogenous processes or indeed exogenously.

The expression of cytochrome *c′* from *R. sphaeroides* 2.4.3 was shown to be increased as oxygen levels were lowered and to be highest under conditions of photosynthetic denitrification (Choi, Grigoryants, Abruna, Scholes, & Shapleigh, 2005). Notably, and in apparent contrast to the previous proposal (Cross et al., 2001), NO reductase activity of cytochrome *c′* itself was not detected. This led to the proposal that cytochrome *c′* may play a role in shuttling NO to the membrane where it could be reduced to N_2O by NOR. This concept unifies the idea of an 'NO sequestration/buffering' role with the apparent NO reductase catalytic activity in cells. To our knowledge, an NO reductase activity in purified cytochrome *c′* has not been demonstrated.

In the proposed 'shuttle' model, NO generated by NiR in the periplasm of denitrifying bacteria binds to ferrous cytochrome *c′*. This complex diffuses to the inner membrane where the protein is oxidized to the ferric state, promoting NO release adjacent to the lipid bilayer, into which it will preferentially diffuse increasing its availability to the integral membrane protein NOR. Choi et al. proposed that this loading of NO into the lipid bilayer would facilitate the cellular response to NO (Choi et al., 2005). This role would likely become important *in vivo* in conditions of low nitrogen oxides availability rather than under more typical laboratory conditions with very high concentrations of these species. However, this result contrasted with that of an earlier study of *R. sphaeroides* IL-106 in which addition of nitrate to cultures actually repressed *cycP* expression.

Further work on cytochrome *c′* from *N. gonorrhoeae* showed that in this human pathogen, the protein is a lipoprotein that is anchored to the outer membrane. A knockout mutant showed that both control and knockout were denitrifying under oxygen-limited conditions in the presence of nitrite, but that the knockout exhibited a longer lag phase prior to exponential growth. In this case, the localization of cytochrome *c′* at the outer membrane was suggested to be consistent with a role in overcoming NO generated by the immune response. Sequestering of NO could also prevent formation of peroxynitrite which could aid the cell in survival within a macrophage or abscess where oxygen availability is low. It is known that both *N. meningitidis* and *N. gonorrhoeae* are able to denitrify and so it is possible to envisage multiple roles for cytochrome *c′* in these organisms involving endogenous NO from denitrification and exogenous NO from the immune response.

4. STRUCTURAL PROPERTIES OF CYTOCHROMES *C′*

The first structural characterization of any cytochrome *c′* was that of *Rhodospirillum molischianum* cytochrome *c′* (RMCP), determined to 1.67 Å resolution in 1980 (Weber et al., 1980), followed by a series of structures of homologues from other organisms (Table 1). While sequence identity between individual cytochromes *c′* is low (~28%; Ambler et al., 1981), the proteins in this family have a remarkably similar monomer fold consisting of a left-handed, antiparallel, elongated 4α-helix bundle (helices A, B, C, D) connected by several short loops. The helices are of differing lengths, comprising 27 (A), 13 (B), 19 (C) and 21 (D) residues in RRCP (Yasui et al., 1992), with small variations of length in other cytochromes *c′*.

The overall monomeric shape is that of a truncated cone, with the prosthetic haem group located within cavities created by the divergence of the helices towards the base (Fig. 2). The short loops connecting helices A–B and C–D lie at the narrow end of the cone, with the longer B–C connecting loop lying over the wider end. The loops are somewhat variable in length, with the B–C loop in particular containing between 17 (AXCP; Dobbs et al., 1996) and 31 (RCCP; Tahirov et al., 1996b) residues. Within each monomer, the haem CXXCH motif is located near to the C-terminus of the protein sequence and towards the base of the truncated cone. The proximal haem pocket is solvent exposed, while the distal pocket is buried within the monomer structure and devoid of water molecules.

Recently, the 1.0 Å resolution crystal structure of cytochrome *c′* from the thermophile *Thermochromatium tepidum* (TTCP) was described, with several adaptations to a higher temperature environment (Hirano et al., 2012). Notably, TTCP denatures at 75.4 °C, some 22 °C higher than its mesophilic homologue CVCP (with which it shares 82% sequence identity), despite the monomer structures being superimposable with an RMSD in Cα positions of only 1.16 Å. Three hydrophobic residues in the distal haem pocket have less bulky side chains in TTCP than in CVCP, allowing one helix to approach the haem more closely. This leads to a more compact tertiary structure consistent with greater protein stability. TTCP is further stabilized by 17 additional main chain hydrogen bonds compared to those present in CVCP. Analysis of the TTCP sequence reveals numerous substitutions of smaller residues compared to those in CVCP and a lower number of Gly residues, both of which would be expected to contribute to protein stability. A later study (Kimura et al., 2015) investigated mutations at two

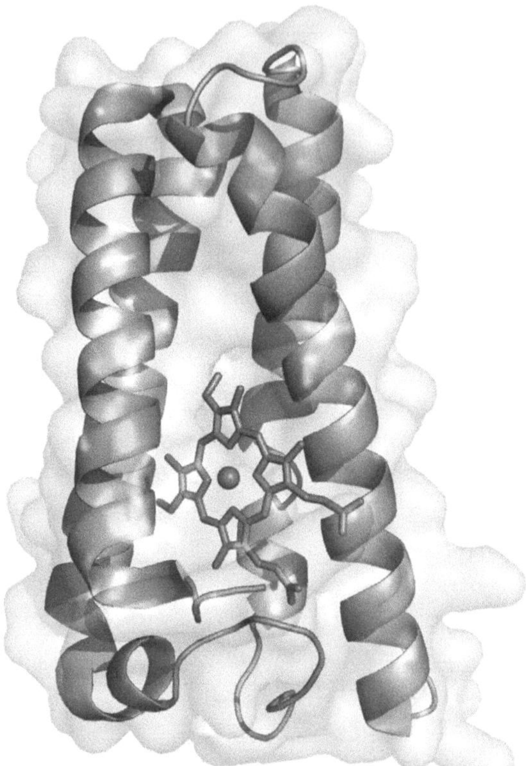

Figure 2 The monomer structure of the AXCP homodimer illustrating the truncated cone morphology and the location of the haem prosthetic group. The molecular surface is also shown.

amino acids, R129 and D131, revealing that these charged residues play important roles in promoting thermal stability and modulating the haem environment in TTCP.

4.1 Oligomeric State and Solvent Channels

4.1.1 Dimer Arrangements and Monomerization

Structurally characterized cytochromes *c*′ have been described as homodimers (ADCP, AXCP, CVCP, RGCP, RRCP, RMCP, TTCP (Finzel et al., 1985; Weber, Howard, Xuong, & Salemme, 1981; Hirano et al., 2012)), monomers (RPCP) or occupying a monomer–dimer equilibrium (RCCP, RSCP (Ramirez et al., 2003)). Where a monomer–dimer equilibrium is present in solution, the available crystal structures represent the dimeric form. The favoured oligomeric state appears to arise from differences in the surface properties of each monomer, the nature of the dimer interface and in particular its hydrophobicity.

The dimeric 'type 1' cytochromes *c′* (according to the definitions proposed by Shibata et al., 1998) exhibit an X-shaped pair of monomers (viewed perpendicular to the non-crystallographic twofold axis), with an extensive interface region involving helices A and B from each monomer (e.g. RMCP structure in Fig. 3). Typical helix crossing angles are −125° (A/A′) and −149° (B/B′) [values from the structure of CVCP (Ren et al., 1993)]. In such dimers, the interface is composed of many hydrophobic residues, while charged amino acids are largely confined to the periphery. Unusually, in CVCP the type 1 homodimer dissociates into monomers upon CO or NO binding to haem. The mechanism of dimer dissociation in CVCP is discussed in detail later.

The 'type 2' cytochromes *c′* are much more diverse in their interface properties than those of type 1 and have a greater tendency towards dissociation of the dimer into monomers. Of the known cytochromes *c′*, only RPCP is strictly monomeric and this is associated with a very different molecular surface to that in the type 1 cytochromes *c′*. Whereas type 1 cytochromes *c′* typically have no more than two or three charged residues at the A–B helix surface, RPCP has no fewer than seven, including several lysines together with three polar residues. The A–B surface of the type 2 RPCP also contains far fewer hydrophobic residues. The overall effect is that the 'interface' region of the RPCP monomer surface is much more hydrophilic which has been proposed to promote a monomeric state.

RCCP exists as a mixture of monomers and dimers and exhibits quite similar surface properties to those of RPCP with seven charged residues at the interface. A further difference is a larger number of Ala residues at the surface

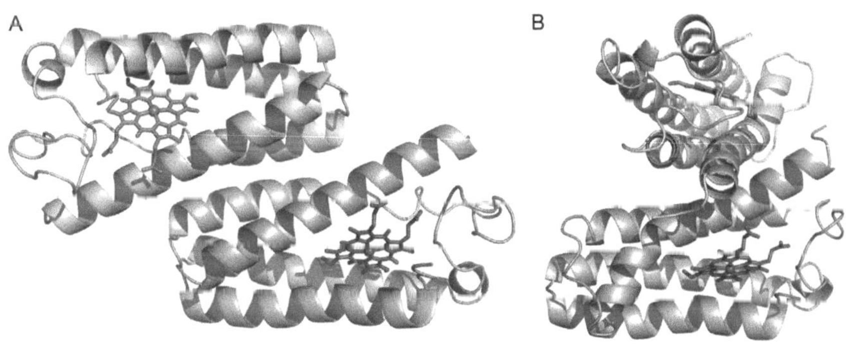

Figure 3 Homodimer arrangements in (A) RCCP and (B) RMCP. Note the X-shaped dimeric arrangement in the type 1 RMCP and the near-parallel arrangement in the type 2 RCCP. (See the colour plate.)

of RPCP and RCCP (Shibata et al., 1998). The two monomers of the dimer lie approximately parallel to each other and are linked by interactions involving helix B (Fig. 3). Analysis of buried surface area in the dimer interface of several cytochrome *c*′ crystal structures revealed that RCCP had the smallest buried area consistent with its propensity to dissociate into monomers.

Other cytochromes *c*′ display intermediate oligomerization properties related to their interface surfaces. The dimeric RRCP has a very hydrophilic dimer interface containing 15 hydrogen bonds that contribute greatly to dimer stability. The monomers lie near-parallel to each other although with a different dimer architecture to that of RCCP. In RSCP, a 'flattened' X-shaped dimer occurs (Ramirez et al., 2003) with dimer interface properties intermediate between those of the type 1 and type 2 cytochromes *c*′. While the buried surface area is closer to that of the type 1 proteins, RSCP possesses 15% more polar residues at the interface than CVCP (Ramirez et al., 2003). As a consequence, RSCP also exists as a monomer–dimer equilibrium.

4.1.2 Solvent Accessible Channels Between the Protein Surface and the Haem Pockets

Cytochromes *c*′ have previously been assigned to two structural groups on the basis of distal pocket residues and channels (Tahirov et al., 1996b; Table 1). Note that the term 'groups' is used here to be distinct from the 'types' classification based on interface properties. In group 1 cytochromes *c*′, including RCCP and CVCP, a deep channel is formed between two helices allowing solvent access to the distal haem pocket. By contrast, this channel is absent in group 2 proteins and solvent is not able to access the distal pocket (Fig. 4).

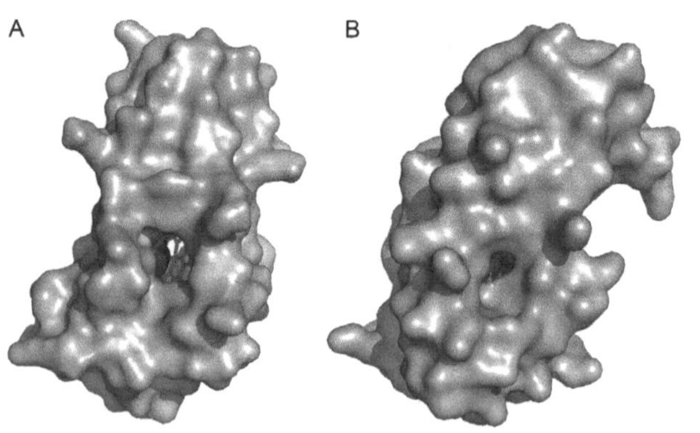

Figure 4 Surface representations of monomers of (A) group 1 RCCP and (B) group 2 RMCP. Note the large access channel to the distal haem pocket in RCCP that is not present in the group 2 cytochromes *c*′. (See the colour plate.)

The channel in CVCP lies between helices A and C and leads from the protein surface to the distal residue Tyr 16. Comparison of this region of the structure in different proteins led to the observation that smaller residues in proteins with an open channel were substituted with bulky residues in cases where the channel was blocked. It was also noted that the group 2 cytochromes *c*′ typically exhibit a three-residue sequence deletion in this region relative to members of group 1 (Ramirez et al., 2003). Group 1 proteins generally have an aromatic residue (Phe or Tyr) positioned over the distal haem face, whereas for group 2 proteins this residue is typically aliphatic (Leu or Met). In addition, all group 1 proteins studied to date are homodimeric, members of group 2 may be homodimeric (e.g. RRCP), monomeric (RPCP) or exist as an equilibrium mixture of the two (e.g. RCCP).

In RCCP, the channel between helices B and C was proposed to allow larger exogenous ligands to access the distal pocket. This assignment was supported by the observation that imidazole was able to readily access and bind to ferric haem in group 1 RCCP but not group 2 AXCP (Andrew et al., 2005), and that alkylisocyanides were able to access and bind to haem iron in group 1 RSCP and CVCP but could not bind to group 2 RMCP, RRCP or RPCP (Kassner, 1991).

A random-acceleration molecular dynamics (RAMD) simulation (Pietra, 2013) examined the migration of NO away from the AXCP haem pockets following dissociation from the starting models of 6cNO distal and 5cNO proximal complexes. NO was shown to primarily escape from the distal pocket via a defined channel, with several residues acting as gates. The simulation of escape from the proximal pocket suggested a main 'pseudo-channel' pathway between helix 7 and solvent—via His 120, Lys 117 and hence to bulk solvent. As expected from structural data, NO is unable to pass directly between the distal and proximal pockets.

4.2 Haem Environment (in the Absence of Exogenous Ligands)

The *c*-type haem in the cytochromes *c*′ has a conserved histidine proximal ligand within the CXXCH motif, close to the C-terminus, and a vacant sixth coordination position on the distal face. As such, the ferric or ferrous haem iron in the absence of exogenous ligands is five-coordinate (5c). The haem moiety makes extensive contacts with protein residues in helices A, B and C in addition to the covalent attachment via two cysteines of the CXXCH motif. The haem propionates are in differing environments, one deeply buried and one more solvent-exposed, but both are stabilized by extensive interactions with the protein. For example in the RMCP crystal structure, Lys 10 of one monomer forms a salt bridge to the more exposed haem propionate from the second monomer of the homodimer. Within each monomer, a conserved Arginine (Arg 12) interacts with the second, more deeply buried propionate group leading to the proposal that Arg 12 is essential for stabilization of the haem moiety. Several additional interactions involving both amino acids and water molecules are also observed, adding to the stabilization of the propionates.

4.2.1 Structure of the Distal Pocket

The distal pocket forms a hydrophobic cage and features an occluding near-haem residue which imposes significant steric restrictions that require conformational rearrangements to allow distal ligands to coordinate (see Section 4.3). The distal occluding residue is either aliphatic (commonly Leu, rarely Met) or aromatic (commonly Phe, rarely Tyr) (Fig. 5). Phe and Tyr distal residues are observed sitting over the Fe with their aromatic rings near-parallel to the haem plane in a position to occlude the binding of ligands. Water molecules are not observed in the distal pocket due to the lack of polar residues. The interior surface of the distal pocket is almost entirely hydrophobic with a high proportion of aromatic residues (Fig. 6). In AXCP, these are Leu 16, Met 19, Pro 55, Trp 56, Phe 59, Phe 79 and Phe 86. There is some variation in the pocket residues between species, but a high level of hydrophobicity is maintained in all cases.

The haem propionates are observed within the distal pocket although they may 'flip' to the proximal side upon the binding of certain ligands (*vide infra*). Superposition of different cytochrome *c′* haem pocket structures shows several consistent features. In the structures of *Rubrivivax gelatinosus* cytochrome *c′* (RGCP), RSCP RCCP and SFCP, the distal Phe lies parallel to haem although with slightly varying positions. Interestingly, Phe positions fall into two groups. In SFCP and RGCP, the Phe sits directly above the Fe atom, while in RCCP and RSCP, Phe is displaced slightly to one side by ~0.7 Å such that it would occlude distal ligand binding to a lesser extent (Fig. 7). The importance of distal steric constraints in defining haem-NO reactivity (including 6cNO vs. 5cNO coordination) is discussed later. In RMCP, the distal Met similarly lies in a position to occlude distal ligand binding. The S atom of Met lies directly above the haem Fe at a distance

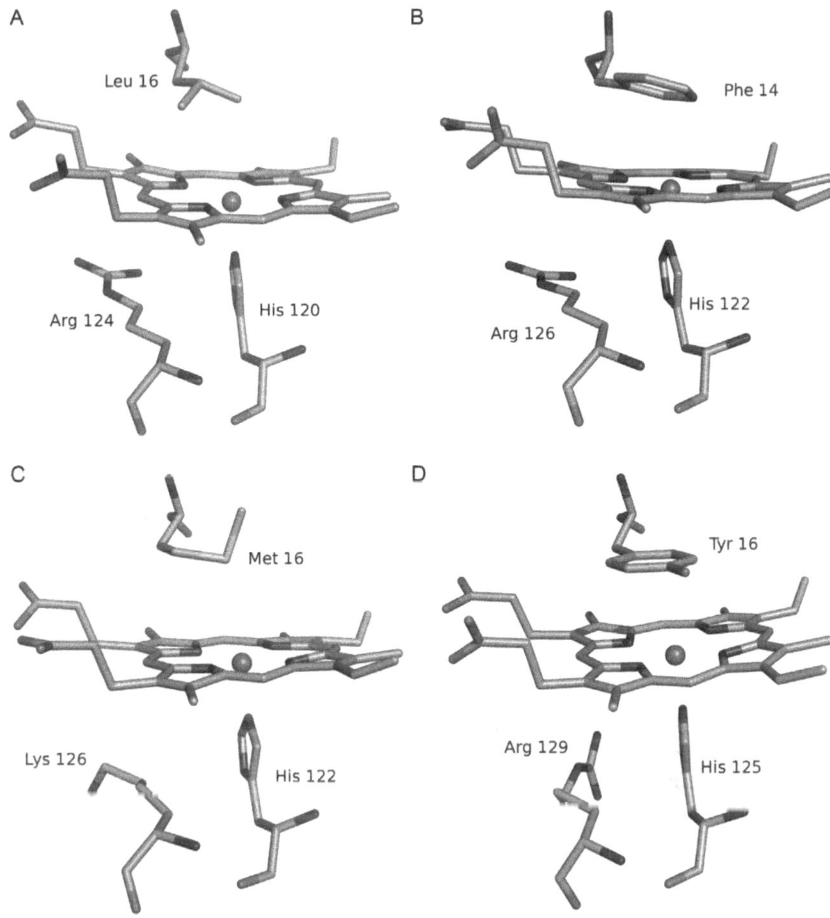

Figure 5 Distal (top) and proximal (bottom) pocket residues in (A) AXCP, (B) RCCP, (C) RMCP and (D) CVCP with occluding residues Leu, Phe, Met and Tyr, respectively. (See the colour plate.)

of 3.7 Å from it. For both Phe and Met occluding residues, large conformational arrangements of the occluding residue in the distal pocket are presumed to be necessary to allow binding of exogenous distal ligands (see later).

For cytochromes *c′* containing a distal Leu residue in the occluding position (e.g. AXCP, RPCP, RRCP), this smaller amino acid side chain is positioned directly over the haem iron in such a way as to provide significant steric hindrance to ligand binding. It is clear that side-chain rearrangement is required in order to allow either linear or bent modes of CO or bent modes of NO binding to form 6c ligand complexes.

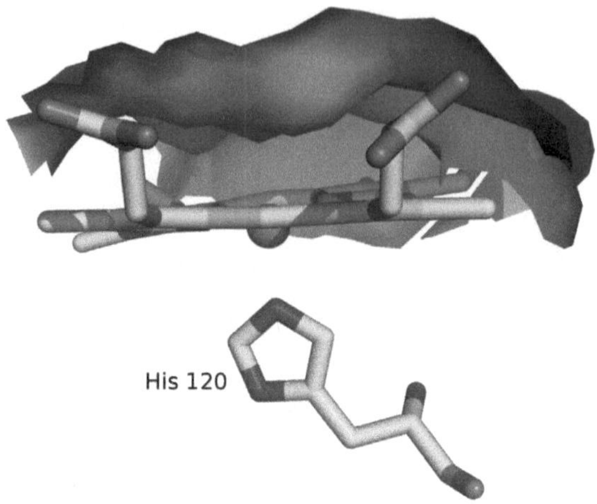

Figure 6 The distal pocket of AXCP. Carbon atoms are shown in green, oxygen in red, nitrogen in blue and sulphur in yellow. (See the colour plate.)

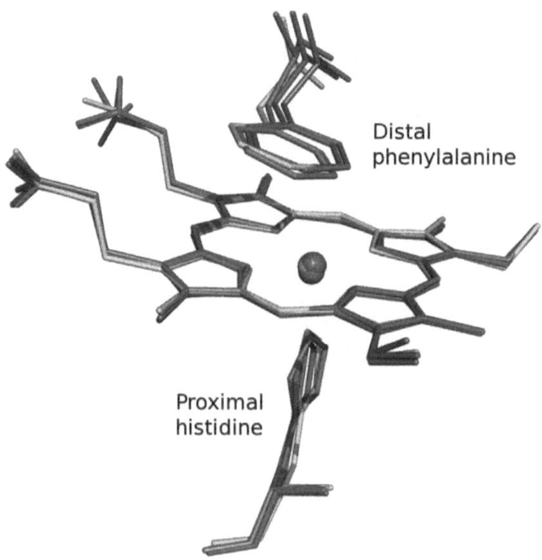

Figure 7 Occlusion of the Fe distal ligand binding site by Phe residues in SFCP (red), RGCP (grey), RCCP (green) and RSCP (blue). Note that the Phe positions are clustered into two groups. (See the colour plate.)

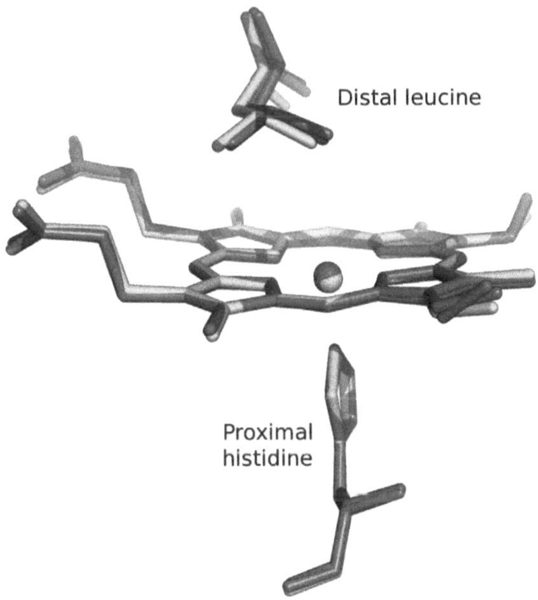

Figure 8 Superposition of the haem environment in AXCP (red), ADCP (blue) and RPCP (green). Note that coordinates are not available for RRCP which also has a distal Leu residue. (See the colour plate.)

A superposition of the haem environments in cytochromes *c*′ with distal Leu residues is shown in Fig. 8. Leu adopts a similar position and conformation in AXCP, ADCP and RPCP.

Where a polar Tyr residue is present in the occluding distal position (CVCP and TTCP), its side chain is positioned such that distal ligand binding must cause conformational changes to this residue, which are proposed to be propagated to the dimer interface causing monomerization in CVCP. In both TTCP and CVCP, the occluding distal tyrosine residue lies such that its O atom is only 2.7–2.8 Å from the equivalent atom of a second tyrosine. This interaction is presumed to be broken on ligand binding to the distal face of haem (*vide infra*). Notably, in these proteins with a polar distal pocket residue, the OH atom of tyrosine forms a H-bond to a water molecule in the pocket.

4.2.2 Structure of the Proximal Haem Pocket

In contrast to the distal haem pocket, the proximal pocket is near to the protein surface and solvent exposed. The haem ligand (His 120 in AXCP) coordinates to the proximal face of haem with a bond length of ~2.1 Å via its

Nε2 atom. The Nδ1 atom of the His ligand has been shown to be bound to a water molecule in some but not all cytochrome c' structures. This variation could be due to differences in pH (which might affect the His protonation state) or it might simply reflect the different resolutions of the structures. The proximal pocket contains several polar residues including a conserved positively charged Arg or Lys residue (Arg 124 in AXCP) that forms a cation–π bond with the bound His ligand and a residue (Asp 121 in AXCP) positioned to be able to form a H-bond to His 120 following its dissociation from haem. Several water molecules are typically observed in the proximal pocket, hydrogen bonded to polar residues.

4.2.3 Redox State-Dependent Changes to the Haem Environment

Relative to the ferric state, the structure of chemically reduced AXCP (Lawson et al., 2000) revealed few changes to the overall structure (with ferric and ferrous structures being superimposable with a root mean square deviation (RMSD) in C_α positions of 0.13 Å). Later structures of ferric and ferrous AXCP at higher resolution confirmed that only minor changes occurred between redox states. However, within the proximal haem pocket, a large flip in the side chain of Arg 124 in the proximal pocket is observed upon reduction of the haem (Fig. 9).

In ferric AXCP, the guanidinium group lies parallel to the haem plane and perpendicular to His 120. On reduction, the side chain rotates so that its guanidinium group is parallel to His 120 and perpendicular to the haem plane. Based on this observation, Arg 124 was proposed to act as an indicator residue for the redox state of the haem (Lawson et al., 2000). Cytochrome c' structures in the absence of exogenous ligands have generally been determined for the as-isolated, nominally ferric oxidation state although it is likely that some degree of X-ray photo-reduction to the ferrous form occurs during structure determination (*vide infra*). A recent combined single-crystal resonance Raman (RR) and crystallographic study of AXCP (Kekilli et al., 2014) investigated the effect of photo-reduction of ferric haem by exposure to laser light and irradiation by X-rays (via generation of solvated electrons from solvent). This analysis showed that where X-ray photoreduction of haem occurs, structural rearrangements including the 'flip' of Arg 124 may be suppressed at the cryogenic temperature (100 K) typically used in macromolecular crystallography experiments. However, the Arg was observed to flip to the 'ferrous' conformation when the crystal was reduced by X-rays at increased temperatures between 160 and 180 K.

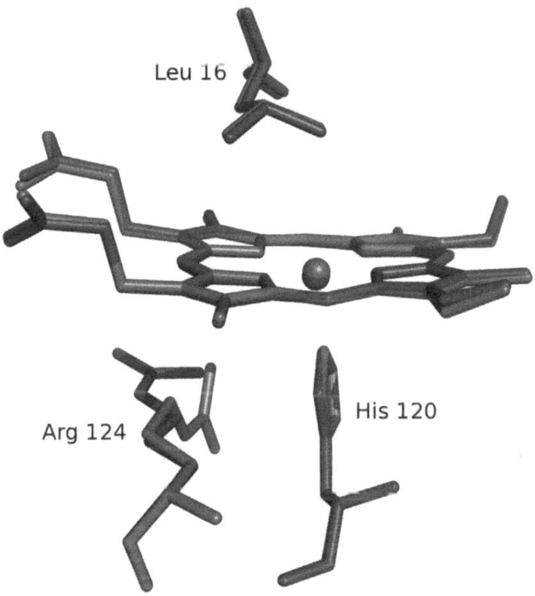

Figure 9 The haem environment in ferric (red) and ferrous (blue) AXCP. (See the colour plate.)

4.3 Crystal Structures of Ligand-Bound Forms

4.3.1 Alkylisocyanide Binding to RCCP

The first structure of a cytochrome *c'* with a ligand bound in the distal pocket was that of RCCP with a bound *N*-butylisocyanide (BIC) ligand (Tahirov et al., 1996a). BIC is representative of alkylisocyanides that were shown to be useful probes of haem site accessibility, being able to bind to haem in CVCP and RSCP with ~1000-fold greater affinity than in RMCP, RPCP and RRCP (Kassner, 1991; Patel, Kassner, Meyer, & Cusanovich, 1989; Rubinow & Kassner, 1984).

In the structure of the RCCP–BIC adduct (Tahirov et al., 1996a), the ligand occupies the channel between helices B and D and binds directly to the haem Fe such that its N-butyl group lies between residue Tyr 13 and the porphyrin. The Tyr 13 aromatic ring rotates to lie close to perpendicular to its position in the ligand-free RCCP structure. Concomitantly, the occluding Phe 14 phenyl ring undergoes a large conformational change, rotating ~63° around its Cβ—Cγ bond and 140° around the Cα—Cβ bond. This in turn forces movement of Glu 11 and one of the haem propionates such that the propionate flips towards the proximal pocket, causing further shifts in Lys 15. A conformational change was also described such that

Arg 126 flips to become perpendicular to its previous position. Later work on AXCP (Lawson et al., 2000) however suggested that the flip of Arg 126 was in fact due to reduction of the haem to the ferrous state and not a consequence of ligand binding. This structure clearly demonstrated the ability of ligand binding in the distal pocket to affect the structure of the proximal haem pocket. Indeed, a distal to proximal propionate flip was also observed in a recent crystal structure of the AXCP–CO complex (*vide infra*) (Antonyuk et al., 2011). The propagation of structural rearrangements on BIC binding to RCCP was used to propose a mechanism for the ligand-associated dimer dissociation of CVCP (Tahirov et al., 1996a) and why this does not occur in other cytochromes *c′*. In this mechanism, CO binding to subunit B of CVCP causes a steric clash between a flipped haem propionate and residue Glu 17 at the dimer interface, leading to a rearrangement of the Glu. This is propagated through interactions with Thr 11 in a manner that could cause subunit dissociation and which is consistent with the observed cooperativity of CO binding to CVCP.

4.3.2 CO- and NO-Bound Structures of AXCP and SFCP

A breakthrough came in 2000 with the high-resolution structures of CO- and NO-bound AXCP, revealing 6c distal CO binding and an unprecedented proximal 5c NO-binding mode (Lawson et al., 2000) with relevance to the activation mechanism of sGC. Subsequent improvements in the resolution of wt AXCP structures to atomic resolution (<1.2 Å), along with the characterization of L16A and L16G variants (Antonyuk et al., 2011), led to further insights into the determinants of ligand binding and discrimination.

In wt AXCP, CO forms a 6cCO complex, with CO bound at the distal face of the haem (Lawson et al., 2000). In the most recent 1.25 Å resolution structure of the AXCP–CO complex (Antonyuk et al., 2011), two alternate positions of CO were observed, one with a near-linear Fe—C—O geometry (Fe—C—O angle 170°), while the other adopted a more 'bent' geometry (Fe—C—O angle 158°) (Fig. 10; Table 2). The linear CO geometry was associated with a significant structural rearrangement of the distal Leu 16 and a flip of the haem 7-propionate from the distal to proximal pocket. This rearrangement was consistent with a ~600 mV drop in haem redox potential on the binding of CO and led to the proposal of a 'loaded spring' mechanism preventing irreversible CO binding to AXCP. Interestingly, the 1.95 Å resolution AXCP–CO structure of Lawson and co-workers showed a second CO molecule in

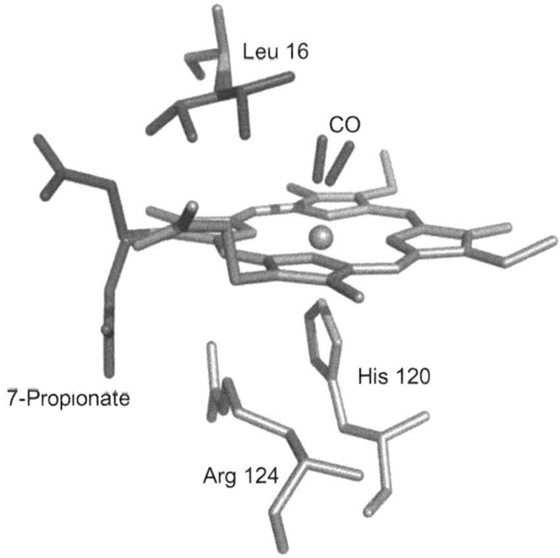

Figure 10 Crystal structure of the CO adduct of AXCP. The linear mode of CO binding (blue; grey in the print version) is associated with a rearrangement of Leu 16 and consequent flip of the haem 7-propionate in comparison to the bent CO mode (red; dark grey in the print version). Invariant atoms are shown in grey.

the haem vicinity (Lawson et al., 2000), bound to the Nδ1 atom of His 120, although this was not evident in the subsequent 1.25 Å structure from the Hasnain group (Antonyuk et al., 2011).

Mutation of the native Leu 16 in AXCP to the smaller hydrophobic residues Gly and Ala led to the removal of essentially all distal steric hindrance arising from this residue, allowing unhindered coordination of gases to the distal haem face (Antonyuk et al., 2011; Garton et al., 2012). The affinity for CO was increased so greatly over that of native AXCP (*vide infra*) that both L16A and L16G variants were isolated as 6cCO adducts. In comparison with native AXCP, where both near-linear (170°) and bent (158°) Fe—C—O geometries were observed, CO binds to L16A and L16G AXCP in a near-linear mode only with Fe—C—O angles of 170° (L16A) and 178° (L16G). No flipping of the haem 7-propionate was observed in either CO complex, consistent with the hypothesis that rearrangement of Leu 16 on CO binding to native AXCP drives propionate flipping as part of a 'loaded spring' mechanism.

Structures of nitric oxide complexes have to date been determined only for AXCP and SFCP (Table 2). In both cases, NO is bound to the proximal face of haem (Lawson et al., 2000; Manole et al., 2015) forming a 5cNO

Table 2 Crystal Structures of Cytochrome *c'*–Ligand Complexes

Organism	Mutation	Ligand	Resolution (Å)	PDB Code	References
Alcaligenes xylosoxidans	Native	CO	1.25	27ld	Antonyuk et al. (2011)
Alcaligenes xylosoxidans	Native	NO	1.19	2xlm	Hough et al. (2011)
Alcaligenes xylosoxidans	L16A	CO	1.03	2yl1	Antonyuk et al. (2011)
Alcaligenes xylosoxidans	L16G	CO	1.05	2ylg	Antonyuk et al. (2011)
Alcaligenes xylosoxidans	R124A	NO	1.07	2xl6	Hough et al. (2011)
Alcaligenes xylosoxidans	R124E	NO	1.14	2xlo	Hough et al. (2011)
Alcaligenes xylosoxidans	R124F	NO	1.24	2xlv	Hough et al. (2011)
Alcaligenes xylosoxidans	R124K	NO	1.12	2xle	Hough et al. (2011)
Alcaligenes xylosoxidans	R124Q	NO	1.17	2xlw	Hough et al. (2011)
Alcaligenes xylosoxidans	D121A	NO	1.45	4d4n	Ghafoor et al. (2015)
Alcaligenes xylosoxidans	D121I	NO	1.30	4d4x	Ghafoor et al. (2015)
Alcaligenes xylosoxidans	D121Q	NO	1.09	5agf	Ghafoor et al. (2015)
Shewanella frigidimarina	Native	NO	1.43	4cx9	Manole et al. (2015)
Rhodobacter capsulatus	Native	BIC	2.40	1nbb	Tahirov et al. (1996a)

Note that the first structures of CO- and NO-bound native AXCP were determined by Lawson and co-workers to lower resolution (Lawson et al., 2000). The highest resolution structures available are given in the table.

complex with the proximal histidine ligand dissociated from haem and rotated by ~110° around its Cα—Cβ bond into a position where it is able to form a H-bond to a sequence-adjacent Asp residue in the proximal pocket. The first structure for the 5cNO complex of AXCP

(Lawson et al., 2000) at 1.35 Å resolution revealed two alternate positions of proximal 5cNO, with Fe—N—O angles of 124° and 132°, respectively. Subsequent structures at 1.19 (Hough et al., 2011) and 1.26 Å (Kekilli et al., 2014) resolutions in contrast showed only a single NO conformer, although rather different NO geometries were observed in structures of site-directed mutants (Hough et al., 2011). In the 1.2 Å resolution structure of wt AXCP, the Fe—N—O angle is ~139° with NO oriented towards Arg 124. Notably, NO does not form any bonding interactions with proximal pocket residues in this structure, while both NO—Arg 124 and His 120—Arg 124 H-bonds were present in the 1.35 Å structure. The origin of these differences remains unresolved.

The structure of the proximal 5cNO complex of SFCP is similar to that of AXCP although a Lys is present rather than the Arg 124 of AXCP. The Fe—N—O angle in SFCP is compressed, with an angle of 128° in one monomer of the dimer and 115° in the second monomer, where its binding is distorted by a nearby glycerol molecule from the cryoprotectant used in crystal preparation. A superposition of the 5cNO-binding modes in AXCP and SFCP is given in Fig. 11.

Mutation of the charged proximal residue Arg 124 of AXCP to Ala and also to Gln, Glu, Phe and Lys (Barbieri, Murphy, Sawers, Eady, & Hasnain, 2008) increased the rate at which the proximal 5cNO species was formed from the initial distal 6cNO complex. Structural analysis of these variants (Hough et al., 2011; Table 3) definitively showed that H-bonding is not required to stabilise the proximal NO ligand. Structures of the R124F and R124K 5cNO complexes showed a single proximal haem-NO conformation similar to that of native AXCP, whereas R124E and R124Q showed two alternate positions of proximal 5cNO with interactions between the mutated side chain and the bound ligand. Structural and kinetic analysis concluded that the role of positively charged Arg 124 in the proximal pocket was, contrary to previous proposals, to slow formation of the proximal 5cNO species by steric and electrostatic interactions with the proximal His 120 haem ligand (Hough et al., 2011).

Interestingly, the atomic resolution crystal structure of the NO complex of the R124A variant revealed two populations of haem (in different positions within the protein) and corresponding partially occupied structures of the haem pockets. In the major conformation (occupancy 0.7), haem adopts a similar but not identical position to that in native AXCP, and two proximal NO-binding orientations are observed, one forming a H-bond with Arg 124 and the second oriented towards Cys 116. In the minor conformation (occupancy = 0.3), the haem is displaced into the proximal pocket and lies

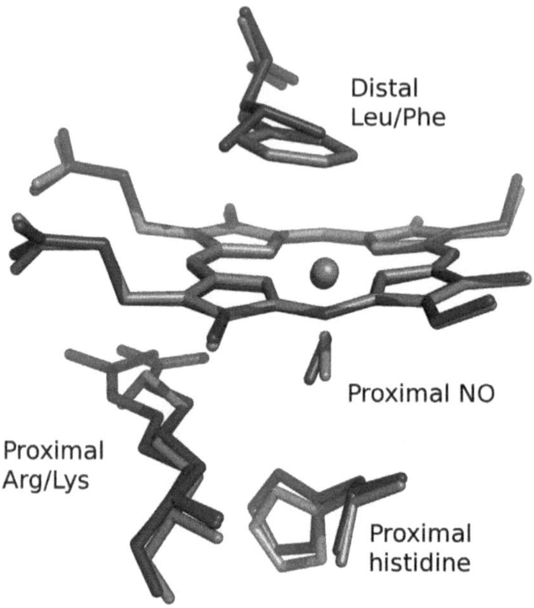

Figure 11 Superposition of the 5cNO complexes in AXCP (red) and SFCP (blue). (See the colour plate.)

Table 3 Crystal Structures of Ferrous Cytochromes *c*′ in the Absence of Exogenous Ligands

Organism	Mutation	Redox State	Resolution (Å)	PDB Code
Alcaligenes xylosoxidans	Native	Fe(II)	1.22	4cip
Alcaligenes xylosoxidans	R124A	Fe(II)	1.14	2xlh
Alcaligenes xylosoxidans	R124E	Fe(II)	1.10	2xm4
Alcaligenes xylosoxidans	R124F	Fe(II)	1.14	2xl8
Alcaligenes xylosoxidans	R124Q	Fe(II)	1.25	2xm0
Alcaligenes xylosoxidans	R124K	Fe(II)	1.40	2xld

All data taken from Hough et al. (2011) and Kekilli et al. (2014).

in a solvent-exposed position such that it is adjacent to Ala 124 and His 120, which now lies parallel to and ~4 Å above the haem plane, occluding the binding of proximal NO. In this case, NO is bound to the distal face of haem in an unusual 5cNO distal complex. As the structure of ferrous R124A without NO is very similar to that of native AXCP, the shifts in haem position are presumed to occur following cleavage of the His 120 to Fe bond after formation of the 6cNO intermediate. With His 120 detached and without the steric bulk and charge of Arg 124, the haem is able to relax into new positions in the proximal pocket. Structural characterization of R124A AXCP was followed by a detailed study of ultrafast NO-rebinding kinetics (Russell et al., 2013). Here, following laser photolysis of the Fe—NO bond, complex rebinding kinetics were observed consistent with the presence of two starting NO complexes in solution in a similar manner to that of the crystal. Processes occurring over relatively long timescales (~300 ns) could correspond to movements of the His with consequent structural rearrangements.

A follow-up study of the AXCP proximal pocket directly addressed the role of residue Asp 121 in stabilizing the haem-dissociated His 120 via a H-bond. Mutation of Asp 121 in AXCP to Ala, Ile or Gln did not prevent formation of proximal NO, suggesting that formation of the His 120 to Asp 121 hydrogen bond is not essential for proximal NO binding to occur (Ghafoor et al., 2015). However, altered NO-binding kinetics and proximal haem pocket geometries pointed to some involvement of Asp 121 in controlling the rate of formation of proximal 5cNO. Small variations in Fe–N–O angles occurred in the different structures, all compressed relative to that in native AXCP.

5. SPECTROSCOPIC PROPERTIES OF CYTOCHROMES *C'*

Spectroscopic characterization of cytochromes *c'* has been carried out primarily with haem-based techniques, including UV–vis, RR and electron paramagnetic resonance (EPR) spectroscopy. By providing 'fingerprints' of the haem cofactor, spectroscopic data can confirm haem identity and integrity, while providing a means of monitoring kinetic measurements. Spectroscopic data can also yield detailed structural information (e.g. haem geometry, bond strengths and electronic properties) complementary to that obtained by X-ray crystallography. UV–vis absorbance and RR spectra can distinguish the Fe coordination number, oxidation state and spin state of cytochrome *c'* haem complexes (Figs. 12 and 13; Tables 4, 5). RR spectroscopy (which detects vibrations of the haem chromophore) can also identify axial ligands (including diatomic gases) in isotopically labelled samples and can provide information on ligand bond strengths, bond angles and chemical environments. EPR spectroscopy provides information on haem complexes with unpaired electrons (e.g. ferric- and ferrous-nitrosyl cytochromes *c'*) (Table 9) and has been used to confirm the haem spin state and coordination number, as well as provide detailed information on the electron spin

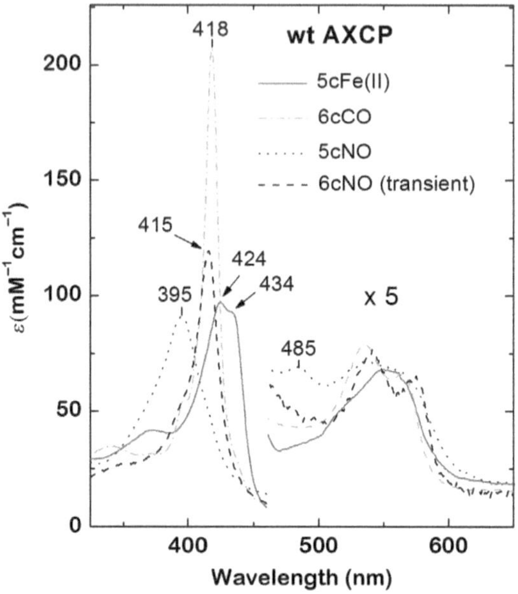

Figure 12 Spectroscopic distinction of AXCP complexes by UV–vis absorbance. (See the colour plate.)

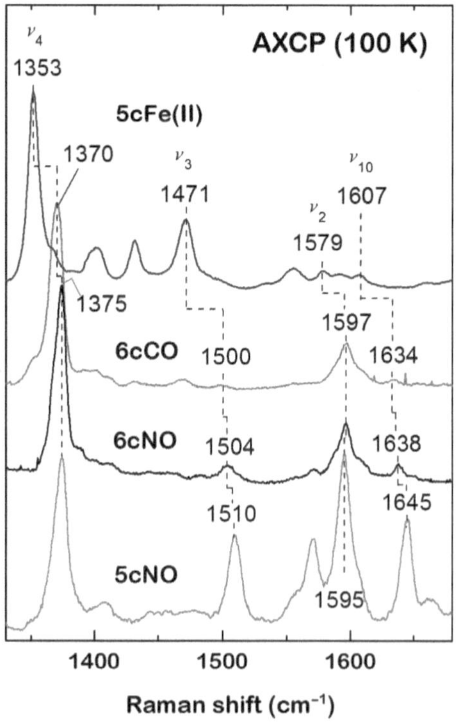

Figure 13 Spectroscopic distinction of AXCP complexes by RR spectroscopy: porphyrin marker frequencies (413-nm excitation) at pH 7.0, 100 K.

Table 4 Representative UV–Vis Spectra of Ferrous and Ferric Cytochrome *c′* Haem Complexes

Haem Complex	Spin State	Cytochrome *c′*	λ_{max} (nm)[a]					References
				Soret	β	α	CT	
Fe(II)								
4c	HS	AXCP		**~412**	*nr*	*nr*		Yoo et al. (2013)
5c	HS	AXCP	372	**424**, 434sh	~550			Andrew et al. (2001)
		AXCP (L16A)		**420**, 432sh	~550			Antonyuk et al. (2011); Garton et al. (2012)
		CVCP		**426**, 337sh	547			Bartsch (1963)
		PDCP		**427**	550			Gilmour, Goodhew, and Pettigrew (1991)
		RCCP		**425**, 435sh	~550			Andrew et al. (2005)
		RPCP		**426**, 435sh	552			Bartsch (1963)
		RRCP		**423**, 423sh	550			Bartsch (1963)
		RSCP		**425**	546			Bartsch (1978); Lee et al. (2009)
		SFCP		**426**, 436sh	560			Manole et al. (2015)
5cNO	LS	AXCP		**395**	538br	564br		Andrew et al. (2001)
		SFCP		**398**				Manole et al. (2015)
6cNO	LS	AXCP[b]		**415**	540	575		Andrew et al. (2002)

Continued

Table 4 Representative UV–Vis Spectra of Ferrous and Ferric Cytochrome c' Haem Complexes—cont'd

Haem Complex	Spin State	Cytochrome c'	λ_{max} (nm)					References
				Soret	β	α	CT	
		AXCP (L16A)		**416**	542	574		Antonyuk et al. (2011)
6cCO	LS	AXCP	341, 396sh	**418**	534	565		Andrew et al. (2001)
		AXCP (L16A)		**418**	535	566		Antonyuk et al. (2011)
		CVCP		**418**	545	565		Bartsch (1963)
		RPCP		**418**	535	570		Bartsch (1963)
		RRCP		**417**	534	564		Bartsch (1963)
		RSCP		**417**	535	561		Lee et al. (2009)
		SFCP		**417**	537	564		Manole et al. (2015)
$6cO_2$	LS	AXCP (L16A)	339	**415**	537	571		Garton et al. (2012)
6cEIC	LS	CVCP		**428**	527	557		Kassner, Rubinow, and Cusanovich (1983)
	LS	RPCP		**427**	527	556		Rubinow and Kassner (1984)
	LS	RRCP		**426**	526	555		Rubinow and Kassner (1984)
	LS	RSCP		**427**	527	556		Rubinow and Kassner (1984)
Fe(III)								
5c	HS	AXCP (pH 11)		374sh, **404**	508	540sh	643	Yoshimura et al. (1985)
5c	QS	AXCP (pH 5.3)		**396br**	500	535sh	643	Yoshimura et al. (1985)
5c	HS	PDCP (pH 7)		**407**	500		638	Gilmour et al. (1991)

Table 4 Representative UV–Vis Spectra of Ferrous and Ferric Cytochrome *c′* Haem Complexes—cont'd

Haem Complex	Spin State	Cytochrome *c′*	λ_{max} (nm)					References
				Soret	β	α	CT	
6cNO	LS	CVCP	354	**418**	530	564		Mayburd and Kassner (2002); Yoshimura et al. (1996)
		RCCP	360sh	**418**	529	562		Yoshimura, Suzuki, Iwasaki, and Takakuwa (1987b)
		RPCP	360sh	**417**	528	562		Yoshimura et al. (1996)
		RRCP		**413**	527	560		Yoshimura et al. (1996)
6cIm	LS	RCCP	354	**409**	529	558sh		Andrew et al. (2005)
6cCN	LS	CVCP		**417**	538			Kassner, Kykta, and Cusanovich (1985)

[a]Bold type denotes most intense peak.
[b]Transient precursor to 5cNO species.
Abbreviations: *nr*, not reported; *sh*, shoulder; *br*, broad absorption; HS, high spin; LS, low spin; QS, quantum mechanical admixture of intermediate spin and high spin.

distribution. As summarized in a previous review (Weiss et al., 2006), the haem environments of cytochromes *c′* have also been characterized by other spectroscopic techniques (including ENDOR, EXAFS, Mössbauer, MCD and NMR), while the fluorescence properties of cytochromes *c′* have been applied to the fabrication of NO-selective biosensors (Barker, Kopelman, Meyer, & Cusanovich, 1998).

Spectroscopic measurements can also reveal how sample and experimental conditions affect haem structure. For example, the spectroscopic features of some cytochrome *c′* complexes exhibit significant temperature dependences, highlighting the importance of comparing spectroscopic (and structural) data under similar conditions. Spectroscopic techniques have also proved useful in comparing cytochrome *c′* haem structures in solution

Table 5 Representative Porphyrin Marker RR Frequencies (cm^{-1}) of Cytochrome *c′* Haem Complexes

Haem Complex	Spin State	Cytochrome *c′*	Temp (K)	ν_4	ν_3	ν_2	ν_{10}	References
Fe(II)								
4c	HS	AXCP	rt	1363	1483	1574	1624	Kruglik et al. (2007)
5c	HS	AXCP	rt	1351	1469	1577	1603	Andrew et al. (2001)
			90	1353	1471	1579	1607	Andrew et al. (2002)
		CVCP	rt	1352	1469	1577		Hobbs et al. (1990)
5cNO	LS	AXCP	rt	1373	1506	1592	1641	Andrew et al. (2001)
			100	1375	1510	1595	1645	Servid et al. (2015)
6cNO	LS	AXCP	90	1375	1504	1596	1638	Andrew et al. (2002)
6cCO	LS	AXCP	rt	1368	1500	1596	1632	Andrew et al. (2001)
			100	1370	1500	1597	1634	Antonyuk et al. (2011)
		AXCP (L16A)	rt	1369		1596	1631	Antonyuk et al. (2011)
			100	1371		1597	1634	Antonyuk et al. (2011)
		CVCP	rt	1371		1591		Hobbs et al. (1990)
6cO_2	LS	AXCP (L16A)	rt	1374	1504	1593	1635	Garton et al. (2012)
6cCN		CVCP	rt	1356	1471	1588		Hobbs et al. (1990)
Fe(III)								
5c	HS	CVCP (pH 10)	rt	1371	1493	1580	1629	Hobbs et al. (1990)
5c	QS	CVCP (pH 7)	rt	1371	1502	1580	1637	Hobbs et al. (1990)
6cCN	LS	CVCP	rt	1372	1508	1589		Hobbs et al. (1990)

Abbreviation; rt, room temperature.

with those in the crystalline state. In particular, a recent study (Kekilli et al., 2014) used single-crystal RR spectroscopy to unambiguously assign redox and ligand states in AXCP crystals used for structure determination. Distinct spectral fingerprints for the ferric-,ferrous-, NO- and CO-bound crystalline states were obtained and used to validate structures in each of these states as well as to provide an effective comparison to solution data. The photo-reduction of ferric haem due to the Raman excitation laser light and/or X-rays (via generation of solvated electrons from solvent) was also studied. A separate study used single-crystal UV–vis absorption spectroscopy to monitor X-ray reduction of the haem in SFCP (Dworkowski, Hough, Pompidor, & Fuchs, 2015). In the following sections, we describe spectroscopic data for cytochromes *c*′ in the absence and presence of exogenous ligands, focusing on structural insights and correlations with X-ray crystallography.

5.1 Haem Spectroscopy in the Absence of Exogenous Ligands

5.1.1 Ferric Cytochromes c′

Purified forms of native cytochromes *c*′ are isolated as 5cFe(III) complexes, which, in light of their unusual spin-state properties, made this state a focus for initial structural and spectroscopic studies. The properties of ferric cytochromes *c*′ have been discussed extensively in a previous review (Weiss et al., 2006). EPR and other magnetic spectroscopies suggest an unusual spin-state equilibrium between a quantum-mechanically admixed high-spin/intermediate-spin (QS) state ($S=5/2$, $3/2$) and a pure high-spin (HS) species ($S=5/2$), the proportion of HS increasing at alkaline pH (p$K_a\sim$7–10) (Maltempo & Moss, 1976; Maltempo, Moss, & Cusanovich, 1974; Moore & Pettigrew, 1990; Weiss et al., 2006). The 5cQS state is very rare in haem proteins: aside from cytochromes *c*′ and model complexes, the only other reported examples are for class III peroxidases and a non-symbiotic plant haemoglobin (Droghetti et al., 2007; Smulevich, Feis, & Howes, 2005). The percentage $S=3/2$ contribution within the 5cQS state can be estimated from spectroscopic data, although the values obtained depend on the spectroscopic technique used (Ikezaki & Nakamura, 2002; Weiss et al., 2006).

EPR spectra of pure 5cHS Fe(III) cytochromes *c*′ are characterized by values of $(g_1+g_2)/2$ between ~5.9 and 6.0 and $g_3\sim2.0$, whereas 5cQS Fe(III) haems are distinguished by smaller values of $(g_1+g_2)/2$ between 4 and 6 (Fujii et al., 1995; Huston et al., 2006; Maltempo et al., 1974; Monkara et al., 1992; Yoshimura et al., 1990). Further discussion of cytochrome *c*′ EPR spectra is provided by the review of Weiss et al. (2006).

UV–vis absorption spectra are also sensitive to the spin state of ferric cytochromes *c*′. At pH values below the spin-state pK_a, the 5cQS state exhibits a broad asymmetric Soret band (λ_{max} ~395–400 nm), along with a β band near 500 nm and a CT1 band near 640 nm (Monkara et al., 1992; Weiss et al., 2006). Conversion to the pure 5cHS Fe(III) state under moderately alkaline conditions (pH 7–11) results in a red-shifted Soret maximum (λ_{max} ~405 nm), the appearance of a pronounced shoulder near ~370–380 nm, and an increase in the intensity of the ~640 nm CT1 band (a specific marker of 5cHS haem) (Weiss et al., 2006).

RR spectra of ferric cytochromes *c*′ obtained with Soret excitation suggest that 5cQS states can be distinguished by the higher frequency of their ν_3 (1500–1502 cm^{-1}) and ν_{10} (~1634–1637 cm^{-1}) modes relative to 5cHS, which exhibit ν_3 (1490–1494 cm^{-1}) and ν_{10} (~1625–1629 cm^{-1}) (Table 6; Hobbs et al., 1990; Huston et al., 2006; Kimura et al., 2015; Othman et al., 1996). This behaviour is illustrated by ferric CVCP at pH values above and below its spin-state pK_a of ~9 (Hobbs et al., 1990). At pH 7.0, the predominantly 5cQS form exhibits porphyrin marker bands: ν_4 (1371 cm^{-1}), ν_3 (1502 cm^{-1}), ν_2 (1580 cm^{-1}) and ν_{10} (1637 cm^{-1}). Raising the pH from 7 to 10 changes the CVCP spin state to 5cHS, with corresponding downshifts in ν_3 (1493 cm^{-1}) and ν_{10} (1629 cm^{-1}), but no change in the ν_4 and ν_2 frequencies. Other ferric cytochromes *c*′ exhibit ν_3 and ν_{10} bands that are doublets, consistent with the presence of both HS and QS forms. For example, RMCP at pH 6.8 exhibits ν_3 (1494, 1501 cm^{-1}) and ν_3 (1627, 1635 cm^{-1}) (Othman et al., 1996), while RSCP at pH 7.2 exhibits ν_3 (1490, 1502 cm^{-1}) and ν_{10} (1628, 1635 cm^{-1}) (Othman et al., 1996). In these cases, the presence of both HS and QS features may indicate that the pH is near the spin-state pK_a. In the case of TTCP (pH 8), the R129A mutation causes the ν_{10} band to change from a 1628, 1635 cm^{-1} doublet to essentially a singlet at 1635 cm^{-1}, consistent with stabilization of the QS (which may be due to removal of the positive charge implicated in stabilizing of the 5cFe(III) state) (Kimura et al., 2015). In the case of the K42E variant of RCCP, RR spectra consistent with a predominantly QS state (ν_3=1500 cm^{-1} and ν_{10}=1635 cm^{-1}) are observed throughout the pH range 4.0–9.5, suggesting that the spin-state pK_a is >10 at room temperature (Huston et al., 2006). Inconsistencies between spin-state pK_a values determined by UV–vis and RR spectroscopy compared to those measured by EPR have been ascribed to differences in sample temperature (Monkara et al., 1992; Moore & Pettigrew, 1990).

Table 6 pH-Dependent RR Features of 5cFe(III) Cytochromes *c′*

	pH	Spin-State pK_a[a]	Spin State[b]	RR Bands (cm^{-1}) ν_3	ν_{10}	References
AXCP	7.2	~7.1	HS +QS	nr	1631[c]	Yoshimura et al. (1985)
CVCP	7.0	~9	QS	1502	1637	Hobbs et al. (1990)
	10.0		HS	1493	1629	
RCCP, wt		8.4				Monkara et al. (1992)
K42E	4.0–9.5	>10	QS	1500	1635	Huston et al. (2006)
RMCP	6.8	nr	HS +QS	1494, 1501	1627, 1635	Othman, Richaud, Verméglio, and Desbois (1996)
RPCP	6.9	nr	QS	1500	1637	Strekas and Spiro (1974)
RRCP	7.3	~8.2	QS	nr	1637	Kitagawa, Ozaki, Kyooku, and Horio (1977)
RSCP	7.2	nr	HS +QS	1490, 1502	1628, 1635	Othman et al. (1996)
TTCP, wt	8.0	nr	HS +QS	nr	1625, 1635	Kimura et al. (2015)
R129A	8.0	nr	QS	nr	1635	Kimura et al. (2015)

[a]Spin-state pK_a determined from ambient temperature UV–vis absorbance measurements (Moore & Pettigrew, 1990).
[b]Predominant spin states consistent with spin-state pK_a, pH and RR frequencies.
[c]Unresolved RR band containing QS and HS species.
nr, not reported.

The structural determinants of 5cQS vs. 5cHS spin states in ferric cytochromes *c′* have been a matter of extensive debate. Formation of a 5cQS haem appears to be associated with a weak axial ligand. It has been suggested that the positive charge of the adjacent Arg/Lys residue weakens the Fe—His bond by shifting electron density away from the iron to the His ligand (Weiss et al., 2006). However, R124 mutations in AXCP were found to have no effect on the Fe–His stretching frequency, albeit in the 5cFe(II) state (Hough et al., 2011). ENDOR measurements of 5cQS Fe(III) RSCP have ascribed the relatively large hyperfine coupling of the His ligand N_δ proton to the absence of H-bonding partners (Usov, Choi,

Shapleigh, & Scholes, 2005). This lack of H-bonding may account for the 5cQS state of cytochromes *c*′, since the neutral His should be a weaker ligand than the proximal His of 5cHS proteins such as Mb, where H-bonding leads to partial imidazolate character (Usov et al., 2005).

It has been proposed that the 5cQS → 5cHS transition in cytochromes *c*′ results from deprotonation of the proximal His ligand (imidazolate being a stronger ligand than imidazole) (Usov et al., 2005; Weber, 1982). Although pH-dependent structural changes have not been investigated in detail by X-ray crystallography, NMR studies of 5cFe(III) cytochromes *c*′ have provided evidence that the spin-state pK_a corresponds to the deprotonation of the proximal His ligand (Weiss et al., 2006). It was further suggested that the positive charge of the adjacent Arg/Lys residue promotes imidazolate formation, thereby lowering the spin-state pK_a values (~7–10) below that of free imidazolate (pK_a 10.5). In this scenario, mutations that remove the positive charge at this position should increase the spin-state pK_a. Consistent with this prediction, UV–vis spectra of R124E and R124Q variants of AXCP indicate that the 5cFe(III) state remains primarily 5cQS throughout pH 5.0–9.0, consistent with a higher spin-state pK_a (>9) relative to wt AXCP (pK_a ~ 7.1), whereas the R124K variant, which retains a positive side chain, displays pH-dependent UV–vis spectra similar to wt protein (Barbieri et al., 2008). The R129A variant of TTCP also appears to stabilise the 5cQS state (Kimura et al., 2015). While these results appear to confirm the correlation of spin-state pK_a with the charge of the residue adjacent to the His ligand, R124F and R124A AXCP exhibit pH dependences similar to wt protein, despite the prediction of an increased spin-state pK_a (Barbieri et al., 2008). Interestingly, the K42E mutation in RCCP, located on the dimer interface, has a marked impact on RCCP haem properties, including stabilizing the 5cQS Fe(III) state, showing the potential for long-range effects to influence the haem (Huston et al., 2006).

5.1.2 Ferrous Cytochromes c′

Reduction of cytochromes *c*′ with dithionite yields the 5cFe(II) state, characterized by a distinctive split Soret absorption (λ_{max} ~ 420–425 nm with a ~435-nm shoulder) together with an αβ envelope near 550 nm (Fig. 12; Table 4). The split Soret band has been attributed to electronic inequivalency in the *x*, *y* directions of the porphyrin chromophore (Strekas & Spiro, 1974; Yoshimura et al., 1985), rather than sample heterogeneity. Consistent with a single 5cFe(II) structure, RR spectra obtained with Soret excitation show a single set of porphyrin marker vibrations (~1350–1610 cm^{-1}) characteristic of a 5cFe(II) haem (Table 7).

Table 7 RR Frequencies (cm^{-1}) of 5cFe(II) Cytochromes c'[a]

	ν_4	ν_3	ν_2	ν_{10}	ν(Fe—His)	References
AXCP						
Native	1351	1469	1577	1603	231	Andrew et al. (2001)
	1351	1469	1577	1603	232	Hough et al. (2011)
	1353	1471	1575	1604		Yoshimura et al. (1985)
	1353	*1471*	*1579*	*1607*		Andrew et al. (2002)
R124A	1352	1470	1576	1604	234	Hough et al. (2011)
R124E	1351	1469	1575	1603	234	Hough et al. (2011)
R124K	1350	1469	1576	1603	234	Hough et al. (2011)
R124Q	1350	1469	1577	1603	235	Hough et al. (2011)
R124F	1351	1470	1575	1603	232	Hough et al. (2011)
L16A	1350	1467	1577	1603	230	Garton et al. (2012)
CVCP						
Native	1352	1469	1577		232	Hobbs et al. (1990)
RCCP						
Native	1351	1470	1575	1604	227	Andrew et al. (2005)
K42E					229	Huston et al. (2006)
RMCP						
Native	1354	1468	1573	1604	228	Othman et al. (1996)
RRCP						
Native	1355	1472		1609		Kitagawa et al. (1977)
RSCP						
Native	1354	1468	1573	1604	231	Othman et al. (1996)
TTCP						
Native	1348			1603	230	Kimura et al. (2015)
R129A					230	Kimura et al. (2015)

[a]Data for room temperature solutions, except for frequencies in italics (solutions at 100 K).

RR measurements on cytochromes c' have observed the Fe—N(His) stretching vibration, ν(Fe—His), which is detected only in the 5cFe(II) state. Ferrous cytochromes c' exhibit relatively high ν(Fe—His) values (~227–235 cm^{-1}) (Table 7), indicating that the proximal Fe—His bond is relatively strong. Compared to other 5cFe(II) haem proteins, the ν_4 band (1348–1355 cm^{-1}) is quite low, while the ν_2 band (1573–1579 cm^{-1}) is quite high. The low ν_4 frequencies (Table 7) are consistent with a strong Fe—His bond as electron density is transferred to the porphyrin π^* orbitals (Yu, 1986).

The high ν(Fe—His) frequencies of 5cFe(II) cytochromes c' relative to most other haem proteins suggest that the proximal ligand has significant imidazolate character. While the ν(Fe—His) values of cytochromes c' are not as high as those of peroxidases (~245 cm^{-1}), Desbois and co-workers proposed that ν(Fe—His) frequencies of 5cFe(II) haem proteins appear to be sensitive to the dihedral angle (ϕ) between the imidazole plane and the nearest N(pyrrole)—Fe—N(pyrrole) axis, downshifting by approximately 0.5 cm^{-1} per degree (Othman et al., 1996). The same study suggested that archetypal imidazolate ligation (e.g. peroxidases with deprotonated N_δ) can be distinguished from partial imidazolate character (e.g. Mb, in which the protonated N_δ is an H-bond donor) by a ~30 cm^{-1} upward displacement of the correlation line. In AXCP crystal structures, the imidazole of His 120 is oriented such that ϕ is ~33° in both the ferrous and ferric states (Lawson et al., 2000). This, together with the ν(Fe—His) frequency of 231 cm^{-1}, places AXCP on the same correlation line as that of peroxidases with imidazolate ligation. A similar conclusion was reached regarding RMCP using the ν(Fe—His) frequency of 228 cm^{-1} together with the ϕ value of ~43° taken from the structure of the ferric state (Othman et al., 1996). It has been proposed that imidazolate coordination in 5cFe(II) cytochromes c' might be stabilized ionically by the positive charge of the adjacent conserved Arg/Lys residue (Dobbs et al., 1996; Moore & Pettigrew, 1990; Othman et al., 1996). However, R124 mutations in AXCP do not decrease ν(Fe—His) (Hough et al., 2011). Hydrogen-bonding between the solvent-exposed His ligand and a water molecule has also been suggested to stabilise imidazolate character (Weiss et al., 2006). The Fe—His bond strength of cytochromes c' is pertinent to the mechanism of proximal 5cNO formation, discussed later in this chapter (Section 6.2). Despite the fact that cytochromes c' (such as AXCP) form 5cNO complexes, their ν(Fe—His) frequencies lie significantly above the proposed value (~210 cm^{-1}) at which haem proteins are predicted to remain in the 6cNO state (Schelvis, Seibold, Cerda, Garavito, & Babcock, 2000). To explain this anomaly, it was proposed that the protonation state of the cytochrome c' proximal His ligand changes from imidazolate in the 5cFe(II) state to neutral imidazole in the 6cNO state, weakening the Fe—His bond beyond that ascribed to the negative *trans* effect of NO (Andrew et al., 2001).

On the other hand, Marti et al. have argued, on the basis of theoretical studies of AXCP, that the His ligand remains neutral throughout, with the high ν(Fe—His) value ascribed to an internal Stark effect from the positive charge of the adjacent Arg/Lys residue (Martí, Capece, Crespo, Doctorovich, & Estrin, 2005). However, subsequent RR studies of 5cFe(II) AXCP Arg 124 variants observed no correlation between the charge of residue 124 and the ν(Fe—His) frequency (Table 7) (Hough et al., 2011). A similar result was observed upon mutation of Arg 129 in TTCP (Kimura et al., 2015). Consequently, the Fe—His bond strength appears to be governed by the conformation of the surrounding polypeptide matrix, rather than the proximal haem environment. Finally, it is noted that RR studies of 5cFe(II) AXCP observed no significant changes in ν(Fe—His) or porphyrin marker band frequencies over the pH range 3.3–8.8 (Andrew et al., 2001). Thus, in contrast to ferric AXCP, whose pH-dependent spin state and spectroscopic properties have been attributed to His-ligand deprotonation (pK_a ~ 7.1) (see Section 5.1.1), the protonation state in ferrous AXCP (whether imidazole or imidazolate) appears constant throughout this pH range.

5.1.3 Four-Coordinate Cytochrome *c′*

Flash photolysis studies of 5cNO AXCP using femtosecond laser pulses have enabled the characterization of a transient four-coordinate ferrous state, 4cFe(II), using UV–vis absorption and RR measurements on the picosecond timescale (Kruglik et al., 2007; Yoo et al., 2013). RR porphyrin marker bands attributed to the 4cFe(II) species (recorded 10 ps after haem-NO photodissociation to minimize contributions from excited states) were observed at ν_4 (1363 cm^{-1}), ν_3 (1483 cm^{-1}), ν_2 (1574 cm^{-1}) and ν_{10} (1624 cm^{-1}). The downshift of the 4cFe(II) core-size marker frequencies (ν_3, ν_2 and ν_{10}) relative to those of the 5cNO complex (Table 5) is indicative of a ~0.05-Å increase in the distance from the haem centre to the pyrrole Nε atoms. The higher intensity of the ν_4 band relative to ν_2 was taken as evidence that the 4c haem is high spin. The absorption spectrum of the 4cFe(II) haem, obtained from difference measurements before and 10 ps after photodissociation, show a Soret λ_{max} at ~412 nm. The blue shift in the 4cFe(II) λ_{max} (~412 nm) relative to 5cFe(II) AXCP (λ_{max} ~424 nm) is similar to the trend observed in a protohaem model compound. Although 4cFe(II) AXCP is only observed using ultrafast photolysis/transient spectroscopy, its spectroscopic signature has facilitated the study of geminate recombination processes involving the proximal 5cNO complex, including reattachment of the His ligand (see Section 6).

5.2 Haem Spectroscopy of Cytochromes *c′* with Exogenous Ligands

5.2.1 Ferrous Complexes with NO

Ferrous haem proteins react with NO to form Fe(II) nitrosyl complexes, denoted $\{FeNO\}^7$ in the Enemark–Feltham notation (Enemark & Feltham, 1974). Depending on the organism and sample conditions, ferrous haem-nitrosyl cytochromes *c′* exhibit 6cNO and/or 5cNO geometries (Yoshimura et al., 1996), the latter involving displacement of the proximal His ligand. For some cytochromes *c′* the 6cNO complex is a transient precursor to a proximal 5cNO product (e.g. AXCP), while others form equilibrium mixtures of varying proportions of 6cNO and 5cNO species (Table 8). Structural determinants of 6cNO vs. 5cNO are discussed in Section 6.

The coordination number of ferrous haem nitrosyls (6cNO vs. 5cNO) can be readily distinguished by UV–vis absorption, EPR and vibrational spectroscopy (RR and FTIR). UV–vis absorption spectra of 6cNO complexes exhibit an intense Soret peak near 415 nm, with well-resolved αβ bands at ~565 and ~540 nm (Fig. 14), whereas 5cNO complexes exhibit a broader Soret peak near 395 nm, together with a weak feature near 480 nm and broad αβ bands near 560 and 540 nm. It should be noted that contamination of NO gas by O_2 during ferrous-nitrosyl preparation can generate higher oxides of nitrogen (including nitrous acid) which can cause sample acidification and conversion to a ferric 6cNO species (evident from a Soret maximum near 418 nm and αβ bands near 565 and 530 nm) (Cole et al., 2008).

Molar absorptivities for the ferrous-nitrosyl 5cNO product of AXCP (~95,000 M^{-1} cm^{-1}, 395 nm) and its transient 6cNO precursor (~120,000 M^{-1} cm^{-1}, 415 nm) (Andrew et al., 2002) were calculated by reference to a previously determined value for its $5cFe^{2+}$ state (~97,000 M^{-1} cm^{-1}, 426 nm) (Cusanovich et al., 1970). By estimating the contributions of 5cNO and 6cNO species to the Soret absorption envelope, the equilibrium 5cNO/6cNO ratios have been compared for a range of cytochromes *c′* (Table 8).

For most cytochromes *c′*, the 5cNO form predominates at equilibrium with excess NO, with the 6cNO ratio being ~0 to ~40% (Table 8). By contrast, the L16A variant of AXCP exists exclusively in the 6cNO state with no

Table 8 Soret Absorption Maxima of Ferrous-Nitrosyl Cytochrome *c′* 6cNO and 5cNO Complexes

		λ_{max} (nm)		
Cytochrome *c′*	**pH**	**6cNO**	**5cNO**	**References**
AXCP, wt	5.3–11.0	415 (transient)	395 (∼100%)	Andrew et al. (2002); Yoshimura et al. (1986)
L16A	7.0–8.9	416 (∼100%)	*n.o.*	Antonyuk et al. (2011); Garton et al. (2012)
R124A	7.0–8.9	*n.o.*	395 (∼100%)	Barbieri et al. (2008); Hough et al. (2011)
R124E	7.0–8.9	*n.o.*	395 (∼100%)	Barbieri et al. (2008); Hough et al. (2011)
R124F	7.0–8.9	416 (transient)	395 (∼100%)	Barbieri et al. (2008); Hough et al. (2011)
R124K	7.0–8.9	416 (transient)	395 (∼100%)	Barbieri et al. (2008); Hough et al. (2011)
R124Q	7.0–8.9	416 (transient)	395 (∼100%)	Barbieri et al. (2008); Hough et al. (2011)
CVCP	7.2	414 (∼30%)	400 (∼70%)	Yoshimura et al. (1996)
RCCP, wt	7.0–7.2	414 (∼40%)	395 (∼60%)	Andrew et al. (2005); Yoshimura, Suzuki, Iwasaki, and Takakuwa (1987a)
K42E	7.0	*n.o.*	399 (∼100%)	Huston et al. (2006)
RPCP	7.2	414 (∼40%)	397 (∼60%)	Yoshimura et al. (1996)
RRCP	7.2	415 (<10%)	393 (>90%)	Yoshimura et al. (1996)
RSCP, wt	7.0	414 (<10%)	397 (>90%)	Lee et al. (2009)
R127A	7.0	*n.o.*	397 (∼100%)	Lee et al. (2009)
SFCP	7.0	*n.o.*	398 (∼100%)	Manole et al. (2015)

Equilibrium populations of each species (shown in parentheses) at room temperature with excess NO are estimated from the relative intensities of overlapping Soret bands, allowing for differences in molar absorptivities.
n.o., not observed.

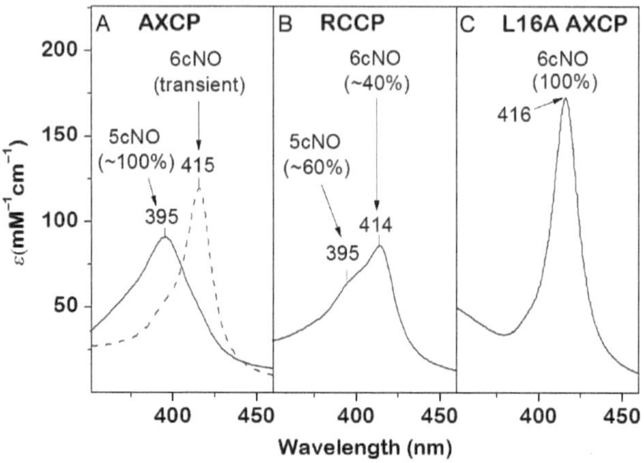

Figure 14 UV–vis absorption spectra for ferrous-nitrosyl complexes of (A) wt AXCP, (B) wt RCCP and (C) L16A AXCP showing the effects of variations in 6cNO:5cNO ratios. Intensities are normalized with respect to the 5cFe(II) states.

evidence of a 5cNO form (Fig. 14; Table 8). While PDCP also appears to form a predominantly 6cNO complex (Moir, 1999), the observation of ferric-nitrosyl absorption peaks upon prolonged NO exposure suggests that the reaction may have been contaminated by higher oxides of nitrogen (*vide supra*) (Cole et al., 2008). Sample temperature can also affect the 5cNO/6cNO ratio in certain cases. For example, the AXCP–NO reaction product has absorbance features typical of a pure 5cNO species (λ_{max} 395 nm) at room temperature, whereas cooling to 77 K causes the emergence of a ~415-nm shoulder indicative of a minor 6cNO population (Yoshimura et al., 1986). Similarly, EPR and RR spectra of AXCP–NO reaction products also show evidence of a minor 6cNO population upon cooling to 77–100 K (*vide infra*) (Servid et al., 2015; Yoshimura et al., 1986). More dramatic changes in 5cNO vs. 6cNO ratios have been observed following mutations in the distal and proximal haem pockets, as well as at the dimer interface (*vide infra*).

EPR spectra of 5cNO and 6cNO complexes are distinguished by their g values and hyperfine patterns. 5cNO complexes exhibit typical g values of ~2.10 (g_{max}), 2.06 (g_{mid}) and 2.01 (g_{min}), the latter showing a well-resolved-three-line hyperfine splitting ($A \sim 45$ MHz) due to the ^{14}N nucleus of NO ($I=1$) (Lehnert, 2008; Lehnert, Berto, Galinato, & Goodrich, 2011; Praneeth, Näther, Peters, & Lehnert, 2006). By contrast, EPR spectra of 6cNO complexes exhibit smaller g values of ~2.08 (g_{max}), 2.00 (g_{mid}) and 1.98 (g_{min}), with resolved hyperfine splitting from the ^{14}NO ligand

($A \sim 60$ MHz) evident on g_{mid} (rather than g_{min}), which is further split by weaker coupling with the *trans*^{14}N-(Im) donor ligand ($A \sim 15$ MHz) (Yoshimura, 1982), resulting in an overall nine-line hyperfine pattern. Single-crystal EPR data of model complexes indicate that only the g axis most closely aligned with the Fe—NO bond (g_{min} for 5cNO complexes and g_{mid} for 6cNO) exhibits a well-resolved hyperfine pattern. The unresolved nature of hyperfine coupling in g values associated with the porphyrin plane is ascribed to precession of the NO ligand around the Fe—NO bond, leading to multiple conformations with slight differences in g values (so-called g strain) (Goodrich, Paulat, Praneeth, & Lehnert, 2010; Lehnert et al., 2011).

Published EPR data for ferrous-nitrosyl cytochromes c' are dominated by 5cNO features (Table 9). For AXCP solutions (pH 7.2, 77 K) rhombic 5cNO spectra are observed with $g_{max} = 2.106$, $g_{med} = 2.058$ and $g_{min} = 2.10$,

Table 9 X-Band EPR Parameters of Cytochrome c' 5cNO Complexes

		5cNO					
Fe(II)NO Cytochrome c'	**Temp (K)**	g_{max}	g_{med}	g_{min}	$A[g_{min}]$ **(MHz)**	**Minor 6cNO**	**References**
AXCP (pH 7.2)	77	2.106	2.058	2.010	45	Yes	Yoshimura et al. (1986)
	293	2.088	2.05	2.016	45	*nr*	Yoshimura et al. (1986)
RCCP (B100)	77	2.111	2.033	2.010	44	Yes	Yoshimura et al. (1987b)
RPCP (ATCC 17001)	77	2.111	2.033	2.010	44	Yes	Yoshimura et al. (1996)
RRCP	77	2.107	2.036	2.010	46	No	Yoshimura et al. (1996)
RSCP (wt)	77	2.12	*nr*	2.010	48	No	Lee et al. (2009)
RSCP (R127A)	77	2.075[a], 2.11[b]	*nr*	2.010	45	No	Lee et al. (2009)
SFCP	10	*nr*	*nr*	2.013	45	No	Manole et al. (2015)

[a]Major species.
[b]Minor species.
nr, not reported.

the latter exhibiting a well-resolved hyperfine pattern, $A=45$ MHz (Yoshimura et al., 1986; Table 9). An additional weak absorption at $g=1.98$ ($A\sim19$ MHz) is consistent with a minor population of 6cNO species. A similar EPR spectrum obtained from denitrifying cell extracts of *Alc. xylosoxidans* has also been attributed to 5cNO AXCP, which in turn suggests that AXCP may have a physiological role in NO metabolism (Yoshimura et al., 1988). Compared to spectra measured at 77 K, lower-resolution spectra of AXCP at room temperature retain the characteristic EPR features of a 5cNO complex, albeit with small changes in *g* values (Table 9). As with AXCP, spectra of RCCP and RPCP solutions at 77 K are also dominated by 5cNO features (Table 9), with additional weak *g* absorptions suggestive of minor 6cNO populations (Yoshimura et al., 1996). Scholes and co-workers investigated the effect of the proximal R127A mutation on the EPR spectrum of 5cNO RSCP (Lee et al., 2009). While the wt RSCP spectrum is consistent with a single 5cNO species with $g_{max}=2.012$, the R127A variant shows evidence of at least two 5cNO species with g_{max} values of 2.075 (major species) and 2.11 (minor species) (Table 9). Another effect of the R127A mutation is to broaden the g_{max} region (an effect ascribed to increased *g* strain), although interestingly, hyperfine coupling in this region is more resolved than in wt RSCP.

RR spectroscopy can selectively probe molecular vibrations of haem cofactors, distinguishing 5cNO vs. 6cNO coordination via the frequencies of porphyrin-based 'marker bands' as well as vibrations of the FeNO unit. While 5cNO and 6cNO cytochromes *c′* display similar ν_4 and ν_2 porphyrin frequencies, the intensity of the ν_2 mode relative to the ν_4 mode is much greater in 5cNO complexes, and the ν_3 and ν_{10} modes of 5cNO complexes are 5–10 cm^{-1} higher than those of their 6cNO counterparts (Fig. 13; Table 10). Definitive distinction of 5cNO vs. 6cNO complexes is provided by the frequencies of the N—O stretch, ν(NO), Fe—NO stretch, ν(Fe—NO) and Fe—N—O bend δ(Fe—N—O) which can be assigned using isotopically labelled NO. For 5cNO cytochromes *c′*, the span of ν(NO) frequencies (~1651–1671 cm^{-1}) (Table 10) is within the range reported for other 5cNO haem proteins and model complexes. Although relatively little RR data are available for 6cNO cytochromes *c′* (Table 10), ν(NO) values for the transient 6cNO complex of AXCP (1624 cm^{-1}) and the 6cNO equilibrium population of RCCP (1625 cm^{-1}) are typical of other 6cNO haem complexes and significantly lower than those of 5cNO species (Table 10).

Table 10 Vibrational Frequencies (cm^{-1}) for 5cNO and 6cNO Cytochromes c'[a]

Cytochrome c'	Temperature (K)	ν_4	ν_3	ν_2	ν_{10}	ν(NO)	ν(FeNO)[b]	References
5cNO complexes								
AXCP, wt	100	1375	1510	1595	1645	1664	531	Servid et al. (2015)
	rt	1373	1506	1592	1641	1661	526	Andrew et al. (2001)
	100	(1378)	(1512)	(1597)	(1648)	*nd*	*nd*	Kekilli et al. (2014)
R124K	100	1375	1509	1594	1644	1657	533	Servid et al. (2015)
R124F	100	1375	1509	1594	1644	1662	529	Servid et al. (2015)
R124Q	100	1375	1510	1595	1645	1662	532	Servid et al. (2015)
R124E	100	1375	1510	1595	1647	1661	533	Servid et al. (2015)
R124A	100	1375	1511	1597	1647	1671	536	Servid et al. (2015)
	rt					~1655[c]		Russell et al. (2013)
SFCP	100	1375	1508	1591	1643	1651	525	Servid et al. (2015)
	rt	1373	1505	1589	1640	1658	519	Servid et al. (2015)
RCCP, wt	90	1375	1509	1596	1646	*nd*	*nd*	Andrew et al. (2005)
	rt	1375	1509	1593	1645	*nd*	*nd*	Andrew et al. (2005)
K42E	rt	1367	1511	1594	1648	1666	523	Huston et al. (2006)

Continued

Table 10 Vibrational Frequencies (cm^{-1}) for 5cNO and 6cNO Cytochromes *c*′—cont'd

Cytochrome *c*′	Temperature (K)	ν_4	ν_3	ν_2	ν_{10}	ν(NO)	ν(FeNO)	References
6cNO complexes								
AXCP	90	1375	1504	1596	1638	1624	579	Andrew et al. (2002)
	rt					1625[d]		George et al. (2001)
RCCP	90	1377	1506	1598	1640	1625	460, 569	Andrew et al. (2005)
RCCP	rt	1375	1503	1593	1635	*no*	458, 562	Andrew et al. (2005)

[a]RR frequencies are for protein solutions at pH 7.0, except for data in parentheses obtained from single crystals at pH 7.5.
[b]Mixed ν(Fe—NO)/δ(Fe—NO) mode: RR spectra of 5cNO cytochromes *c*′ detect a single vibration in the 519–536 cm^{-1} region with predominant ν(Fe—NO) character. For 6cNO complexes, two possible modes are observed, designated $\nu(FeNO)_I$ (~458–460 cm^{-1}) and $\nu(FeNO)_{II}$ (562–579 cm^{-1}).
[c]Flash-photolysis FTIR data.
[d]Stopped-flow FTIR data at pD 9.4.
Abbreviations: rt, room temperature; *nr*, not reported; *nd*, not determined; *no*, not observed.

Unlike 6cCO complexes (which prefer a linear Fe—C—O geometry), the inherently bent FeNO geometries of ferrous nitrosyls lead to significant mixing of the Fe—NO stretching and Fe—N—O bending modes. For 5cNO cytochromes *c*′, an RR band with predominant ν(Fe—NO) character is observed between 519 and 536 cm^{-1} (Table 10), although the corresponding δ(Fe—N—O) mode, expected near 390 cm^{-1} based on nuclear resonance vibrational spectroscopy (NRVS) of model complexes, is not observed. By contrast, 6cNO cytochromes *c*′ exhibit two RR bands with mixed Fe—NO bending/stretching character: a weak band in the 450–460 cm^{-1} range, denoted $\nu(FeNO)_I$, and a more intense band the 560–580 cm^{-1} range, denoted $\nu(FeNO)_{II}$ (Table 10). The assignment of these mixed-mode RR bands in 6cNO complexes has been controversial (Benko & Yu, 1983), although recent NRVS data on model complexes suggest that the higher frequency $\nu(FeNO)_{II}$ band has more Fe—NO bending character, while the lower frequency $\nu(FeNO)_I$ band is the predominant Fe—NO stretch (Hunt & Lehnert, 2015; Lehnert et al., 2010; Paulat et al., 2008; Zeng et al., 2005).

Trends in ν(NO) and ν(FeNO) frequencies can provide additional structural information on the Fe(II)NO unit and the surrounding environment. Common to all haem-XO complexes (X=C, N, O), the frequencies of ν(XO) and ν(FeXO) vibrations are sensitive to the extent of Fe(II)→XO(π^*) backbonding, which in turn may be influenced by the haem-XO environment (Spiro, Soldatova, & Balakrishnan, 2013). For example, a positive charge or H-bond donor close to the XO ligand may enhance Fe(II)→XO(π^*) backbonding, strengthening the Fe—XO bond and boosting the ν(FeXO) frequency, while weakening the X—O bond and lowering ν(XO). For example, compared to other 6cCO proteins, the ν(FeCO) and ν(CO) frequencies of cytochromes *c*′ lie midway along the 6cCO backbonding trendline, consistent with the non-polar character (hydrophobic cage) of the distal pocket (Andrew et al., 2001, 2005). By contrast, the effect of the haem protein environment on 5cNO and 6cNO vibrations is more complicated (and less defined), in part because of vibrational mixing between the ν(Fe—NO) and δ(Fe—N—O) modes. As well as the influence of Fe(II)→XO(π^*) backbonding, recent studies of Fe(II)NO porphyrins point to a positive correlation of ν(FeNO) vs. ν(NO) frequencies arising from variations in σ donation from the NO(π^*) orbital to the Fe(II) (d_{z^2}) orbital (Lehnert, Scheidt, & Wolf, 2014). DFT calculations also predict an intricate dependence of ν(FeNO) and ν(NO) frequencies on the Fe—N—O bond angle, the value of which may be distorted from its equilibrium value of ~140° by steric crowding or by H-bonding interactions (Ibrahim, Derbyshire, Soldatova, Marletta, & Spiro, 2010).

The recent X-ray crystal structures of 5cNO complexes of AXCP (native and R124 variants) and SFCP have enabled the influence of the proximal haem-NO environment on 5cNO RR frequencies to be experimentally tested (Servid et al., 2015). RR measurements were carried out on the proximal 5cNO complex of SFCP which has a Lys 126 residue near the proximal NO-binding site, as well as for AXCP variants in which the proximal Arg 124 (near the NO ligand) was mutated to Ala, Gln, Glu, Lys and Phe (Servid et al., 2015). Overall, the RR data for 5cNO cytochromes *c′* highlight the complex structure–vibrational relationships exhibited by Fe(II)NO haem proteins. Most of the cytochrome *c′* RR frequencies lie close to the 5cNO backbonding trendline of model porphyrins, with the exceptions of R124A AXCP and SFCP, which are displaced significantly above and below, respectively (Fig. 15). Surprisingly, the positive charge of the AXCP Arg 124 residue (nearest the proximal NO ligand) appears to have little impact on haem-NO bonding, since R124F, R124Q and R124E mutations have only minor influences on ν(NO) and ν(FeNO) frequencies. On the other hand, the R124K mutation causes a 6-cm^{-1} decrease in ν(NO) and a 2-cm^{-1} increase in ν(FeNO) frequency, suggesting that the positively charged Lys side chain is more able to enhance Fe(II) → NO(π^*) backbonding.

It is notable that the ν(FeNO) and ν(NO) RR bandwidths of R124Q, R124E and R124A variants (whose crystal structures reveal multiple haem-NO conformers) are similar to those of the other 5cNO cytochromes *c′* with single haem-NO conformations in their crystal structures (Hough et al., 2011; Servid et al., 2015). This suggests either that (i) different haem-NO conformers within each protein exhibit similar RR frequencies, or (ii) only one detectable haem-NO conformer is present in solution. In the case of SFCP, deviations from the backbonding trendline might arise from its relatively small Fe—N—O angle (128°) (Manole et al., 2015) and/or changes in NO(π^*) → Fe^{2+}(d_{z^2}) σ-bonding (Lehnert et al., 2014). Interpretation of R124A AXCP vibrational data is complicated by the existence of distal and proximal 5cNO populations in its crystal structure (Hough et al., 2011) (Section 4.3.2). The singlet nature of its ν(NO) and ν(FeNO) RR bands suggests that these populations give rise to similar haem-NO frequencies, or that only one detectable

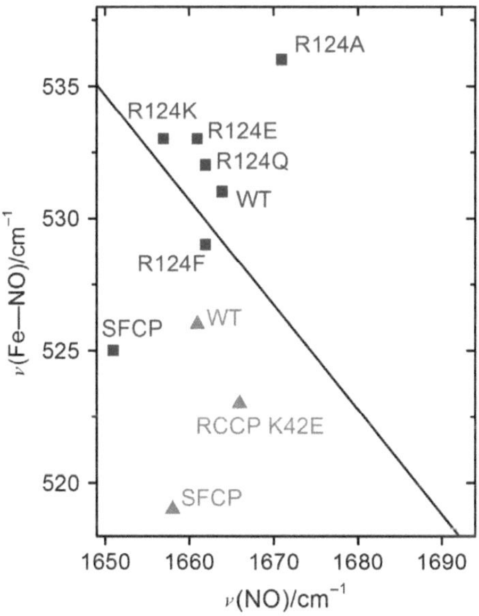

Figure 15 Correlation of ν(FeNO) and ν(NO) RR frequencies for 5cNO cytochromes *c'* at 100 K (squares) and room temperature (triangles). Protein labels refer to AXCP unless otherwise stated. Superimposed is a plot of the experimental backbonding line previously determined in 5cNO model complexes (Vogel et al., 1999) .

5cNO population (proximal or distal) is present in solution. Interestingly, the ν(NO) frequency of R124A AXCP determined by RR spectroscopy (1671 cm^{-1}) (Servid et al., 2015) is much higher than that observed in recent flash photolysis FTIR measurements (~1655 cm^{-1}) (Russell et al., 2013). The cause of this discrepancy is puzzling, but may relate to different sample conditions. For example, RR frequencies of 5cNO cytochromes *c'* were found to be quite temperature sensitive (Servid et al., 2015), consistent with a high degree of conformational flexibility in the haem-NO environment.

Although trends in vibrational frequencies have yet to be investigated in detail for 6cNO cytochromes *c'*, the unusually high $\nu(FeNO)_{II}$ RR frequency of 6cNO AXCP (579 cm^{-1}) has been tentatively ascribed to steric compression of the Fe—N—O angle (Andrew et al., 2002), although this remains to be confirmed by X-ray crystallography.

5.2.2 Ferrous Complexes with Carbon Monoxide

Ferrous cytochromes *c'* bind CO at the distal site to form 6cCO complexes that are not susceptible to His-ligand dissociation (as are 6cNO complexes) or autoxidation (as are $6cO_2$). Their UV–vis absorbance spectra are characterized by an intense Soret peak near 418 nm, together with αβ bands near 565 and 535 nm (Fig. 12; Table 4), and porphyrin marker RR frequencies are consistent with 6cLS Fe(II) (Fig. 13; Table 5). Additional structural information is available from vibrations within the FeCO unit. At room temperature, wt AXCP exhibits a C—O stretch, ν(CO) at 1966 cm^{-1}, Fe—CO stretch, ν(Fe—CO) at 491 cm^{-1} and Fe—C—O bend, δ(Fe—C—O) at 572 cm^{-1} (Table 11). Similar vibrational frequencies (±4 cm^{-1}) are observed for the other cytochromes c', although FTIR measurements of CVCP at pH 7.0 reveal a broad ν(CO) absorption envelope centred at 1978 cm^{-1} (Hobbs et al., 1990). The relative positions of the ν(FeCO) and ν(CO) frequencies on the backbonding trendline for 6cCO proteins are typical of CO ligands in non-polar environments (Phillips, Teodoro, Li, Smith, & Olson, 1999; Spiro et al., 2013), in line with crystallographic data.

Whereas room temperature RR and kinetic studies of the wt AXCP 6cCO complex detected only a single Fe—C—O species, the crystal structure (obtained at 100 K) shows two Fe—C—O conformers (Section 4.3.2).

Table 11 Vibrational Frequencies (cm^{-1}) of 6cCO Cytochromes c'[a]

Cytochrome c'	Temperature (K)	ν(FeCO)	δ(Fe—C—O)	ν(CO)	$\delta(C_\beta C_c C_d)$	References
AXCP	rt	491	572	1966	384	Andrew et al. (2001)
	100	482, 493	570, 580	1965	385	Antonyuk et al. (2011)
AXCP, L16A	rt	487	575	1964	385	Antonyuk et al. (2011)
	100	491	577	1965	386	Antonyuk et al. (2011)
CVCP	rt	492	565	1978		Hobbs et al. (1990)
RCCP	rt	494	573	1966	384	Andrew et al. (2005)

[a]All data are from RR spectra at pH 7.0, apart from the ν(CO) frequency of CVCP which is the apparent maximum of a broad FTIR absorption band.

Additional RR measurements of the wt AXCP 6cCO complex at 100 K (to match the crystallographic temperature) revealed two partially resolved ν(FeCO) modes at ~493 and ~482 cm^{-1}, together with two overlapping δ(FeCO) modes at ~580 and ~570 cm^{-1}. The ν(FeCO) component at ~493 cm^{-1} was attributed to the near-linear Fe—C—O conformer in the wt AXCP crystal structure, since this frequency is close to that of L16A AXCP (491 cm^{-1} at 100 K) which exhibits only a near-linear Fe—C—O geometry in its crystal structure. However, a detailed investigation of the effect of Fe—C—O angle on cytochrome *c'* RR vibrations has yet to be carried out. The observation of a single propionate $\delta(C_\beta C_c C_d)$ bending mode near 385 cm^{-1} in all 6cCO AXCP RR spectra (Table 11) suggests that this vibration is not significantly affected by the propionate conformation.

5.2.3 Ferrous Complexes with Dioxygen

Although native ferrous cytochromes *c'* autoxidise upon exposure to O_2 (without a detectable haem-O_2 complex), spectroscopic measurements have recently characterized a 6cO_2 species in the L16A variant of AXCP (Garton et al., 2012). UV–vis absorbance measurements of oxy L16A show a Soret band at 415 nm, with αβ bands at 571 and 537 nm, similar to other 6cO_2 proteins such as myoglobin (Mb) (Table 4) (Antonini & Brunori, 1971). RR spectra, obtained with 413-nm excitation, reveal porphyrin marker bands consistent with a Fe(II) 6cLS complex: ν_4 (1374 cm^{-1}), ν_3 (1504 cm^{-1}), ν_2 (1593 cm^{-1}), ν_{10} (1635 cm^{-1}) (Table 5). In the low-frequency region, RR spectra confirm the presence of the 6cO_2 complex from an $^{18}O_2$-sensitive vibration at 572 cm^{-1}, which is assigned to the Fe—O_2 stretch, ν(Fe—O_2). Similar to most 6cO_2 RR spectra (Mukai, Savard, Ouellet, Guertin, & Yeh, 2002), the ν(O—O) vibration of 6cO_2 L16A (expected in the ~1100 cm^{-1} region) was not enhanced.

5.2.4 Ferrous Complexes with Alkyl Isocyanides

Alkyl isocyanides (RNC) are neutral ligands that behave as electron-rich analogues of CO. The spectroscopic properties of 6c complexes of ferrous cytochromes *c'* with ethyl isocyanide (EIC) have been characterized in CVCP, RMCP, RPCP, RRCP and RSCP using UV–vis absorption (Patel et al., 1989; Rubinow & Kassner, 1984). For example, upon binding of EIC to RRCP, the split Soret band of the 5cFe(II) state (423, 432sh nm) changes to a single peak at 426 nm with a twofold increase in intensity, while the broad αβ envelope (~550 nm) resolves into separate peaks at 555 and 526 nm (Rubinow & Kassner, 1984; Table 4). Similar absorption trends were reported for EIC binding to other cytochromes *c′*, as well as for a range of additional alkyl isocyanide complexes of RMCP (Patel et al., 1989). The absorption properties of 6cEIC cytochromes *c′* are similar to analogous complexes reported for Mb and hemoglobin (Hb). A crystal structure of the *n*-butylisocyanide complex of CVCP, showing the coordination of the distal isocyano ligand, has been reported (Tahirov et al., 1996a; Section 4.3.1). Compared to the 6cEIC complexes of CVCP and RSCP, the higher molar absorptivities and smaller bandwidths of the RSCP, RRCP and RMCP complexes have been ascribed to more conformationally restricted distal haem environments (Rubinow & Kassner, 1984).

5.2.5 Ferric Complexes with Exogenous Ligands

Nitric oxide is unique among diatomic gases in being able to coordinate to haem in the Fe(III) as well as the Fe(II) state. Ferric 6cNO complexes, denoted $\{FeNO\}^6$ in the Enemark–Feltham notation (Enemark & Feltham, 1974), have been identified in a range of cytochromes *c′* based on their UV–vis absorbance spectra (Mayburd & Kassner, 2002; Yoshimura et al., 1987b, 1996) which resemble those of the isoelectronic ferrous 6cCO complexes. Compared to their ferrous 6cNO counterparts, ferric 6cNO complexes typically exhibit a red-shifted Soret band near 418 nm, as well as blue-shifted αβ bands near 565 and 530 nm (Table 4). Only the ferrous-nitrosyl state is observed when NO is added to ferric AXCP (Yoshimura et al., 1986), consistent with slow NO binding to the ferric AXCP haem, followed by rapid reductive nitrosylation due to its relatively high reduction potential.

Apart from NO, only a few other examples of Fe(III) cytochrome *c′* complexes with exogenous ligands have been reported. In the case of RCCP, imidazole binding to the Fe(III) state yields absorption bands at 409, 529 and 558 nm, typical of a 6c LS complex (Andrew et al., 2005; Table 4). Cyanide (CN^-) is the sole anionic ligand reported to bind to cytochrome *c′* (and only for CVCP). After several hours incubation with 0.1 M cyanide, Fe(III) CVCP forms a 6cCN complex as judged by the shift in the haem Soret band to 417 nm and the appearance of a broad αβ envelope at 538 nm (Kassner et al., 1985), resembling the absorption features of the cyano complexes of ferric Mb and Hb (Antonini & Brunori, 1971). Hobbs et al. reported porphyrin marker RR bands for ferric 6cCN CVCP: ν_4 (1372 cm^{-1}),

ν_3 (1508 cm^{-1}), ν_2 (1589 cm^{-1}), consistent with a 6c Fe(III) LS haem (Table 5), although no actual spectra were presented (Hobbs et al., 1990). The same authors also reported the formation of a ferrous 6cCN CVCP complex with porphyrin marker RR bands: ν_4 (1356 cm^{-1}), ν_3 (1471 cm^{-1}), ν_2 (1588 cm^{-1}). In this case, the relatively high ν_2 frequency is typical of a low spin Fe(II), although as with the ferric complex, no spectra were shown.

6. STRUCTURE–REACTIVITY RELATIONSHIPS IN CYTOCHROMES *C'*

The reactivities of native and variant cytochromes *c'* have been studied using a variety of kinetic, thermodynamic and electrochemical techniques. Trends in ligand-binding rate constants and binding affinities, together with other physicochemical parameters, have provided insights into reaction mechanisms as well as structure–reactivity relationships. In this section, we review kinetic studies on cytochrome *c'*, focusing on structural determinants of reversible ligand-binding reactions at the distal and proximal haem sites.

6.1 Distal Haem Coordination

Unlike other gas-binding haem proteins that bind a variety of anionic and neutral ligands, the hydrophobic and sterically crowded distal pockets of cytochrome *c'* favour small neutral ligands that do not require H-bond stabilization. Only two neutral physiological ligands (NO and CO) are able to bind to the distal sites of all cytochromes *c'*, although some proteins can also bind imidazole and alkylisocyanides. The only example of distal anion binding to cytochrome *c'* is the cyanide complex of CVCP. Mutations of haem pocket residues can also increase the range of ligands able to coordinate to the haem.

6.1.1 Reactivity of Cytochromes c' with Diatomic Gases

6.1.1.1 Ferrous 6cXO Complexes (X=N, C, O)

The conserved non-polar occluding residue in the distal pockets of native cytochromes *c'* imposes steric restrictions on the coordination of diatomic gases. Compared to Mb, haem-gas affinities of cytochrome *c'* 6cNO and 6cCO complexes are ~10^4-fold lower, while reaction with O_2 leads to autoxidation without a detectable 6cO_2 complex (Tables 12–14). Removal of distal steric constraints in AXCP via the Leu 16→Ala mutation boosts distal haem affinities for NO and CO by a remarkable 10^6–10^8-fold,

Table 12 Kinetic Parameters for 6c Fe(II) Haem Complexes with O_2

Haem	$k_{on}(O_2)$ (M^{-1} s^{-1})	$k_{off}(O_2)$ (s^{-1})	$K_d(O_2)$ (M)	$K_d(O_2)/K_d(CO)$	Distal H-Bond Donor	References
AXCP	~5000 (predicted)	~5000 (predicted)	~1–10 (predicted)[a]		None	Garton et al. (2012)
AXCP (L16A)	3.5×10^6	0.17	4.9×10^{-8}	1.5×10^4	None	Garton et al. (2012)
Mb (wt)	1.7×10^7	15	8.8×10^{-7}	24	His	Carver et al. (1992); Quillin et al. (1995)
Mb (H64L)	9.8×10^7	4100	4.2×10^{-5}	4.5×10^4	None	Springer, Sligar, Olson, and Phillips (1994)
sGC			~1–3 (predicted)[a]		None	Boon, Huang, and Marletta (2005)
Ns H-NOX			1.3×10^{-2}	9.3×10^3	None	Tsai et al. (2010)
Vc H-NOX			$\sim 1.3 \times 10^{-3}$ (predicted)[a]			Wu, Liu, Berka, and Tsai (2013)
Chelated protohaem	6.2×10^7	4200	6.7×10^{-5}	2.9×10^4	None	Tani et al. (2003)

[a] $K_d(O_2)$ estimated from K_d values for CO and NO and the sliding scale rule (Tsai, Berka, et al., 2012; Tsai, Martin, et al., 2012).

Table 13 Kinetic Parameters for 6c Fe(II) Haem Complexes with CO and Ethylisocyanide (EIC)

Haem	6cCO k_{on} (M^{-1} s^{-1})	6cCO k_{off} (s^{-1})	6cCO K_d[a] (M)	6cEIC k_{on} (M^{-1} s^{-1})	6cEIC k_{off} (s^{-1})[b]	6cEIC K_d (M)[c]	K_{dCO}/K_{dEIC}	References
AXCP (wt)	101	0.028	2.8×10^{-4}					Antonyuk et al. (2011)
	92	0.0125	1.4×10^{-4}					Cusanovich and Gibson (1973)
AXCP (L16A)	1.1×10^{6}	3.7×10^{-6}	3.4×10^{-12}					Antonyuk et al. (2011)
AXCP (L16G)	5.0×10^{5}	7.4×10^{-5}	1.5×10^{-11}					Antonyuk et al. (2011)
CVCP	140	0.0018	1.3×10^{-5}	32	0.0097	3.0×10^{-4}	23	Cusanovich and Gibson (1973); Kassner et al. (1983)
RCCP	142							Cusanovich and Gibson (1973)
RGCP	11	0.0053	4.8×10^{-4}					Cusanovich and Gibson (1973)
RMCP	440	0.010[b]	2.3×10^{-5}[c]	5	1.5	0.30	1.3×10^{4}	Doyle, Weber, and Gill (1985); Kassner (1991); Patel et al. (1989)
RPCP	334	0.0002	6.0×10^{-7}			0.45	7.5×10^{5}	Cusanovich and Gibson (1973); Rubinow and Kassner (1984)
RRCP	8.4	0.0051	6.1×10^{-4}			0.83	1.3×10^{3}	Cusanovich and Gibson (1973); Rubinow and Kassner (1984)

Continued

Table 13 Kinetic Parameters for 6c Fe(II) Haem Complexes with CO and Ethylisocyanide (EIC)—cont'd

	6cCO			6cEIC				
Haem	k_{on}(M^{-1} s^{-1})	k_{off} (s^{-1})	K_d(M)	k_{on}(M^{-1} s^{-1})	k_{off} (s^{-1})	K_d(M)	K_{dCO}/ K_{dEIC}	**References**
RSCP	2620/150		~10^{-5}–10^{-4}			2.6×10^{-4}	2.6–26	Cusanovich and Gibson (1973); Rubinow and Kassner (1984)
SFCP	54	0.015	2.8×10^{-4}					Manole et al. (2015)
Mb (wt)	5.1×10^5	0.019	3.7×10^{-8}	5.4×10^4	0.31	5.8×10^{-6}	160	Olson, Mims, and Reisberg (1982); Springer et al. (1994)
Mb (H64L)	2.6×10^7	0.024	9.2×10^{-10}					Springer et al. (1994)
sGC	4×10^4	10.7	2.6×10^{-4}					Boon et al. (2005), Martin, Berka, Bogatenkova, Murad, and Tsai (2006), Stone and Marletta (1996)
Ns H-NOX	3×10^6	3.6	1.4×10^{-6}					Tsai et al. (2010)
Vc H-NOX	1.1×10^6	0.85	7.7×10^{-7}					Wu et al. (2013)
Chelated protohaem	1.1×10^7	0.025	2.3×10^{-9}	1.9×10^8	1.2	6×10^{-9}	2.6	Olson, McKinnie, Mims, and White (1983); Tani et al. (2003)

[a] K_d values calculated from the ratio k_{off}/k_{on}, except for RMCP.
[b] k_{off} value calculated from the product $K_d \cdot k_{on}$.
[c] K_d values determined by titration.

Table 14 Kinetic Parameters for 6cNO and 5cNO Complexes

Haem	6cNO			5cNO			References
	k_{6on} ($\mu M^{-1}\,s^{-1}$)	k_{6off} (s^{-1})	K_d (μM)	k_{6-5}[a] ($\mu M^{-1}\,s^{-1}$)	k_{5on} ($\mu M^{-1}\,s^{-1}$)	k_{5off} (s^{-1})	
AXCP							
Native	0.043	0.006	0.14	0.011		*nd*[b]	Pixton et al. (2009)
R124K	0.065	*nd*		0.026		*nd*	Hough et al. (2011)
R124Q	0.040	*nd*		0.29		*nd*	Hough et al. (2011)
R124F	0.061	*nd*		0.59		*nd*	Hough et al. (2011)
R124A				$k_{6-5} \gg k_{6on}$[c]	0.032[c]	*nd*	Hough et al. (2011)
R124E				$k_{6-5} \gg k_{6on}$[c]	0.079[c]	*nd*	Hough et al. (2011)
D121A	0.029	*nd*		0.056		*nd*	Ghafoor et al. (2015)
D121I	0.033	*nd*		0.06		*nd*	Ghafoor et al. (2015)
D121Q	0.029	*nd*		0.06		*nd*	Ghafoor et al. (2015)
L16A	2.9	2×10^{-7}	7×10^{-8}	–	–	–	Garton et al. (2012)
SFCP				$k_{6-5} \gg k_{6on}$[c]	0.069[c]	9.2×10^{-5}	Manole et al. (2015)
NMCP				$k_{6-5} \gg k_{6on}$[c]	0.0019[c]	*nd*	Cole et al. (2008)

Continued

Table 14 Kinetic Parameters for 6cNO and 5cNO Complexes—cont'd

Haem	6cNO				5cNO		References
	k_{6on} ($\mu M^{-1}\,s^{-1}$)	k_{6off} (s^{-1})	K_d (μM)	k_{6-5} ($\mu M^{-1}\,s^{-1}$)	k_{5on} ($\mu M^{-1}\,s^{-1}$)	k_{5off} (s^{-1})	
Mb	22	1×10^{-4}	4.5×10^{-6}	*n.o.*			Quillin et al. (1995)
sGC	450	27	0.06	~1			Tsai, Berka, Sharina, and Martin (2011); Zhao et al. (1999)
*Vc*H-NOX	1100	0.3	2.7×10^{-4}	26		0.01 (16%) 0.001 (84%)	Wu et al. (2013)

[a]Rate-determining step for 6cNO → 5cNO conversion involving the formation of a putative dinitrosyl species.
[b]Observed rate is dependent on the concentration of dithionite NO scavenger.
[c]The rate-determining step of the monophasic Fe^{2+} → 5cNO reaction is assumed to be 6cNO formation followed by rapid conversion to the 5cNO state such that 6cNO → 5cNO conversion (k_{6-5}) is not observed (i.e. $k_{6-5} \gg k_{6on}$).
nd, not determined.

stabilises haem-O_2 coordination (without a distal H-bond donor) and abolishes 6cNO→5cNO conversion (Tables 12–14; Antonyuk et al., 2011; Garton et al., 2012). A similar increase in CO affinity was observed for the L16G variant. The fact that the L16A and L16G variants are isolated as 6cCO complexes is attributed to their picomolar haem-CO affinity, which could sequester low levels of CO produced by the *E. coli* expression culture (Antonyuk et al., 2011).

Kinetic studies reveal that the increased haem-gas affinities of the L16A (G) variants result from higher k_{on} values, coupled with lower k_{off} values (Tables 12–14). For example, k_{on} for 6cCO AXCP increases from 101 $M^{-1}\,s^{-1}$ (wt) to $1.1 \times 10^6\ M^{-1}\,s^{-1}$(L16A), while k_{off} decreases from $6.0 \times 10^{-3}\ s^{-1}$ (wt) to $2 \times 10^{-7}\ s^{-1}$(L16A). For the 6cNO AXCP complex, k_{on} increases from $4.3 \times 10^4\ M^{-1}\,s^{-1}$ (wt) to $2.9 \times 10^6\ M^{-1}\,s^{-1}$(L16A), while k_{off} decreases from $6.0 \times 10^{-3}\ s^{-1}$ (wt) to $2 \times 10^{-7}\ s^{-1}$(L16A) (Tables 13 and 14). Although a 6cO_2 complex is not observed for wt AXCP, the haem-O_2 affinity of L16A AXCP is actually higher than that of Mb due to a comparable k_{on} (3.5×10^6) combined with a significantly lower k_{off} ($0.17\ s^{-1}$) (Table 12). Whereas the relatively low K_{dO2}/K_{dCO} ~24 for Mb signifies H-bonding to the O_2 ligand, the much higher K_{dO_2}/K_{dCO} ~15,000 for L16A AXCP is consistent with an apolar distal haem environment. Assuming that the K_{dO_2}/K_{dCO} ratios of L16A and wt AXCP are similar, the wt AXCP K_{dCO} value (~0.28 mM) points to an extremely weak haem-O_2 affinity ($K_{dO_2} \geq 1$ M) with the sliding scale rule predicting values of k_{on} (~5000 $M^{-1}\,s^{-1}$) and k_{off} (~5000 s^{-1}) (Garton et al., 2012; Tsai, Berka, et al., 2012; Tsai, Martin, et al., 2012). The weak haem-O_2 affinity of wt AXCP explains the lack of a detectable haem-O_2 complex (solubility of 1 atm. O_2 ~ 1.3 mM) and is consistent with computational studies suggesting an unusually low O_2-binding energy (Martí et al., 2006).

6.1.1.2 Structural Determinants of Distal k_{on} Values

To account for the high on rates in L16A AXCP, Tsai and co-workers have proposed that the staggered conformation of the proximal histidine ring relative to the N(pyrrole)—Fe—N(pyrrole) axis (a structural factor associated with unusually high k_{on} values in soybean leghaemoglobin) enhances the intrinsic rates of the Fe—XO bond formation (Tsai, Berka, et al., 2012; Tsai, Martin, et al., 2012). While this effect should operate in both L16A and wt AXCP, k_{on} values are much lower in wt AXCP due to steric hindrance from the Leu 16 side chain. The ~10^4 decrease in the k_{on}(CO) value of wt AXCP relative to L16A has been attributed to the energy costs of conformational changes required for CO binding at the distal site (see Section 4) (Antonyuk et al., 2011). The recent X-ray crystal structures of 5cFe(II) and 6cCO wt AXCP show that steric clashes between Leu 16 and a near-linear (170°) Fe—C—O moiety force a 120° rotation of the Leu 16 Cα—Cβ bond together with a distal to proximal flip of the 7-propionate (Fig. 10). A second conformer with a bent Fe—C—O geometry (158°) is associated with smaller movements of Leu 16 and no propionate flip. Relative to the non-hindered L16A CO complex, where such conformational changes are not observed, the energetic cost of the CO-induced Leu propionate movements is reflected in a 600 mV (57.9 kJ mol^{-1}) decrease in haem potential, a value supported by DFT calculations (Antonyuk et al., 2011). No crystallographic or potentiometric data are yet available for 6cNO AXCP complexes. However, the fact that the k_{on}(NO) value for wt AXCP is ~10^2-fold lower than for L16A suggests that conformational changes triggered by the presence of Leu 16 also inhibit NO binding to wt AXCP, albeit to a lesser extent than for CO (which may be due to the inherently bent Fe—N—O geometry).

6.1.1.3 Structural Determinants of Distal k_{off} Values

Whereas removal of distal steric constraints in AXCP boosts k_{on} values, the opposite trend is observed for k_{off}. In other words, the native Leu16 residue of AXCP inhibits distal coordination of diatomic gases by decreasing k_{on} as well as increasing k_{off}. A possible explanation for why k_{off} values for wt AXCP 6cCO and 6cNO complexes are ~10^4 higher than those of L16A is provided by the 'loaded spring' mechanism (Fig. 16). In this model, which is based on recent crystallographic and potentiometric data for AXCP–CO binding (Antonyuk et al., 2011), distal steric constraints lead to distorted 6cXO structures with increased energies, slowing ligand binding (as the 'spring is primed') and speeding up ligand release (as the 'spring is released'). As well as accounting for the lower k_{on} values and higher k_{off} values of wt AXCP relative to the L16A variant (Tables 12–14), a loaded spring state for 6cXO complexes might also explain why distal steric constraints promote distal 6cNO → 5cNO conversion (*vide infra*). However, the loaded spring hypothesis has yet to be tested beyond the CO complexes of wt and L16A AXCP. Furthermore, other factors such as geminate recombination processes within the haem pocket are also likely to be important in determining k_{off} values (Russell et al., 2013; Yoo et al., 2013).

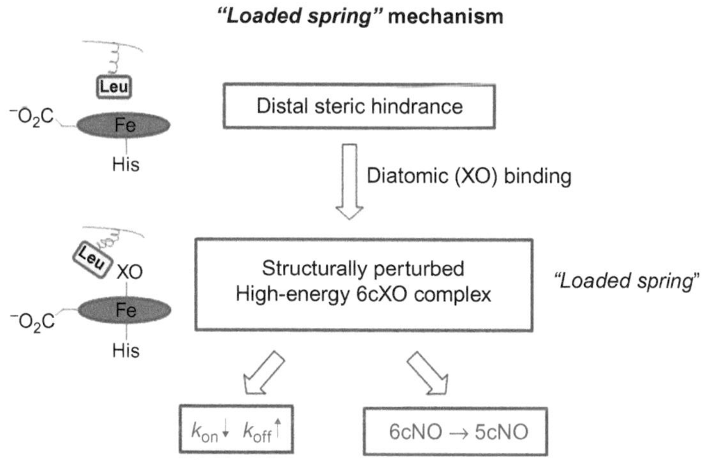

Figure 16 'Loaded spring' mechanism: diatomic XO binding to the crowded wt AXCP distal site yields a structurally perturbed high-energy 6cXO complex, altering the haem reactivity relative to the non-hindered Leu16Ala variant.

6.1.1.4 Ligand-Induced Dimer Dissociation

Intriguingly, the distal binding of CO (ferrous 6cCO complex), NO (ferric 6cNO complex) or CN^- (ferric 6cCN complex) to CVCP, which has a Tyr occluding residue in the distal haem pocket, has been shown to induce a dimer→monomer transition (de Groot et al., 2007; Doyle et al., 1986; Evers & Merkx, 2005; Mayburd & Kassner, 2002; Motie et al., 1991). By contrast, dimer dissociation does not appear to occur with the 5cNO CVCP complex (Mayburd & Kassner, 2002). Clues to the mechanism of dissociation were found from the 1.8 Å crystal structure of CVCP (Ren et al., 1993). Tyr 16 in the distal pocket stacks parallel to the haem in such a position that a conformational rearrangement of the residue would be necessary to allow ligand binding to the distal side of haem. CVCP forms an X-shaped dimer involving residues from helices A and B with a largely hydrophobic interface and held together by shape complementarity in contrast to other cytochrome c' dimers which feature more cross-dimer H-bond interactions. It was proposed that conformational change to Tyr 16, located within helix A, would be propagated to aromatic residues also within this helix that form the hydrophobic core of the dimer interface. Changes to these residues would drive the dissociation into monomers.

6.1.1.5 6cNO Complexes of Ferric Cytochromes c′

Ferric 6cNO haem complexes have been identified in CVCP, RCCP, RPCP and RRCP on the basis of their UV–vis absorption bands at pH

7.2 (Table 4; Mayburd & Kassner, 2002; Yoshimura et al., 1987b, 1996). By contrast, ferric AXCP reacts with NO to yield a ferrous-nitrosyl (5cNO) product without any observable ferric 6cNO species. This was attributed to reductive nitrosylation, facilitated by the higher reduction potential of AXCP (+90 to +132 mV) relative to the other four cytochromes c' (+3 to +105 mV) (Yoshimura et al., 1996). Although no rate constants have been reported, the formation of ferric 6cNO cytochromes c' requires at least 10 min to reach equilibrium in the presence of 0.03–2 mM NO (Mayburd & Kassner, 2002; Yoshimura et al., 1996)—significantly longer than for the formation of ferrous 6cNO AXCP (<5 s under similar conditions). This difference in reactivity is not due to solvent ligation since both the Fe(III) and Fe(II) redox states are pentacoordinate. Instead, slower NO binding to the Fe(III) state of cytochromes c' may stem from increased steric constraints due to the preference for a linear (rather than bent) Fe(III)—N—O geometry. Studies of ferric 6cNO formation in CVCP showed that the Y16F mutation appeared to significantly increase the ferric haem-NO affinity (Mayburd & Kassner, 2002). It was postulated that the higher NO affinity of the Y16F variant might arise from the loss of a H-bond between Tyr 16 and Trp 58 which might facilitate the distal pocket rearrangements required to bind NO.

In the presence of ~1 atm. NO, UV–vis spectra indicate that CVCP is fully converted to the ferric 6cNO state, whereas the content of 6cNO in the other cytochromes c' is estimated as 93% (RCCP), 87% (RPCP) and 50% (RRCP), with the remainder present as 5c Fe(III) haem (Yoshimura et al., 1996). It was proposed that the haem-NO affinity of ferric cytochromes c' is inversely correlated with the $S=3/2$ content of the 5c Fe(III) state, which in turn may be due to increased distances between the Fe and mean haem plane in the $S=3/2$ state (Yoshimura et al., 1996).

6.1.2 Reactivity of Cytochromes c′ with Bulky Distal Ligands

6.1.2.1 Alkyl Isocyanides

Ethyl isocyanide, C_2H_5NC (EIC), is a bulky analogue of CO that can bind to ferrous haem. Haem-EIC affinities have been measured for CVCP, RMCP, RPCP, RRCP and RSCP, while k_{on} values have also been determined for CVCP and RMCP (Table 13; Kassner et al., 1983; Patel et al., 1989; Rubinow & Kassner, 1984). Consistent with significant distal steric hindrance to haem-EIC coordination in cytochromes c', k_{on} values are ~10^7–10^8-fold lower than for unhindered protohaem (Olson et al., 1983) and ~10^3–10^4-fold lower than for Mb (Olson et al., 1982; Table 13).

The overall EIC affinities of cytochromes *c′* fall into two categories: those with K_{dEIC} values $\sim 3 \times 10^{-4}$ M (CVCP and RSCP) and the other with K_{dEIC} values in the range 0.3–0.8 M (RMCP, RPCP, RRCP) (Table 13). Cytochromes *c′* with higher EIC affinities (CVCP and RSCP) belong to the group 1 structural designation (Table 1) (characterized by a solvent channel from the protein surface to the distal pocket), whereas those with lower EIC affinities are belong to the group 2 structural class (Table 1; Kassner, 1991). The K_{dEIC}/K_{dCO} ratios of CVCP and RSCP resemble those of unrestricted protohaem, whereas the much greater K_{dEIC}/K_{dCO} ratios of RMCP, RPCP and RRCP suggest that distal haem coordination in these three cytochromes *c′* is much more inhibited by the increased steric bulk of EIC. The variation in binding affinities between the two categories of cytochromes *c′* diminishes as the size of ligand decreases (e.g. CO), consistent with smaller ligands experiencing less steric hindrance (Tables 12–14). Consistent with a more restricted haem-EIC conformation, Soret absorption bands of RMCP, RPCP and RRCP 6cEIC complexes exhibit smaller half bandwidths and higher intensities than those of CVCP and RSCP (Patel et al., 1989; Rubinow & Kassner, 1984). Moreover, the affinity of RMCP for EIC ($K_d \sim 0.30$ M) is less than that of methyl isocyanide ($K_d \sim 0.11$ M). Reactivity trends for larger alkyl groups are complicated by hydrophobic effects (Patel et al., 1989). Despite significantly different EIC affinities, it is notable that k_{on} values for CVCP (32 M^{-1} s^{-1}) and RMCP (5 M^{-1} s^{-1}) are quite similar (Table 13). While no other kinetic data have been reported for 6cEIC cytochromes *c′*, similar k_{on} values for CVCP and RMCP suggest that variations in K_{d6EIC} values may stem primarily from differences in k_{off} (rather than k_{on}).

6.1.2.2 Imidazole

Imidazole has also been used as a probe of steric accessibility in a variety of haem proteins (Mansy, Olson, Gonzalez, & Gilles-Gonzalez, 1998). Ferric RCCP forms a 6c haem-imidazole complex (6cIm) ($K_d \sim 43$ mM), whereas no imidazole binding was observed in the Fe(II) state (Andrew et al., 2005) By contrast, Fe(III) AXCP shows no detectable imidazole binding for imidazole concentrations up to 1 M. The higher haem-Im affinity of RCCP was taken as evidence of a more sterically accessible distal site, which in turn has been correlated with a higher distal 6cNO: proximal 5cNO equilibrium ratio than in AXCP (Section 4; Andrew et al., 2005). The K_d value for the 6cIm complex of RCCP is similar to that of Mb ($K_d \sim 28$ mM)

(Mansy et al., 1998), although comparison of Fe(III) haem reactivities is complicated by the presence of a distal solvent ligand in met Mb.

6.1.3 Reactivity of Cytochromes c′ with Anionic Distal Ligands

The only example of distal anion binding to cytochrome *c*′ is the cyanide complex of CVCP (Kassner et al., 1985; Motie, Kassner, Meyer, & Cusanovich, 1990). At neutral pH, the binding affinity of the ferric CVCP 6cCN complex ($K_d \sim 7.7 \times 10^{-3}$ M) is $\sim 10^4$-fold lower than that of Mb ($K_d \sim 7.1 \times 10^{-7}$ M) (Antonini & Brunori, 1971; Motie et al., 1990). Kinetic measurements attribute the lower cyanide affinity of CVCP to an unusually low on rate ($k_{on} = 0.0023$ M^{-1} s^{-1}), which is five orders of magnitude lower than that of Mb ($k_{on} = 170$ M^{-1} s^{-1}) (Antonini & Brunori, 1971). The slower reactivity of CVCP with cyanide may be due in part to unfavourable steric interactions with the linear Fe(III)—CN unit, as well as to the hydrophobic distal haem environment. At higher pH values, Fe(III) CVCP exhibits a non-linear dependence of observed rate constants on cyanide concentration, attributed to the formation of a H-bonded complex between the cyanide anion and the distal occluding Tyr residue prior to Fe(III)—CN bond formation (Motie et al., 1990).

6.1.4 Conformational Control of Distal Ligand Binding in SFCP

Bimolecular binding of NO or CO to SFCP appears monophasic when measured by stopped-flow kinetics, but biphasic when measured using flash photolysis (Manole et al., 2015). Stopped-flow data reveal conversion from the 5c ferrous state to a proximal 5cNO species without a detectable 6cNO intermediate, as previously observed in AXCP variants with accelerated distal to proximal conversion (Hough et al., 2011). The k_{on} value for SFCP (6.89×10^3 M^{-1} s^{-1}) is significantly lower than that of AXCP consistent with a higher level of distal steric hindrance. By contrast, flash photolysis data reveal fast (1.42×10^5 M^{-1} s^{-1}) and slow (3.82×10^3 M^{-1} s^{-1}) phases. Similarly, CO binds to the distal face of SFCP to form a 6cCO complex with kinetic parameters from stopped-flow data of $k_{on} = 53.6$ M^{-1} s^{-1} and $k_{off} = 1.46 \times 10^{-2}$ s^{-1}. For this ligand, flash photolysis data again reveal fast and slow phases of bimolecular CO rebinding. These data strongly suggest a conformational element in the mechanism, with only one of two conformations of the distal Phe residue permitting distal ligand binding (Phe accepting), while a second conformation blocks binding (Phe occluding).

The 'accepting' conformation was proposed to be adopted by Phe in only ~1% of the protein molecules at equilibrium, thus strongly affecting the observed kinetics. Whether such kinetics are in effect in other cytochromes c' with aromatic distal ligands remains uncertain as few have been characterized by fast kinetic methods.

6.2 Proximal 5cNO Formation

Stopped-flow absorption measurements of NO binding to AXCP reveal a biphasic reaction in which 6cNO formation ($k_{6on} = 4.3 \times 10^4$ M^{-1} s^{-1}) is followed by its conversion to a 5cNO product ($k_{6\text{–}5} = 1.1 \times 10^4$ M^{-1} s^{-1}) (Andrew et al., 2002; Pixton et al., 2009). The NO dependences of both phases suggest that proximal 5cNO formation requires two NO molecules and a series of distal haem-NO intermediates, including 6cNO and putative distal 5cNO and dinitrosyl species (Fig. 17; Andrew et al., 2002; Hough et al., 2011; Pixton et al., 2009). Measurements of the entropy of activation ($\Delta S^{\ddagger}$), enthalpy of activation ($\Delta H^{\ddagger}$) and volume of activation ($\Delta V^{\ddagger}$) provide more detail on the mechanisms of distal and proximal NO-binding reactions of AXCP (Pixton et al., 2009). These activation parameters describe changes in the properties of reactants on going from the ground state to the transition state and therefore relate to the elementary step that is rate determining. The near-zero $\Delta S^{\ddagger}$ and a positive $\Delta V^{\ddagger}$ (+14.1 cm^3 mol^{-1}) of distal 6cNO formation is consistent with movement of Leu 16 residue to allow NO binding to the crowded distal site. For the 6cNO → 5cNO reaction, the large

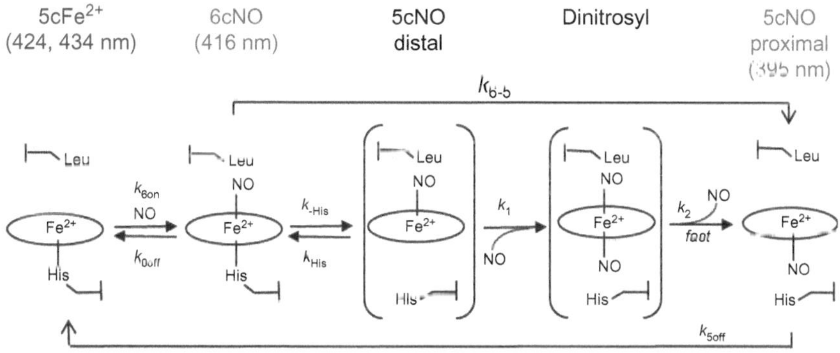

Figure 17 Mechanism for proximal 5cNO formation in AXCP. Absorption maxima of observed haem species are shown in parentheses.

positive $\Delta S^{\ddagger}$ (+103 J K^{-1} mol^{-1}) and $\Delta V^{\ddagger}$ (+24.1 cm^3 mol^{-1}) suggest that the putative dinitrosyl intermediate forms via a dissociative mechanism in which the endogenous His ligand dissociates *prior to the attack* of the second NO molecule on the proximal haem face (Fig. 17). The NO dependence of the 6cNO → 5cNO reaction is ascribed to proximal NO binding competing with recombination of the proximal His ligand.

This unexpected distal → proximal haem-NO conversion has expanded the repertoire of haem-NO reaction mechanisms in proteins (Pixton et al., 2009), with distal vs. proximal 5cNO coordination emerging as a strategy for selective NO response. Indeed, very recently, a proximal 5cNO binding mode was confirmed in the crystal structure of H-NOX from *S. oneidensis* (*So* H-NOX) (Herzik et al., 2014). Mechanistic studies also support distal → proximal 5cNO conversion in sGC (Martin et al., 2012; Tsai, Berka, et al., 2012; Tsai, Martin, et al., 2012), the pro-apoptotic cytochrome *c*/cardiolipin complex (Silkstone, Kapetanaki, Husu, Vos, & Wilson, 2010) and cystathionine β-synthase (Vicente et al., 2014)

6.2.1 Determinants of Proximal 5cNO Formation

Because of the negative σ-*trans* effect exhibited by ferrous haem-NO complexes, the (His) ligand opposite to NO is weakened, promoting Fe–His scission and 6cNO → 5cNO conversion (Goodrich et al., 2010; Hunt & Lehnert, 2015; Lehnert et al., 2010; Traylor & Sharma, 1992; Tsai, 1994). Despite this, the majority of ferrous-nitrosyl haem proteins remain 6cNO due to the 'cage effect' of the surrounding protein structure. Nevertheless, 5cNO formation is observed in some haem proteins, including cytochromes *c*′, as well as T-state Hb α subunits (Chan, Kavanaugh, Rogers, & Arnone, 2004), the CO-sensing transcriptional activator, CooA (Aono, 2003), the O_2-sensing histidine kinase, FixL (Rodgers, Lukat-Rodgers, & Tang, 2000), the circadian-coupled transcriptional regulator, CLOCK (Lukat-Rodgers, Correia, Botuyan, Mer, & Rodgers, 2010), the dissimilative nitrate respiration regulator, DNR (Cutruzzolà et al., 2014), cystathionine β synthase (Vicente et al., 2014), the cytochrome *c*/cardiolipin complex (Silkstone et al., 2010), sGC (Zhao et al., 1999) and certain bacterial haem nitric oxide/oxygen-binding (H-NOX) gas sensors (Boon et al., 2006; Herzik et al., 2014; Tsai, Berka, et al., 2012; Tsai, Martin, et al., 2012; Wu et al., 2013).

Formation of a 5cNO (rather than a 6cNO) complex constitutes a mechanism for controlling haem-NO reactivity and response (Tsai, Berka, et al., 2012; Tsai, Martin, et al., 2012), while allosteric effects triggered by 5cNO formation are believed to underpin the NO-sensing and signal transduction mechanisms of sGC and other haem-NO sensors (Underbakke et al., 2014; Yoo, Lamarre, Martin, Rappaport, & Negrerie, 2015). While the location of the NO ligand (distal or proximal) has yet to be established for most 5cNO proteins, the observation of proximal 5cNO coordination in crystal structures of cytochromes *c′* and *So* H—NOX has expanded the repertoire of haem-NO reaction mechanisms, including the factors responsible for 6cNO → 5cNO conversion. Although the protein architectures of cytochromes *c′* differ from those of other 5cNO proteins, the factors controlling proximal 5cNO formation provide insights into haem-NO-sensing strategies. Determinants of proximal 5cNO formation in cytochromes *c′* have been investigated via structural, spectroscopic and kinetic studies of native and variant proteins.

6.2.1.1 Fe—His Bond Scission

Cleavage of the Fe—His bond is a requirement for 5cNO formation, occurring subsequent to 6cNO formation. Activation parameters for NO binding to AXCP suggest that Fe—His bond scission is the rate-determining step for 6cNO → 5cNO conversion (k_{6-5} in Fig. 17), although the rate of this process also depends on NO concentration since the binding of a second (proximal) NO must compete with His reattachment to form the proximal 5cNO species (Fig. 17; Pixton et al., 2009). The conventional view is that 6cNO → 5cNO conversion in haem proteins requires an *inherently weak Fe—His bond* to enable the negative haem-NO *trans* effect to dissociate the proximal His ligand (Scheidt & Ellison, 1999; Traylor & Sharma, 1992). However, cytochromes *c′* have relatively high ν(Fe—His), RR frequencies (~230 cm^{-1}) in their 5cFe(II) state that are typical of haem nitrosyls that remain 6cNO due to a strong Fe—N(His) bond (see Section 5.1.2) (Andrew et al., 2001; Garton et al., 2012). However, it should be borne in mind that ν(Fe—His) values measured by RR spectroscopy are for the 5cFe(II) state, rather than for 6cNO. This leaves open the possibility that conformational changes (occurring upon NO binding to sterically constrained cytochrome *c′* distal sites) act in concert with the negative *trans* effect to further weaken the Fe—His bond of the 6cNO state to facilitate 5cNO formation.

Because the proximal His ligand is part of the CXXCH motif required for *c*-haem attachment by the *E. coli* cytochrome *c* maturation (Ccm) apparatus (Allen, Leach, & Ferguson, 2005), the role of the His ligand has not been probed directly by mutation (His-ligand variants of holocytochromes

c' cannot be expressed). However, mutagenesis studies of AXCP and RSCP have shown that replacement of the basic Arg/Lys residue adjacent to the His ligand promotes 5cNO formation (Barbieri et al., 2008; Hough et al., 2011; Lee et al., 2009). Stopped-flow studies of NO binding to AXCP Arg 124 variants show that removal of positive charge at position 124 leads to faster 6cNO → 5cNO formation ($k_{6\text{–}5}$) (Table 14). This is apparent in absorbance time traces at 415 nm (Fig. 18) which show less buildup of transient 6cNO species as $k_{6\text{–}5}$ increases relative to $k_{6\text{on}}$ (Table 14). In other words, the native proximal Arg residue in AXCP prolongs the lifetime of the 6cNO complex by lowering the rate of Fe—His bond cleavage. At the same time, RR spectra of R124 AXCP variants in their 5cFe(II) states show only minor variations in the ν(Fe—His) frequency (232–235 cm^{-1}) (Table 7), implying that the changes in $k_{6\text{–}5}$ do not stem from altered Fe—His bond strengths (Hough et al., 2011). It was suggested that Fe—His bond cleavage in AXCP is hindered by basic residues due to a pi–cation interaction between the imidazole ring and the adjacent Arg/Lys side chain (Hough et al., 2011). However, in the case of SFCP (which has a proximal

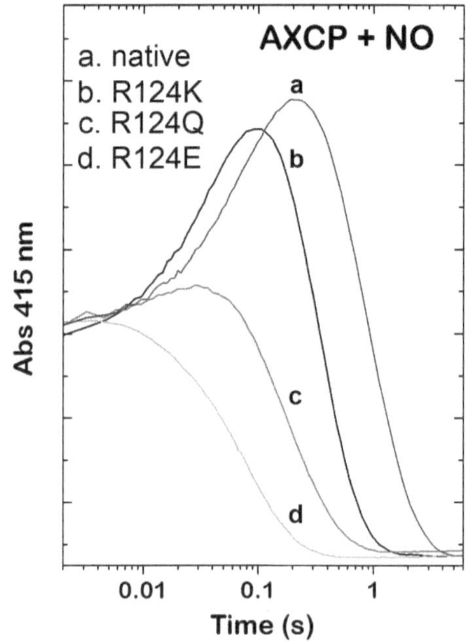

Figure 18 Absorption time traces of the reaction of 5cFe(II) AXCP with NO (190 μM), showing the effect of R124 mutations.

Lys residue), 5cNO formation occurs without a detectable 6cNO intermediate, suggesting that k_{6-5} is not lowered to the same extent as in AXCP (Manole et al., 2015).

Theoretical QM/MM studies by Martí et al. suggested that 5cNO formation in AXCP is driven by an energy minimum for the dissociated His ligand within the proximal pocket (Martí et al., 2005). However, recent studies showed that proximal 5cNO coordination in AXCP is retained upon mutation of Asp 121 which H-bonds to the displaced His ligand in the proximal 5cNO complex (Ghafoor et al., 2015).

6.2.1.2 Rapid Proximal NO Binding

While Fe–His cleavage is the rate-determining step in distal 6cNO → proximal 5cNO conversion (k_{6-5}) (Fig. 17), the value of k_{6-5} also depends on NO concentration since binding of a second (proximal) NO must compete with His reattachment to form the proximal 5cNO species (Fig. 17; Pixton et al., 2009). Accordingly, structural features that enhance the rate of bimolecular NO binding at the proximal site should be a requisite for proximal 5cNO formation. The proximal haem pockets of cytochromes c' are solvent exposed, allowing rapid proximal NO binding upon dissociation of the His ligand to promote proximal 5cNO formation. On the other hand, the formation of a distal 5cNO population in R124A AXCP has been attributed in part to the conformation of the detached His ligand which may hinder bimolecular binding of NO to the proximal site (Hough et al., 2011).

6.2.1.3 Steric Hindrance of Distal NO Binding

According to the reaction mechanism in Fig. 17, NO binding to the proximal site yields a transient dinitrosyl which then decays via the preferential release of the distal (rather than proximal) NO ligand to form the proximal 5cNO product. Indeed, QM–MM calculations on the putative dinitrosyl precursor predict that proximal 5cNO formation is favoured due to a lower energy barrier for release of distal NO relative to proximal NO (Martí et al., 2005). Accordingly, structural features that promote distal NO release, while disfavouring distal NO binding, should promote proximal 5cNO formation. By trapping the proximal 5cNO species in this manner, a 5cNO product can be formed even if Fe—His scission is relatively unfavourable. Although the dinitrosyl species itself has not been characterized, reactivity trends for other 6c distal complexes of cytochromes c' strongly suggest that steric constraints (from the distal occluding residue) do indeed lower the distal ligand affinity by increasing $k_{6\text{off}}$ while decreasing $k_{6\text{on}}$.

The higher imidazole affinity of ferric RCCP ($K_d \sim 43$ mM) relative to AXCP (no observable imidazole binding at concentrations < 1 M) indicates that RCCP has less steric hindrance to distal ligand binding, which in turn correlates with RCCP forming a significant population (~40%) of 6cNO product (Andrew et al., 2005). In RCCP (and from difference spectra also in RSCP (Choi et al., 2005)), an equilibrium mixture of 6c distal and 5c proximal modes is known, while in SFCP only the 5c proximal form is observed in static experiments. As described in Section 4, superposition of crystal structures of ferric cytochromes *c′* with distal Phe residues shows that in SFCP and RSCP, the Phe residue is closer to the haem Fe than in RCCP and RSCP. It is conceivable that a lower level of steric hindrance in RCCP could promote retention of the distal 6cNO form.

More striking are the effects of the distal Leu 16→Ala mutation in AXCP, which traps the 6cNO complex and boosts distal haem-NO affinity by a factor of $\sim 10^8$ (due to a 10^4-fold decrease in $k_{6\text{off}}$ and a 10^4-increase in $k_{6\text{on}}$) (Table 14; Garton et al., 2012). Similarly, $\sim 10^8$-fold increases in distal ligand affinity are observed for the L16A 6cCO (Antonyuk et al., 2011) and $6cO_2$ complexes (in the latter case, the extremely low O_2 affinity predicted for wt AXCP, $K_d \sim 1$ M, precludes its observation) (Garton et al., 2012). A $\sim 10^8$-fold increase in CO affinity was also reported for the L16G variant (Antonyuk et al., 2011). The behaviour of the L16A and L16G variants confirmed earlier predictions by Martí et al., based on QM–MM calculations, that Leu 16 should have a destabilizing effect on distal NO and O_2 coordination (Martí et al., 2005, 2006). According to the recently proposed 'loaded spring' mechanism (see Section 6.1.1) (Antonyuk et al., 2011), distal steric constraints in wt AXCP increase the energy state of 6cXO complexes relative to L16A, slowing ligand binding (as the 'spring is primed') and speeding up ligand release (as the 'spring is released'). The rate of distal ligand release may also be affected by the efficiency of geminate rebinding within the distal pocket.

The only other AXCP distal variant to have been studied is F59Y (Barbieri et al., 2008). This residue is part of the hydrophobic surface of the distal pocket and mutation to a polar Tyr sought to clarify the role of this residue in modulating NO binding to the distal face of haem. While no structural data are available, the F59Y mutant exhibited a faster formation of the 5cNO product such that the 6cNO distal intermediate was not observed in an NO titration. This result implies that the polar Tyr residue provides additional steric hindrance to distal NO binding or that its polar nature acts to discourage NO binding to Fe. A brief report of UV–vis properties of the distal F14A variant of RCCP (Cole et al., 2008) indicated that the ferrous-nitrosyl complex has λ_{max} values similar to those of native RCCP, suggesting that this mutation does not influence the 6cNO/5cNO equilibrium ratio. Curiously, F14A RCCP exhibits an atypical ferrous cytochrome c' absorption spectrum characteristic of bis axial Fe(II) haem coordination, suggesting that an undefined distal ligand is present in the absence of NO. However, no other data or discussion was presented.

6.2.1.4 Stability of the Proximal 5cNO Complex

It was proposed that proximal 5cNO complexes of AXCP and RSCP might be stabilized by H-bonding to the conserved Arg/Lys that lies adjacent to the NO ligand (Lawson et al., 2000; Lee et al., 2009). However, subsequent high-resolution crystal structures of wt AXCP and a selection of Arg 124 variants revealed that H-bonding is not required for proximal 5cNO coordination (see Section 4.3.2) (Hough et al., 2011). Recent crystallographic studies of 5cNO complexes of AXCP Asp 121 variants also showed that the H-bond between Asp 121 and the displaced His ligand is also not required for 5cNO formation (Ghafoor et al., 2015). Although measurement of the k_{off} value for 5cNO AXCP is complicated by side reactions with the dithionite NO scavenger, flash photolysis studies of 5cNO AXCP show that for proximal NO to be released it must overcome geminate recombination that is both ultrafast ($\tau \sim 7$ ps) and highly efficient (~99%) (Kruglik et al., 2007; Yoo et al., 2013). Molecular dynamics calculations simulated the trajectory of dissociated NO within the proximal pocket over a 50 ps timescale (Kruglik et al., 2007). It was found that NO adopted a stable position parallel to haem and interacted with Cys 119, His 120 and Arg 124, with the largest energy of interaction (6.78 kcal mol^{-1}) being with Arg 124. From this analysis, it was proposed that Arg 124 provides both a steric and electrostatic barrier to ligand escape from the pocket following photolysis of the Fe–NO bond. The simulation suggested that His 120 could not rebind to Fe while NO remained within the proximal pocket. Consistent with this proposal, a subsequent ultrafast laser flash photolysis kinetic analysis of the AXCP R124A variant showed a greatly increased escape of photodissociated proximal NO in the variant (Russell et al., 2013).

A 'kinetic trap mechanism' for the release of proximal NO was proposed in which rapid reattachment of the His ligand acts as a one-way gate for releasing proximal NO by precluding bimolecular rebinding of proximal NO after it exits the haem pocket (Andrew et al., 2003). Subsequent flash photolysis studies on the picosecond timescale detected the reattachment of the His ligand to 4c haem ($\tau \sim 100$ ps) in photolysed 5cNO AXCP, supporting the feasibility of the His kinetic trap mechanism as a means of preventing irreversible proximal 5cNO formation (Yoo et al., 2013).

6.2.1.5 Long-Range Effects

A K42E variant of RCCP was generated (Huston et al., 2006) to remove a salt bridge at the dimer interface (K42-E39). While the K42E variant showed near-identical CO-binding properties to those of native RCCP, the equilibrium mixture of 6cNO and 5cNO species that exists in native RCCP was converted to an all-5cNO state in the K42E variant. This is a remarkable result given that the haem Fe and residue K42 are some 19 Å apart and argues for long-range communication through the RCCP structure via transmission of conformational changes from the mutation site. It was noted that K42 lies in close proximity to the deep channel between helices B and C that has been proposed to allow ligand access to the distal haem pocket.

7. RELEVANCE OF CYTOCHROME *C′* TO OTHER PROTEINS, INCLUDING HAEM-BASED GAS SENSORS

Proximal 5cNO binding to haem was first identified in AXCP (Andrew et al., 2001; Lawson et al., 2000) and parallels immediately drawn to the multistep NO-activation mechanism of sGC. In addition to AXCP, proximal 5cNO complexes have recently been structurally characterized in SFCP (*vide infra*; Manole et al., 2015), as well as in *So* H-NOX (Herzik et al., 2014). Although a crystal structure of NO binding to sGC has not yet been reported, sequential stopped-flow and freeze-quench EPR experiments with isotopically labelled NO have confirmed both the presence of a proximal 5cNO complex and the occurrence of distal 6cNO → proximal 5cNO conversion with excess NO (Martin et al., 2012; Tsai, Berka, et al., 2012; Tsai, Martin, et al., 2012). Kinetic data have also revealed NO-dependent 6cNO → 5cNO conversion (consistent with proximal 5cNO formation via a transient dinitrosyl) in H-NOX from *Vibrio cholerae* (*Vc* H-NOX) (Wu et al., 2013), the cardiolipin–cytochrome *c* complex (Hough, Silkstone, Worrall, & Wilson, 2014; Silkstone et al., 2010) and cystathionine β-synthase (Vicente et al., 2014) although we note that structural data are not yet available for these complexes.

It is instructive to compare the recent 1.65 Å resolution crystal structure of the *So* H-NOX proximal 5cNO complex (Herzik et al., 2014) with the structures of proximal NO complexes in cytochromes *c*′. Both cytochrome *c*′ and *So* H-NOX have non-polar distal haem pockets made up of many hydrophobic residues. In the structure of 5cNO *So* H-NOX, the proximal His ligand has been displaced into the proximal pocket to now lie ~8.5 Å from the Fe, in a manner analogous to that occurring in cytochrome *c*′. NO at the proximal face is oriented between the A and B haem pyrroles with an Fe—N bond length of 1.8 Å and Fe—N—O angle of 124°, comparable to that of the 5cNO AXCP and SFCP complexes.

However, there are also many structural differences between the cytochrome *c*′ and *So* H-NOX proximal NO complexes. Whereas the proximal pockets of cytochromes *c*′ are solvent exposed and contain many polar or charged residues, the proximal pocket in *So* H-NOX is largely hydrophobic with the closest water molecule to NO being some 6.6 Å away. A superposition of the two haem environments is given in Fig. 19.

In *So* H-NOX, significant conformational changes occur on NO binding. In the ferrous form of the protein, prior to NO binding, nearby protein residues force the haem into a distorted geometry with a weak Fe–His interaction. Upon scission of the Fe—His bond and formation of the 5cNO proximal complex, the haem is able to relax into a near-planar geometry that

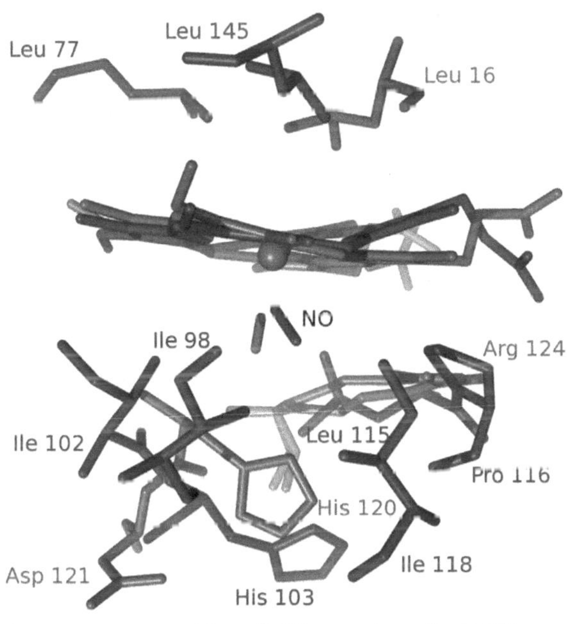

Figure 19 Least squares superposition of all haem atoms for 5cNO complexes in AXCP (blue; grey in the print version) and *So* H-NOX (red; dark grey in the print version).

is proposed to represent the active form of the sensor. Notably in *So* H-NOX, CO binds to the distal haem face with only a small movement of Leu 77 (analogous to Leu 16 in AXCP) occurring and with a near-linear Fe—C—O geometry (176°). This result suggests that, in contrast to cytochromes *c'*, distal steric crowding may not be a determining factor for proximal NO formation in *So* H-NOX. It is noted that k_{on} values for the 6cNO intermediates of the related *VcH*-NOX and sGC proteins are relatively high (Table 14), consistent with a low degree of distal steric hindrance. On the other hand, k_{off} values for the distal 6cNO species in these proteins are also relatively high, such that the putative dinitrosyl species may decompose by preferential release of distal NO in a manner analogous to that proposed for cytochrome *c'*. It is further noted that sGC exhibits an usually high k_{off} value (10.7 s^{-1}) for its distal 6cCO complex (Table 13). It has been proposed that in sGC, CO release is enhanced by a relatively weak Fe—CO bond (evidenced by the low Fe–CO stretching frequency of 473 cm^{-1}), which in turn might be caused by proximal strain and/or extensive H-bonding to the haem propionates (Ibrahim, Derbyshire, Marletta, & Spiro, 2010; Makino et al., 2011).

Despite many efforts, no structure is yet available for the haem domain of sGC itself and so the structural features that may promote proximal NO binding and the conformational changes that transmit the downstream signal following scission of the Fe—His bond remain unclear. Examination of the haem pockets in the other proteins where proximal 5cNO complexes have been proposed reveals a low level of structural similarity. In mitochondrial cytochrome *c*, the proximal haem ligand is involved in a H-bond to the carbonyl of Pro 30, which would limit its ability to dissociate from haem to allow proximal NO binding (Hough et al., 2014; Rajagopal et al., 2013). A large conformational change in the cardiolipin–cytochrome *c* complex must therefore occur in order to facilitate this proximal mode of NO binding. Future crystallographic and kinetic studies should more light on the structural determinants and ramifications of proximal haem-NO coordination in cytochromes *c'* and other haem proteins.

ACKNOWLEDGEMENTS

The authors gratefully acknowledge the many students, group members and collaborators who contributed to these studies over many years, as well as the financial support provided by National Science Foundation Grants MCB-0417152,MCB-0745035 and MCB-1411963 (to C.R.A.) and Royal Society Research Project Grant 120094 (to M.A.H.)

REFERENCES

Allen, J. W. A., Leach, N., & Ferguson, S. J. (2005). The histidine of the c-type cytochrome CXXCH haem-binding motif is essential for haem attachment by the Escherichia coli cytochrome c maturation (Ccm) apparatus. *Biochemical Journal, 389*, 587–592.

Ambler, R. P. (1991). Sequence variability in bacterial cytochromes c. *Biochimica et Biophysica Acta, 1058*(1), 42–47.

Ambler, R. P., Bartsch, R. G., Daniel, M., Kamen, M. D., McLellan, L., Meyer, T. E., et al. (1981). Amino acid sequences of bacterial cytochromes *c'* and *c-556*. *Proceedings of the National Academy of Sciences of the United States of America, 78*, 6854–6857.

Andrew, C. R., George, S. J., Lawson, D. M., & Eady, R. R. (2002). Six- to five-coordinate heme-nitrosyl conversion in cytochrome c' and its relevance to guanylate cyclase. *Biochemistry, 41*(7), 2353–2360.

Andrew, C. R., Green, E. L., Lawson, D. M., & Eady, R. R. (2001). Resonance Raman studies of cytochrome c' support the binding of NO and CO to opposite sides of the heme: Implications for ligand discrimination in heme-based sensors. *Biochemistry, 40*(13), 4115–4122.

Andrew, C. R., Kemper, L. J., Busche, T. L., Tiwari, A. M., Kecskes, M. C., Stafford, J. M., et al. (2005). Accessibility of the distal heme face, rather than Fe-His bond strength, determines the heme-nitrosyl coordination number of cytochromes c': Evidence from spectroscopic studies. *Biochemistry, 44*(24), 8664–8672.

Andrew, C. R., Rodgers, K. R., & Eady, R. R. (2003). A novel kinetic trap for NO release from cytochrome c': A possible mechanism for NO release from activated soluble guanylate cyclase. *Journal of the American Chemical Society, 125*(32), 9548–9549.

Antonini, E., & Brunori, M. (1971). *Hemoglobin and myoglobin in their reactions with ligands*, (Vol. 21)Amsterdam: North-Holland.

Antonyuk, S. V., Rustage, N., Petersen, C. A., Arnst, J. L., Heyes, D. J., Sharma, R., et al. (2011). Carbon monoxide poisoning is prevented by the energy costs of conformational changes in gas-binding haemproteins. *Proceedings of the National Academy of Sciences of the United States of America, 108*(38), 15780–15785.

Aono, S. (2003). Biochemical and biophysical properties of the CO-sensing transcriptional activator CooA. *Accounts of Chemical Research, 36*(11), 825–831.

Arslan, E., Schilz, H., Zufferey, R., Künzler, P., & Thöny-Meyer, L. (1998). Overproduction of the *Bradyrhizobium japonicum c*-type cytochrome subunits of the *cbb_3*-oxidase in *Escherichia coli*. *Biochemical and Biophysical Research Communications, 251*, 744–747.

Ascenzi, P., Santucci, R., Coletta, M., & Polticelli, F. (2010). Cytochromes: Reactivity of the "dark side" of the heme. *Biophysical Chemistry, 152*, 21–27.

Barbieri, S., Murphy, L. M., Sawers, R. G., Eady, R. R., & Hasnain, S. S. (2008). Modulation of NO binding to cytochrome c' by distal and proximal haem pocket residues. *Journal of Biological Inorganic Chemistry, 13*(4), 531–540.

Barker, S. L., Kopelman, R., Meyer, T. E., & Cusanovich, M. A. (1998). Fiber-optic nitric oxide-selective biosensors and nanosensors. *Analytical Chemistry, 70*, 971–976.

Bartsch, R. G. (1963). Non heme iron proteins and Chromatium iron protein. In H. Gest, A. San Pietro, & L. P. Vernon (Eds.), *Bacterial photosynthesis* (pp. 315–326). Yellow Springs, OH: Antioch Press.

Bartsch, R. G. (1978). Cytochromes. In R. K. Clayton & W. R. Sistrom (Eds.), *The photosynthetic bacteria* (pp. 249–279). New York: Plenum Press.

Benini, S., Rypniewski, W. R., Wilson, K. S., & Ciurli, S. (2008). High resolution crystal structure of *Rubrivivax gelatinosus* cytochrome c'. *Journal of Inorganic Biochemistry, 102*(5–6), 1322–1328.

Benko, B., & Yu, N. T. (1983). Resonance Raman studies of nitric oxide binding to ferric and ferrous hemoproteins: Detection of Fe(III)–NO stretching, Fe(III)–N–O bending, and Fe(II)–N–O bending vibrations. *Proceedings of the National Academy of Sciences of the United States of America, 80*(22), 7042–7046.

Berry, M. J., George, S. J., Thomson, A. J., Santos, H., & Turner, D. L. (1990). Cytochrome c″ isolated from *Methylophilus methylotrophus.* An example of bis-histidine-co-ordinated Fe3+ haem, with near-perpendicular orientation of the ligands. *Biochemical Journal, 270*, 413–417.

Boon, E. M., Davis, J. H., Tran, R., Karow, D. S., Huang, S. H., Pan, D., et al. (2006). Nitric oxide binding to prokaryotic homologs of the soluble guanylate cyclase beta1 H-NOX domain. *Journal of Biological Chemistry, 281*(31), 21892–21902.

Boon, E. M., Huang, S. H., & Marletta, M. A. (2005). A molecular basis for NO selectivity in soluble guanylate cyclase. *Nature Chemical Biology, 1*(1), 53–59.

Carver, T. E., Brantley, R. E. J., Singleton, E. W., Arduini, R. M., Quillin, M. L., Phillips, G. N., et al. (1992). A novel site-directed mutant of myoglobin with an unusually high O_2 affinity and low autooxidation rate. *Journal of Biological Chemistry, 267*, 14443–14450.

Chan, N.-L., Kavanaugh, J. S., Rogers, P. H., & Arnone, A. (2004). Crystallographic analysis of the interaction of nitric oxide with quaternary-T human hemoglobin. *Biochemistry, 43*, 118–132.

Choi, P. S., Grigoryants, V. M., Abruna, H. D., Scholes, C. P., & Shapleigh, J. P. (2005). Regulation and function of cytochrome c' in *Rhodobacter sphaeroides 2.4.3. Journal of Bacteriology, 187*(12), 4077–4085.

Cole, L. J., Huston, W. M., & Moir, J. W. (2008). Delivery of nitric oxide for analysis of the function of cytochrome c'. *Methods in Enzymology, 436*, 21–33.

Cross, R., Aish, J., Paston, S. J., Poole, R. K., & Moir, J. W. B. (2000). Cytochrome *c'* from *Rhodobacter capsulatus* confers increased resistance to nitric oxide. *Journal of Bacteriology, 182*, 1442–1447.

Cross, R., Lloyd, D., Poole, R. K., & Moir, J. W. B. (2001). Enzymatic removal of nitric oxide catalyzed by cytochrome *c'* in *Rhodobacter capsulatus. Journal of Bacteriology, 183*, 3050–3054.

Cusanovich, M. A. (1971). Molecular weights of some cytochromes *cc'*. *Biochimica et Biophysica Acta, 236*, 238–241.

Cusanovich, M. A., & Gibson, Q. H. (1973). Anomalous ligand binding by a class of high spin c-type cytochromes. *Journal of Biological Chemistry, 248*(3), 822–834.

Cusanovich, M. A., Tedro, S. M., & Kamen, M. D. (1970). *Pseudomonas denitrificans* cytochrome *cc'*. *Archives of Biochemistry and Biophysics, 141*, 557–570.

Cutruzzolà, F., Arcovito, A., Giardina, G., dell Longa, S., D'Angelo, P., & Rinaldo, S. (2014). Distal–proximal crosstalk in the heme binding pocket of the NO sensor DNR. *Biometals, 27*, 763–773.

de Groot, M. T., Evers, T. H., Merkx, M., & Koper, M. T. M. (2007). Electron transfer and ligand binding to cytochrome c' immobilized on self-assembled monolayers. *Langmuir, 23*, 729–737.

Dobbs, A. J., Anderson, B. F., Faber, H. R., & Baker, E. N. (1996). Three-dimensional structure of cytochrome *c'* from two *Alcaligenes* species and the implications for four-helix bundle structures. *Acta Crystallographica. Section D, Biological Crystallography, 52*, 356–368.

Doyle, M. L., Gill, S. J., & Cusanovich, M. A. (1986). Ligand controlled dissociation of *Chromatium vinosum* cytochrome *c'*. *Biochemistry, 25*, 2509–2516.

Doyle, M. L., Weber, P. C., & Gill, S. J. (1985). Carbon monoxide binding to *Rhodospirillum molischianum* ferrocytochrome c'. *Biochemistry, 24*, 1987–1991.

Droghetti, E., Howes, B. D., Feis, A., Dominici, P., Fittipaldi, M., & Smulevich, G. (2007). The quantum mechanically mixed-spin state in a non-symbiotic plant hemoglobin: The effect of distal mutation on AHb1 from *Arabidopsis thaliana*. *Journal of Inorganic Biochemistry, 101*, 1812–1819.

Dworkowski, F. S., Hough, M. A., Pompidor, G., & Fuchs, M. R. (2015). Challenges and solutions for the analysis of in situ, in crystallo micro-spectrophotometric data. *Acta Crystallographica. Section D, Biological Crystallography, 71*(Pt 1), 27–35.

Eady, R. R., & Hasnain, S. S. (2003). Denitrification. In L. T. W. Que (Ed.), *Comprehensive coordination chemistry II: Bio-coordination chemistry: Vol. 8.* (pp. 759–786). Oxford: Elsevier.

Elmore, B. O., Bergmann, D. J., Klotz, M. G., & Hooper, A. B. (2007). Cytochromes P460 and c'-beta; a new family of high-spin cytochromes c. *FEBS Letters, 581*, 911–916.

Enemark, J. H., & Feltham, R. D. (1974). Principles of structure, bonding, and reactivity for metal nitrosyl complexes. *Coordination Chemistry Reviews, 13*(4), 339–406.

Evers, T. H., & Merkx, M. (2005). Successful recombinant production of *Allochromatium vinosum* cytochrome c' requires coexpression of cmm genes in heme-rich *Escherichia coli JCB712*. *Biochemical and Biophysical Research Communications, 327*(3), 668–674.

Evers, T. H., van Dongen, J. L., Meijer, E. W., & Merkx, M. (2007). Ligand-induced monomerization of Allochromatium vinosum cytochrome c' studied using native mass spectrometry and fluorescence resonance energy transfer. *Journal of Biological Inorganic Chemistry, 12*, 919–928.

Finzel, B. C., Weber, P. C., Hardman, K. D., & Salemme, F. R. (1985). Structure of ferricytochrome c' from *Rhodospirillum molischianum* at 1.67 Å resolution. *Journal of Molecular Biology, 186*, 627–643.

Fujii, S., Yoshimura, T., Kamada, H., Yamaguchi, K., Suzuki, S., Shidara, S., et al. (1995). Electron paramagnetic resonance studies of ferric cytochrome c' from photosynthetic bacteria. *Biochimica et Biophysica Acta, 1251*, 161–169.

Garton, E. M., Pixton, D. A., Petersen, C. A., Eady, R. R., Hasnain, S. S., & Andrew, C. R. (2012). A distal pocket Leu residue inhibits the binding of O_2 and NO at the distal heme site of cytochrome *c'*. *Journal of the American Chemical Society, 134*, 1461–1463.

George, S. J., Andrew, C. R., Lawson, D. M., Thorneley, R. N. F., & Eady, R. R. (2001). Stopped-flow infrared spectroscopy reveals a six-coordinate intermediate in the formation of the proximally bound five-coordinate NO adduct of cytochrome *c'*. *Journal of the American Chemical Society, 123*, 9683–9684.

Ghafoor, D. D., Kekilli, D., Abdullah, G. H., Dworkowski, F. S., Hassan, H. G., Wilson, M. T., et al. (2015). Hydrogen bonding of the dissociated histidine ligand is not required for formation of a proximal NO adduct in cytochrome c'. *Journal of Biological Inorganic Chemistry, 20*(6), 949–956.

Gilmour, R., Goodhew, C. F., & Pettigrew, G. W. (1991). Cytochrome c' from *Paracoccus denitrificans*. *Biochimica et Biophysica Acta, 1059*, 233–238.

Goodrich, L. E., Paulat, F., Praneeth, V. K., & Lehnert, N. (2010). Electronic structure of heme-nitrosyls and its significance for nitric oxide reactivity, sensing, transport, and toxicity in biological systems. *Inorganic Chemistry, 49*, 6293–6316.

Harris, R. L., Barbieri, S., Paraskevopoulos, K., Murphy, L. M., Hasnain, S. S., & Sawers, R. G. (2010). Characterization of cycP gene expression in *Achromobacter xylosoxidans* NCIMB 11015 and high-level heterologous synthesis of cytochrome *c'* in *Escherichia coli*. *Journal of Molecular Microbiology and Biotechnology, 18*, 102–108.

Herzik, M. A. J., Jonnalagadda, R., Kuriyan, J., & Marletta, M. A. (2014). Structural insights into the role of iron-histidine bond cleavage in nitric oxide-induced activation of H-NOX gas sensor proteins. *Proceedings of the National Academy of Sciences of the United States of America, 111*(40), E4156–E4164.

Hirano, Y., Kimura, Y., Suzuki, H., Miki, K., & Wang, Z. Y. (2012). Structure analysis and comparative characterization of the cytochrome c' and flavocytochrome c from thermophilic purple photosynthetic bacterium *Thermochromatium tepidum*. *Biochemistry*, *51*(33), 6556–6567.

Hobbs, J. D., Larsen, R. W., Meyer, T. E., Hazzard, J. H., Cusanovich, M. A., & Ondrias, M. R. (1990). Resonance Raman characterization of *Chromatium vinosum* cytochrome *c*'. Effect of pH and comparison of equilibrium and photolyzed carbon monoxide species. *Biochemistry*, *29*(17), 4166–4174.

Hough, M. A., Antonyuk, S. V., Barbieri, S., Rustage, N., McKay, A. L., Servid, A. E., et al. (2011). Distal to proximal NO conversion in hemoproteins: The role of the proximal pocket. *Journal of Molecular Biology*, *405*, 395–409.

Hough, M. A., Silkstone, G., Worrall, J. A. R., & Wilson, M. T. (2014). NO binding to the proapoptotic cytochrome *c*-cardiolipin complex. *Vitamins and Hormones*, *96*, 193–209.

Hunt, J. P., & Lehnert, N. (2015). Heme-nitrosyls: Electronic structure implications for function in biology. *Accounts of Chemical Research*, *48*, 2117–2125.

Huston, W. M., Andrew, C. R., Servid, A. E., McKay, A. L., Leech, A. P., Butler, C. S., et al. (2006). Heterologous overexpression and purification of cytochrome c' from *Rhodobacter capsulatus* and a mutant (K42E) in the dimerization region. Mutation does not alter oligomerization but impacts the heme iron spin state and nitric oxide binding properties. *Biochemistry*, *45*(14), 4388–4395.

Huston, W. M., Lowe, E. C., Butler, C. S., & Moir, J. W. B. (2005). Purification and characterization of cytochrome *c*′ from *Neisseria meningitidis*. *Biochemical Society Transactions*, *33*, 187–189.

Ibrahim, M., Derbyshire, E. R., Marletta, M. A., & Spiro, T. G. (2010a). Probing soluble guanylate cyclase activation by CO and YC-1 using resonance Raman spectroscopy. *Biochemistry*, *49*, 3815–3823.

Ibrahim, M., Derbyshire, E. R., Soldatova, A. V., Marletta, M. A., & Spiro, T. G. (2010b). Soluble guanylate cyclase is activated differently by excess NO and by YC-1: Resonance Raman spectroscopic evidence. *Biochemistry*, *49*, 4864–4871.

Ikezaki, A., & Nakamura, M. (2002). Models for cytochromes c': Spin states of mono(imidazole)-ligated (meso-tetramesitylporphyrinato)iron(III) complexes as studied by UV–vis, 13C NMR, 1H NMR, and EPR spectroscopy. *Inorganic Chemistry*, *41*, 6225–6236.

Iwasaki, H., Yoshimura, T., Suzuki, S., & Shidara, S. (1991). Spectral properties of *Achromobacter xylosoxidans* cytochromes *c*' and their NO complexes. *Biochimica et Biophysica Acta*, *1058*, 79–82.

Kassner, R. J. (1991). Ligand binding properties of cytochromes *c*'. *Biochimica et Biophysica Acta*, *1058*, 8–12.

Kassner, R. J., Kykta, M. G., & Cusanovich, M. A. (1985). Binding of cyanide to cytochrome c' from *Chromatium vinosum*. *Biochimica et Biophysica Acta*, *831*(1), 155–158.

Kassner, R. J., Rubinow, S. C., & Cusanovich, M. A. (1983). Binding of ethyl iso cyanide to cytochrome c' from *Chromatium vinosum*. *Biochimica et Biophysica Acta*, *743*(2), 195–199.

Kekilli, D., Dworkowski, F. S., Pompidor, G., Fuchs, M. R., Andrew, C. R., Antonyuk, S., et al. (2014). Fingerprinting redox and ligand states in haemprotein crystal structures using resonance Raman spectroscopy. *Acta Crystallographica. Section D, Biological Crystallography*, *70*(Pt 5), 1289–1296.

Kimura, Y., Kasuga, S., Unno, M., Furusawa, T., Osoegawa, S., Sasaki, Y., et al. (2015). The roles of C-terminal residues on the thermal stability and local heme environment of cytochrome c' from the thermophilic purple sulfur bacterium *Thermochromatium tepidum*. *Photosynthesis Research*, *124*(1), 19–29.

Kitagawa, T., Ozaki, Y., Kyooku, Y., & Horio, T. (1977). Resonance Raman study of the pH-dependent and detergent-induced structural alterations in the heme moiety of *Rhodospirillum rubrum* cytochrome *c*'. *Biochimica et Biophysica Acta*, *495*, 1–11.

Kruglik, S. G., Lambry, J.-C., Cianetti, S., Martin, J.-L., Eady, R. R., Andrew, C. R., et al. (2007). Molecular basis for nitric oxide dynamics and affinity with *Alcaligenes xylosoxidans* cytochrome *c'*. *Journal of Biological Chemistry*, *282*, 5053–5062.

Lawson, D. M., Stevenson, C. E. M., Andrew, C. R., & Eady, R. R. (2000). Unprecedented proximal binding of nitric oxide to heme: Implications for guanylate cyclase. *The EMBO Journal*, *19*(21), 5661–5671.

Lawson, D. M., Stevenson, C. E. M., Andrew, C. R., George, S. J., & Eady, R. R. (2003). A two-faced molecule offers NO explanation: The proximal binding of nitric oxide to haem. *Biochemical Society Transactions*, *31*, 553–557.

Lee, B., Usov, O. M., Grigoryants, V. M., Myers, W. K., Shapleigh, J. P., & Scholes, C. P. (2009). The role of arginine-127 at the proximal NO-binding site in determining the electronic structure and function of 5-coordinate NO-heme in cytochrome c' of *Rhodobacter sphaeroides*. *Biochemistry*, *48*(38), 8985–8993.

Lehnert, N. (2008). EPR and low-temperature MCD spectroscopy of ferrous heme nitrosyls. In A. Ghosh (Ed.), *The smallest biomolecules. Diatomics and their interactions with heme proteins* (pp. 147–171). Amsterdam: Elsevier.

Lehnert, N., Berto, T. C., Galinato, M. G. I., & Goodrich, L. E. (2011). The role of heme-nitrosyls in the biosynthesis, transport, sensing, and detoxification of nitric oxide (NO) in biological systems: Enzymes and model complexes. In K. M. Kadish, K. M. Smith, & R. Guilard (Eds.), *Handbook of porphyrin science: Vol. 14*. (pp. 1–247): New Jersey: World Scientific.

Lehnert, N., Sage, J. T., Silvernail, N., Scheidt, W. R., Alp, E. E., Sturhahn, W., et al. (2010). Oriented single-crystal nuclear resonance vibrational spectroscopy of [Fe(TPP)(MI)(NO)]: Qualitative assessment of the *trans* effect of NO. *Inorganic Chemistry*, *49*, 7197–7215.

Lehnert, N., Scheidt, W. R., & Wolf, M. W. (2014). Structure and bonding in heme-nitrosyl complexes and implications for biology. *Structure and Bonding*, *154*, 155–224.

Li, B.-R., Anderson, R., Mowat, C. G., Miles, C. S., Reid, G. A., & Chapman, S. K. (2008). Rhodobacter sphaeroides haem protein: A novel cytochrome with nitric oxide dioxygenase activity. *Biochemical Society Transactions*, *36*, 992–995.

Lukat-Rodgers, G. S., Correia, C., Botuyan, M. V., Mer, G., & Rodgers, K. R. (2010). Heme-based sensing by the mammalian circadian protein CLOCK. *Inorganic Chemistry*, *49*(14), 6349–6365.

Makino, R., Park, S., Obayashi, E., Iizuka, T., Hori, H., & Shiro, Y. (2011). Oxygen binding and redox properties of the heme in soluble guanylate cyclase. *Journal of Biological Inorganic Chemistry*, *286*, 15678–15687.

Maltempo, M. M., & Moss, T. H. (1976). The spin 3/2, 5/2 state and quantum spin mixtures in heme proteins. *Quarterly Review of Biophysics*, *9*, 181–215.

Maltempo, M. M., Moss, T. H., & Cusanovich, M. A. (1974). Magnetic studies on the changes in the iron environment in *Chromatium* ferricytochrome *c'*. *Biochimica et Biophysica Acta*, *342*, 290–305.

Manole, A., Kekilli, D., Svistunenko, D. A., Wilson, M. T., Dobbin, P. S., & Hough, M. A. (2015). Conformational control of the binding of diatomic gases to cytochrome c'. *Journal of Biological Inorganic Chemistry*, *20*(4), 675–686.

Mansy, S. S., Olson, J. S., Gonzalez, G., & Gilles-Gonzalez, M. A. (1998). Imidazole is a sensitive probe of steric hindrance in the distal pockets of oxygen-binding heme proteins. *Biochemistry*, *37*, 12452–12457.

Martí, M. A., Capece, L., Crespo, A., Doctorovich, F., & Estrin, D. A. (2005). Nitric oxide interaction with cytochrome *c'* and its relevance to guanylate cyclase. Why does the iron histidine bond break? *Journal of the American Chemical Society*, *127*, 7721–7728.

Martí, M. A., Crespo, A., Capece, L., Boechi, L., Bikiel, D. E., Scherlis, D. A., et al. (2006). Dioxygen affinity in heme proteins investigated by computer simulation. *Journal of Inorganic Biochemistry*, *100*, 761–770.

Martin, E., Berka, V., Bogatenkova, E., Murad, F., & Tsai, A. L. (2006). Ligand selectivity of soluble guanylyl cyclase: Effect of the hydrogen-bonding tyrosine in the distal heme pocket on binding of oxygen, nitric oxide, and carbon monoxide. *Journal of Biological Chemistry*, *281*(38), 27836–27845.

Martin, E., Berka, V., Sharina, I., & Tsai, A. L. (2012). Mechanism of binding of NO to soluble guanylyl cyclase: Implication for the second NO binding to the heme proximal site. *Biochemistry*, *51*(13), 2737–2746.

Mayburd, A. L., & Kassner, R. L. (2002). Mechanism and biological role of nitric oxide binding to cytochrome *c'*. *Biochemistry*, *41*, 11582–11591.

McGuirl, M. A., Lee, J. C., Lyubovitsky, J. G., Thanyakoop, C., Richards, J. H., Gray, H. B., et al. (2003). Cloning, heterologous expression, and characterization of recombinant class II cytochromes c from *Rhodopseudomonas palustris*. *Biochimica et Biophysica Acta*, *1619*(1), 23–28.

Meyer, T. E., & Kamen, M. D. (1982). New perspectives on *c*-type cytochromes. *Advances in Protein Chemistry*, *35*, 105–212.

Moir, J. W. (1999). Cytochrome c' from *Paracoccus denitrificans*: Spectroscopic studies consistent with a role for the protein in nitric oxide metabolism. *Biochimica et Biophysica Acta*, *1430*(1), 65–72.

Monkara, F., Bingham, S. J., Kadir, F. H. A., McEwan, A. G., Thomson, A. J., Thurgood, A. G. P., et al. (1992). Spectroscopic studies of *Rhodobacter capsulatus* cytochrome *c'* in the isolated state and in intact cells. *Biochimica et Biophysica Acta*, *1100*, 184–188.

Moore, G. R., & Pettigrew, G. W. (1990). *Cytochromes c: Evolutionary, structural and physicochemical aspects*. Berlin: Springer-Verlag.

Motie, M., Kassner, R. J., Meyer, T. E., & Cusanovich, M. A. (1990). Kinetics of cyanide binding to *Chromatium vinosum* ferricytochrome *c'*. *Biochemistry*, *29*, 1932–1936.

Motie, M., Kassner, R. J., Meyer, T. E., & Cusanovich, M. A. (1991). Cyanide-linked dimer-monomer equilibrium of *Chromatium vinosum* ferric cytochrome *c'*. *Biochimica et Biophysica Acta*, *1076*, 97–102.

Mukai, M., Savard, P.-Y., Ouellet, H., Guertin, M., & Yeh, S. R. (2002). Unique ligand-protein interactions in a new truncated hemoglobin from *Mycobacterium tuberculosis*. *Biochemistry*, *41*, 3897–3905.

Olson, J. S., McKinnie, R. E., Mims, M. P., & White, D. K. (1983). Mechanisms of ligand binding to pentacoordinate protoheme. *Journal of the American Chemical Society*, *105*, 1522–1527.

Olson, J. S., Mims, M. P., & Reisberg, P. I. (1982). Kinetic mechanisms of ligand binding to heme proteins. In C. Ho, W. A. Eaton, J. P. Collman, Q. H. Gibson, J. S. Leigh, Jr., E. Margoliash, K. Moffat, & W. R. Scheidt (Eds.), *Hemoglobin and oxygen binding* (pp. 393–398). New York: Elsevier.

Othman, S., Richaud, P., Verméglio, A., & Desbois, A. (1996). Evidence for a proximal histidine interaction in the structure of cytochromes *c'* in solution: A resonance Raman study. *Biochemistry*, *35*, 9224–9234.

Patel, M. J., Kassner, R. J., Meyer, T. E., & Cusanovich, M. A. (1989). Steric and hydrophobic effects in alkyl isocyanide binding to *Rhodospirillum molischianum* cytochrome *c'*. *Biochemistry*, *28*, 2140–2144.

Paulat, F., Berto, T. C., George, S. D., Goodrich, L., Praneeth, V. K. K., Sulok, C. D., et al. (2008). Vibrational assignments of six-coordinate ferrous heme nitrosyls: New insight from nuclear resonance vibrational spectroscopy. *Inorganic Chemistry*, *47*, 11449–11451.

Phillips, G. N., Jr., Teodoro, M. L., Li, T., Smith, B., & Olson, J. S. (1999). Bound CO is a molecular probe of electrostatic potential in the distal pocket of myoglobin. *Journal of Physical Chemistry B*, *103*, 8817–8829.

Pietra, F. (2013). Gates and binding pockets for nitric oxide with cytochrome c', according to molecular dynamics. *Chemistry & Biodiversity*, *10*(9), 1574–1588. http://dx.doi.org/10.1002/cbdv.201300164.

Pixton, D. A., Petersen, C. A., Franke, A., van Eldik, R., Garton, E. M., & Andrew, C. R. (2009). Activation parameters for heme-NO binding in *Alcaligenes xylosoxidans* cytochrome c': The putative dinitrosyl intermediate forms via a dissociative mechanism. *Journal of the American Chemical Society*, *131*(13), 4846–4853.

Pletneva, E. V., Zhao, Z., Kimura, T., Petrova, K. V., Gray, H. B., & Winkler, J. R. (2007). Probing the cytochrome c' folding landscape. *Journal of Inorganic Biochemistry*, *101*(11–12), 1768–1775.

Praneeth, V. K., Näther, C., Peters, G., & Lehnert, N. (2006). Spectroscopic properties and electronic structure of five- and six-coordinate iron(II) porphyrin NO complexes: Effect of the axial N-donor ligand. *Inorganic Chemistry*, *45*(7), 2795–2811.

Quillin, M. L., Li, T., Olson, J. S., Phillips, G. N., Jr., Dou, Y., Ikeda-Saito, M., et al. (1995). Structural and functional effects of apolar mutations of the distal valine in myoglobin. *Journal of Molecular Biology*, *245*(4), 416–436.

Quintas, P. O., Catarino, T., Todorovic, S., & Turner, D. L. (2011). Highly selective ligand binding by *Methylophilus methylotrophus* cytochrome *c''*. *Biochemistry*, *50*, 5624–5632.

Rajagopal, B. S., Edzuma, A. N., Hough, M. A., Blundell, K. L., Kagan, V. E., Kapralov, A. A., et al. (2013). The hydrogen-peroxide-induced radical behaviour in human cytochrome c-phospholipid complexes: Implications for the enhanced pro-apoptotic activity of the G41S mutant. *Biochemical Journal*, *456*(3), 441–452.

Ramirez, L. M., Axelrod, H. L., Herron, S. R., Rupp, B., Allen, J. P., & Kantardjieff, K. A. (2003). High resolution crystal structure of ferricytochrome c' from *Rhodobacter sphaeroides*. *Journal of Chemical Crystallography*, *33*(5–6), 413–424.

Ren, Z., Meyer, T., & McRee, D. E. (1993). Atomic-structure of a cytochrome c' with an unusual ligand-controlled dimer dissociation at 1.8 Å resolution. *Journal of Molecular Biology*, *234*, 433–445.

Rodgers, K. R., Lukat-Rodgers, G. S., & Tang, L. (2000). Nitrosyl adducts of FixL as probes of heme environment. *Journal of Biological Inorganic Chemistry*, *5*, 642–654.

Romão, M. J., & Archer, M. (2001). Cytochrome c'. In R. Wieghardt, R. Huber, A. Poulos, & A. Messerschmidt (Eds.), *Handbook of metalloproteins* (pp. 44–54). Chichester, England: Wiley-VCH.

Rubinow, S. C., & Kassner, R. J. (1984). Cytochromes *c'* in their reaction with ethyl isocyanide. *Biochemistry*, *23*, 2590–2595.

Russell, H. J., Harman, S. J., Heyes, D. J., Hough, M. A., Greetham, G. M., Towrie, M., et al. (2013). Modulation of ligand-heme reactivity by binding pocket residues demonstrated in cytochrome c' over the femtosecond-second temporal range. *FEBS Journal*, *280*, 6070–6082.

Scheidt, W. R., & Ellison, M. K. (1999). The synthetic and structural chemistry of heme derivatives with nitric oxide ligands. *Accounts of Chemical Research*, *32*, 350–359.

Schelvis, J. P. M., Seibold, S. A., Cerda, J. F., Garavito, R. M., & Babcock, G. T. (2000). Interaction of nitric oxide with prostaglandin endoperoxide H synthase-1: Implications for Fe-His bond cleavage in heme proteins. *Journal of Physical Chemistry B*, *104*, 10844–10850.

Servid, A. E., McKay, A. L., Davis, C. A., Garton, E. M., Manole, A., Dobbin, P. S., et al. (2015). Resonance Raman spectra of five-coordinate heme-nitrosyl cytochromes c': Effect of the proximal heme-NO environment. *Biochemistry*, *54*, 3320–3327.

Shibata, N., Satoaki, I., Misaki, S., Meyer, T. E., Bartsch, R. G., Cusanovich, M. A., et al. (1998). Basis for monomer stabilization in *Rhodopseudomonas palustris* cytochrome *c'* derived from the crystal structure. *Journal of Molecular Biology*, *284*, 751–760.

Silkstone, G., Kapetanaki, S. M., Husu, I., Vos, M. H., & Wilson, M. T. (2010). Nitric oxide binds to the proximal heme coordination site of the ferrocytochrome c/cardiolipin complex: Formation mechanism and dynamics. *Journal of Biological Chemistry, 285*(26), 19785–19792.

Smulevich, G., Feis, A., & Howes, B. D. (2005). Fifteen years of Raman spectroscopy of engineered heme containing peroxidases: What have we learned? *Accounts of Chemical Research, 38*, 433–440.

Spiro, T. G., Soldatova, A. V., & Balakrishnan, G. (2013). CO, NO and O2 as vibrational probes of heme protein interactions. *Coordination Chemistry Reviews, 257*, 511–527.

Springer, B. A., Sligar, S. G., Olson, J. S., & Phillips, G. N., Jr. (1994). Mechanisms of ligand recognition in myoglobin. *Chemical Reviews, 94*, 699–714.

Stone, J. R., & Marletta, M. A. (1996). Spectral and kinetic studies on the activation of soluble guanylate cyclase by nitric oxide. *Biochemistry, 35*(4), 1093–1099.

Strekas, T. C., & Spiro, T. G. (1974). Resonance-Raman evidence for anomalous heme structures in cytochrome *c*' from *Rhodopseudomonas palustris*. *Biochimica et Biophysica Acta, 351*, 237–245.

Tahirov, T. H., Misaki, S., Meyer, T. E., Cusanovich, M. A., Higuchi, Y., & Yasuoka, N. (1996a). Concerted movement of side chains in the haem vicinity observed on ligand binding in cytochrome c' from Rhodobacter capsulatus. *Nature Structural Biology, 3*(5), 459–464.

Tahirov, T. H., Misaki, S., Meyer, T. E., Cusanovich, M. A., Higuchi, Y., & Yasuoka, N. (1996b). High-resolution crystal structures of two polymorphs of cytochrome *c*' from the purple phototrophic bacterium *Rhodobacter capsulatus*. *Journal of Molecular Biology, 259*, 467–479.

Tani, F., Matsu-ura, M., Ariyama, K., Setoyama, T., Shimada, T., Kobayashi, S., et al. (2003). Iron twin-coronet porphyrins as models of myoglobin and hemoglobin: Amphibious electrostatic effects of overhanging hydroxyl groups for successful CO/O2 discrimination. *Chemistry - A European Journal, 9*, 862–870.

Traylor, T. G., & Sharma, V. S. (1992). Why NO? *Biochemistry, 31*, 2847–2849.

Tsai, A.-L. (1994). How does NO activate hemeproteins? *FEBS Letters, 341*, 141–145.

Tsai, A.-L., Berka, V., Martin, F., Ma, X., van den Akker, F., Fabian, M., et al. (2010). Is Nostoc H-NOX a NO sensor or redox switch? *Biochemistry, 49*, 6587–6599.

Tsai, A. L., Berka, V., Martin, E., & Olson, J. S. (2012). A "sliding scale rule" for selectivity among NO, CO, and O2 by heme protein sensors. *Biochemistry, 51*, 172–186.

Tsai, A.-L., Berka, V., Sharina, I., & Martin, E. (2011). Dynamic ligand exchange in soluble guanylyl cyclase (sGC): Implications for sGC regulation and desensitization. *Journal of Biological Chemistry, 286*, 43182–43192.

Tsai, A. L., Martin, E., Berka, V., & Olson, J. S. (2012). How do heme-protein sensors exclude oxygen? Lessons learned from cytochrome c', Nostoc puntiforme heme nitric oxide/oxygen-binding domain, and soluble guanylyl cyclase. *Antioxidants & Redox Signaling, 17*(9), 1246–1263.

Tsan, P., Hus, J.-C., Caffrey, M., Marion, D., & Blackledge, M. (2000). Rotational diffusion anisotropy and local backbone dynamics of carbon monoxide-bound *Rhodobacter capsulatus* cytochrome c′. *Journal of the American Chemical Society, 122*, 5603–5612.

Turner, S. M., Moir, J. W. B., Griffiths, L., Overton, T. W., Smith, H., & Cole, J. A. (2005). Mutational and biochemical analysis of cytochrome *c*′, a nitric oxide-binding lipoprotein important for adaptation of *Neisseria gonorrhoeae* to oxygen-limited growth. *Biochemical Journal, 388*, 545–553.

Underbakke, E. S., Iavarone, A. T., Chalmers, M. J., Pascal, B. D., Novick, S., Griffin, P. R., et al. (2014). Nitric oxide-induced conformational changes in soluble guanylate cyclase. *Structure, 22*, 602–611.

Usov, O. M., Choi, P. S.-T., Shapleigh, J. P., & Scholes, C. P. (2005). ENDOR investigation of the liganding environment of mixed-spin ferric cytochrome c'. *Journal of the American Chemical Society*, *127*, 9485–9494.
Usov, O. M., Choi, P. S., Shapleigh, J. P., & Scholes, C. P. (2006). ENDOR of NO-ligated cytochrome c'. *Journal of the American Chemical Society*, *128*(15), 5021–5032.
Vernon, L. P., & Kamen, M. D. (1954). Hematin compounds in photosynthetic bacteria. *Journal of Biological Chemistry*, *211*, 643–662.
Vicente, J. B., Colaço, H. G., Mendes, M. I., Sarti, P., Leandro, P., & Giuffrè, A. (2014). NO* binds human cystathionine ß-synthase quickly and tightly. *Journal of Biological Chemistry*, *289*, 8579–8587.
Vogel, K. M., Kozlowski, P. M., Zgierski, M. Z., & Spiro, T. G. (1999). Determinants of the FeNO (X = C, N, O) vibrational frequencies in heme adducts from experiment and density functional theory. *Journal of the American Chemical Society*, *121*, 9915–9921.
Watmough, N. J., Butland, G., Cheesman, M. R., Moir, J. W. B., Richardson, D. J., & Spiro, S. (1999). Nitric oxide in bacteria: Synthesis and consumption. *Biochimica et Biophysica Acta*, *1411*, 456–474.
Weber, P. C. (1982). Correlations between structural and spectroscopic properties of the high-spin heme protein cytochrome c'. *Biochemistry*, *21*, 5116–5119.
Weber, P. C., Bartsch, R. G., Cusanovich, M. A., Hamlin, R. C., Howard, A., Jordan, S. R., et al. (1980). Structure of cytochrome *c'*: A dimeric, high-spin haem protein. *Nature*, *286*, 302–304.
Weber, P. C., Howard, A., Xuong, N. H., & Salemme, F. R. (1981). Crystallographic structure of *Rhodospirillum molischianum* ferricytochrome *c'* at 2.5 Å resolution. *Journal of Molecular Biology*, *153*, 399–424.
Weiss, R., Gold, A., & Terner, J. (2006). Cytochromes c': Biological models for the S = 3/2, 5/2 spin-state admixture? *Chemical Reviews*, *106*(6), 2550–2579.
Wu, G., Liu, W., Berka, V., & Tsai, A.-L. (2013). The selectivity of *Vibrio cholerae* H-NOX for gaseous ligands follows the "sliding scale rule" hypothesis. Ligand interactions with both ferrous and ferric Vc H-NOX. *Biochemistry*, *52*, 9432–9446.
Yasui, M., Harada, S., Kai, Y., Kasai, N., Kusunoki, M., & Matsuura, Y. (1992). Three-dimensional structure of *ferricytochrome c'* from *Rhodospirillum rubrum* at 2.8 Å resolution. *Journal of Biochemistry*, *111*, 317–324.
Yoo, B. K., Lamarre, I., Martin, J. L., Andrew, C. R., & Negrerie, M. (2013). Picosecond binding of the His ligand to four-coordinate heme in cytochrome c': A one-way gate for releasing proximal NO. *Journal of the American Chemical Society*, *135*(8), 3248–3254.
Yoo, B.-Y., Lamarre, I., Martin, J.-L., Rappaport, F., & Negrerie, M. (2015). Motion of proximal histidine and structural allosteric transition in soluble guanylate cyclase. *Proceedings of the National Academy of Sciences of the United States of America*, *112*, E1697–E1704.
Yoshimura, T. (1982). Electron paramagnetic resonance study of nitrosylprotoheme dimethyl ester complexes with aliphatic nitrogenous bases: Characterization of the axial ligand trans to the nitrosyl group in nitrosylhemoproteins. *Archives of Biochemistry and Biophysics*, *216*(2), 625–630.
Yoshimura, T., Fujii, S., Kamada, H., Yamaguchi, K., Suzuki, S., Shidara, S., et al. (1996). Spectroscopic characterization of nitrosylheme in nitric oxide complexes of ferric and ferrous cytochrome *c'* from photosynthetic bacteria. *Biochimica et Biophysica Acta*, *1292*, 39–46.
Yoshimura, T., Iwasaki, H., Shidara, S., Suzuki, S., Nakahara, A., & Matsubura, T. (1988). Nitric oxide complex of cytochrome *c'* in cells of denitrifying bacteria. *Journal of Biochemistry*, *103*, 1016–1019.
Yoshimura, T., Shidara, S., Ozaki, T., & Kamada, H. (1993). 5-Coordinated nitrosylhemoprotein in the whole cells of denitrifying bacterium, *Achromobacter xylosoxidans* NCIB 11015. *Archives of Microbiology*, *160*, 498–500.

Yoshimura, T., Suzuki, S., Iwasaki, H., & Takakuwa, S. (1987a). Spectral properties of cytochrome *c'* from *Rhodopseudomonas capsulatus* B100 and its CO complex. *Biochemical and Biophysical Research Communications, 144*, 224–231.

Yoshimura, T., Suzuki, S., Iwasaki, H., & Takakuwa, S. (1987b). Spectral properties of nitric oxide complex from cytochrome *c'* from *Rhodopseudomonas capsulata* B100. *Biochemical and Biophysical Research Communications, 145*, 868–875.

Yoshimura, T., Suzuki, S., Kohzuma, T., Iwasaki, H., & Shidara, S. (1990). EPR spectra of ferric cytochromes c' from five strains of *Achromobacter xylosoxidans* at low temperature and their temperature dependence. *Biochemical and Biophysical Research Communications, 169*, 1235–1241.

Yoshimura, T., Suzuki, S., Nakahara, A., Iwasaki, H., Masuko, M., & Matsubara, T. (1985). Identification of heme axial ligands of cytochrome c' from *Alcaligenes sp. N.C.I.B. 11015. Biochimica et Biophysica Acta, 831*, 267–274.

Yoshimura, T., Suzuki, S., Nakahara, A., Iwasaki, H., Masuko, M., & Matsubara, T. (1986). Spectral properties of nitric oxide complexes of cytochrome *c'* from *Alcaligenes* sp. NCIB 11015. *Biochemistry, 25*, 2436–2442.

Yu, N. T. (1986). Resonance Raman studies of ligand binding. *Methods in Enzymology, 130*, 350–409.

Zahn, J. A., Arciero, D. M., Hooper, A. B., & Dispirito, A. A. (1996). Cytochrome *c'* of *Methylococcus Bath. European Journal of Biochemistry, 240*, 684–691.

Zeng, W., Silvernail, N. J., Wharton, D. C., Georgiev, G. Y., Leu, B. M., Scheidt, W. R., et al. (2005). Direct probe of iron vibrations elucidates NO activation of heme proteins. *Journal of the American Chemical Society, 127*, 11200–11201.

Zhao, Y., Brandish, P. E., Ballou, D. P., & Marletta, M. A. (1999). A molecular basis for nitric oxide sensing by soluble guanylate cyclase. *Proceedings of the National Academy of Sciences of the United States of America, 96*(26), 14753–14758.

CHAPTER TWO

Bridging Theory and Experiment to Address Structural Properties of Truncated Haemoglobins: Insights from *Thermobifida fusca* HbO

Barry D. Howes*, Leonardo Boechi[†], Alberto Boffi[‡], Dario E. Estrin[§], Giulietta Smulevich*[,1]

*Dipartimento di Chimica "Ugo Schiff", Università di Firenze, Sesto Fiorentino, Italy
[†]Instituto de Cálculo, Facultad de Ciencias Exactas y Naturales, Universidad de Buenos Aires, Ciudad Universitaria, Buenos Aires, Argentina
[‡]Dipartimento di Scienze Biochimiche, Università "Sapienza", Rome, Italy
[§]Departamento de Química Inorgánica, Analítica y Química Física and Inquimae-Conicet, Facultad de Ciencias Exactas y Naturales, Universidad de Buenos Aires, Buenos Aires, Argentina
[1]Corresponding author: e-mail address: giulietta.smulevich@unifi.it

Contents

Abstract

In this chapter, we will discuss the paradigmatic case of *Thermobifida fusca* (Tf-trHb) HbO in its ferrous and ferric states and its behaviour towards a battery of possible ligands. This

Advances in Microbial Physiology, Volume 67
ISSN 0065-2911
http://dx.doi.org/10.1016/bs.ampbs.2015.08.002

choice was dictated by the fact that it has been one of the most extensively studied truncated haemoglobins, both in terms of spectroscopic and molecular dynamics studies. Tf-trHb typifies the structural properties of group II trHbs, as the active site is characterized by a highly polar distal environment in which TrpG8, TyrCD1, and TyrB10 provide three potential H-bond donors in the distal cavity capable of stabilizing the incoming ligands. The role of these residues in key topological positions, and their interplay with the iron-bound ligands, has been addressed in studies carried out on the CO, F^-, OH^-, CN^-, and HS^- adducts formed with the wild-type protein and a combinatorial set of mutants, in which the distal polar residues, TrpG8, TyrCD1, and TyrB10, have been singly, doubly, or triply replaced by a Phe residue. In this context, such a complete analysis provides an excellent benchmark for the investigation of the relationship between protein structure and function, allowing one to translate physicochemical properties of the active site into the observed functional behaviour. Tf-trHb will be compared with other members of the group II trHbs and, more generally, with members of the other trHb subgroups.

ABBREVIATIONS

5c 5 coordinate
6c 6 coordinate
Bs-HbO group II trHb from *Bacillus subtilis*
CCP cytochrome *c* peroxidase
CCP(MI) CCP expressed in *E. coli* containing Met-Ile at the N-terminus
Cj-trHbP group III trHb from *Campylobacter jejuni*
CT1 charge transfer transition $[a_{2u}(\pi) \rightarrow e_g(d\pi)]$
CT2 charge transfer transition $[a_{2u}(\pi)–a_{1g}(d_{z2})]$
DHP dehaloperoxidase
GCSs globin-coupled sensors
HbA tetrameric human Hb
HbI, HbII, HbIII haemoglobin I, II, and III from *Lucina pectinata*
hh horse heart
HRPC horseradish peroxidase isoenzyme C
HS high spin
LS low spin
Mb myoglobin
MD molecular dynamics
Mt-tHbN group I trHb from *Mycobacterium tuberculosis*
Mt-tHbO group II trHb from *Mycobacterium tuberculosis*
Pgbs protoglobins
RR resonance Raman
sw sperm whale
Tf-tHbO group II trHb from *Thermobifida fusca*
trHbs truncated haemoglobins

1. INTRODUCTION

1.1 Microbial Haemoglobins

The first observation of haemoglobins (Hbs) in microorganisms was made by Warburg in the 1930s and thoroughly reported 20 years later by Keilin, who spectroscopically assigned oxygen-dependent Hb-like absorption bands in moulds, fungi, and protozoa (Keilin, 1953). Many years later, Hbs were also demonstrated to be present in bacteria (Wakabayashi, Matsubara, & Webster, 1986) and most recently even in archaea (Freitas et al., 2004). Phylogenetic analysis of protein sequences has indicated that globins are present in all kingdoms of life and can be collected within a single superfamily that contains three main lineages: (i) truncated haemoglobins (trHbs), (ii) protoglobins (Pgbs) and related globin-coupled sensors (GCSs), and (iii) flavohaemoglobins and related single-domain globins (Frey & Kallio, 2003; Vinogradov, Bailly, et al., 2013; Vinogradov & Moens, 2008; Wajcman, Kiger, & Marden, 2009; Wittenberg, Bolognesi, Wittenberg, & Guertin, 2002). Among these, classical vertebrate Hbs represent just a cluster within the last group of the globin superfamily. According to this view, O_2 transport, the hallmark of Hb function, appears as a recently emerged specialized trait associated with the evolution of metazoans (Pesce, Bolognesi, & Nardini, 2013). A survey of globin occurrence in bacterial and archaeal genomes revealed that over half of the bacterial and approximately one-fifth of archaeal genomes contain genes encoding for globins belonging to at least one of the three lineages (Vinogradov, Tinajero-Trejo, Poole, & Hoogewijs, 2013). A physiological role comprising nitric oxide detoxification via NO dioxygenase or denitrosylase activities and/or the sensing of gaseous ligands (oxygen, hydrogen sulphide, cyanide, etc.) in environmental niches has been suggested but it has been demonstrated only in a very few cases, notably for a few GCSs (Martínková, Kitanishi, & Shimizu, 2013). As such, the physiological function of most microbial globins still remains unknown. Structural biology data offer another multifaceted view of the globin superfamily and of microbial globins in particular. Members of the globin superfamily can be assigned to two main structural classes, one based on a 2-on-2 α-helical sandwich (trHbs) and the other showing the classical 3-on-3 α-helical motif (Pgbs and single-domain globins). The differences among the two diverse Hb folds are exemplified in

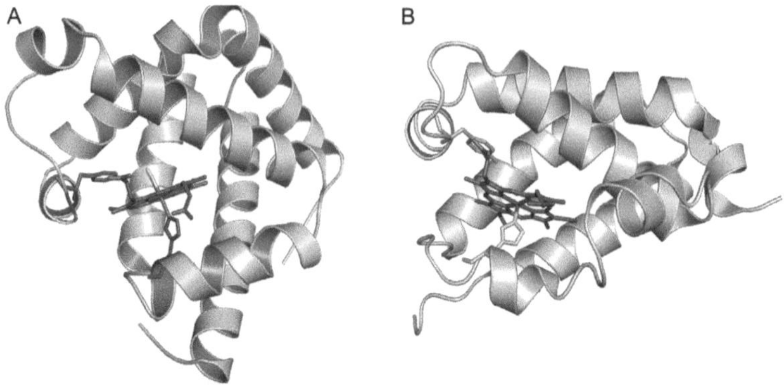

Figure 1 Comparison of the overall fold of a vertebrate globin and a truncated globin. (A) The eight-helix assembly of horse heart myoglobin (PDB code: 1MBN) is shown in magenta (dark grey in the print version); haem, with the HisF8 proximal ligand and PheCD1 is depicted in green (light grey in the print version). (B) The six-helix assembly of *Thermobifida fusca* truncated haemoglobin (PDB code: 2BMM) is shown in cyan (dark grey in the print version), haem, with HisF8 and TyrCD1 in pink (grey in the print version). The figure was prepared using Chimera software (UCSF Chimera).

Fig. 1 and highlight that, at variance with the eight-helix arrangement of classical globins, trHbs achieve a simplified 2-on-2 fold characterized by a very short or absent A-helix, a brief CE inter-helical region, the F-helix occurring as a loop with the B, E, G, and H regions surrounding the haem group. Thus, features common to all microbial Hbs are restricted to the presence of a proximal histidine at the F8 position that coordinates to the haem iron, and a tyrosine or phenylalanine residue at the CD1 position in the distal pocket, necessary for stabilizing the haem moiety in the pocket.

1.2 Truncated Haemoglobins

TrHbs are widely distributed in bacteria, archaea, and plants and are also found in some unicellular eukaryotes. They display less than 20% overall identity with respect to the 3-on-3 globins and can be conveniently classified into three groups, designated I (trHbN), II (trHbO), and III (trHbP). The overall sequence identity between trHbs from different groups may be as low as 20%, though it may be much higher within a given group (80%). Some bacteria display multiple trHbs belonging to different groups, thus suggesting the presence of an ancestor gene (possibly trHbO) that underwent duplications and transfer events during the evolutionary pathway (Pesce et al., 2013). From the structural point of view, each group displays specific features at the basis of distal pocket structure and polarity, as well as ligand stabilization and ligand entry/exit mechanisms. TrHbs share seven key residues in the haem pocket at topologically relevant positions, namely B9, B10, CD1, E7, E11, E14, and F8. Nevertheless, besides HisF8, only PheB9 and TyrB10 are highly conserved and are thus thought to be of critical importance for ligand binding. Other topologically important positions in the haem pocket are not conserved and may vary in size and charge, thus changing completely the dielectric properties of the distal pocket. Such a profound diversity between trHbs might be associated with diverse functional properties concerned with ligand recognition and/or ligand affinity. However, it is noted that physiologically relevant activities such as gaseous ligand sensors, nitric oxide detoxification, and long-term ligand storage are presently not fully clarified. Nevertheless, *in vitro* experimental evidence does indicate diverse ligand-binding properties and complex ligand entry/exit dynamics that are characteristics of each group. Notable examples concern the unusually high affinity for cyanide in trHbP from *Campylobacter jejuni* (Cj-trHbP) (group III), the genuine peroxidase activity of trHbs from *Thermobifida fusca* (Tf-tHbO) and *Mycobacterium tuberculosis* (Mt-tHbO) (group II), or the distinctive NO dioxygenase capabilities identified in trHbNs (group I) (Arya et al., 2013; Ouellet, Milani, et al., 2007; Ouellet, Ranguelova, et al., 2007; Tinajero-Trejo & Shepherd, 2013; Torge et al., 2009).

2. AN OVERVIEW OF RESONANCE RAMAN SPECTROSCOPY OF HAEM PROTEINS

Over the past several decades, resonance Raman (RR) spectroscopy has been extensively applied to the study of structure, function, folding, and dynamics of haem proteins. In fact, it has a number of characteristics that have been advantageous in its application in the field of haem proteins (i) it requires only very small quantities of protein (samples are normally ~50 μL, 30–50 μM); (ii) it can be used easily for proteins in aqueous media, hence, reflecting physiological conditions, since the Raman water spectrum is very weak; and (iii) it is able to enhance only the vibrational modes of a chosen chromophore in proteins with multiple chromophores. There are many excellent reviews on the application of RR to haem proteins (Spiro, 1988), and recently, Yeh and co-workers have published two comprehensive reviews dedicated to trHbs (Egawa & Yeh, 2005; Lu, Egawa, Mukai, Poole, & Yeh, 2008).

In general, the combined analysis of electronic absorption and RR spectra enables some general conclusions to be obtained. The electronic absorption spectrum is characterized by the $\pi \rightarrow \pi^*$ electronic transitions in the Soret (380–440 nm) and visible Q band (500–600 nm) regions deriving from the haem chromophoric group (Gouterman, 1978). In addition, high-spin (HS) ferric haem proteins show also weak bands assigned to porphyrin to iron charge transfer transitions $[a_{2u}(\pi)\text{–}a_{1g}(d_{z2})]$ (CT2) and $[a_{2u}(\pi) \rightarrow e_g(d\pi)]$ (CT1) in the 450–470 and 600–650 nm regions, respectively (Adar, 1978). When haem proteins are excited with a laser wavelength coincident with these electronic transitions, only the Raman spectrum of the haem moiety is enhanced (RR effect). In particular, excitation in the Soret and Q bands leads to the intensification of the haem vibrational modes, whereas by excitation in the CT bands Fe–ligand stretching modes are intensified (see below). Therefore, the haem active site can be probed and information on the structure–function relationship obtained.

2.1 Core-Size Marker Bands

In the high-frequency (1300–1700 cm^{-1}) spectral region, the skeletal modes, the so-called core-size marker bands, have been found to be sensitive to important aspects of porphyrin geometry and electronic structure. Many RR studies on haem proteins and model compounds have shown that their frequencies are inversely correlated with the size of the porphyrin core and, therefore, depend on the spin and coordination states of the Fe atom (Spiro & Li, 1988). Table 1 reports the expected frequencies for a number of more important core-size marker bands, determined by their sensitivity to

Table 1 Resonance Raman Frequencies (cm^{-1}) and Assignments for the Core-Size Marker Bands of Ferric and Ferrous Haem Proteins (Choi et al., 1982)[a]

	Spin State				
	Ferric			Ferrous	
Mode	**6cHS**	**5cHS**	**6cLS**	**5cHS**	**6cLS**
ν_4	1370	1373	1373	1357	1359
ν_3	1480	1491	1502	1471	1493
ν_{11}	1545	1553	1562	1547	1539
ν_2	1559	1570	1579	1562	1584
ν_{37}	1580	1591	1602	1550	1583
ν_{10}	1610	1626	1640	1604	1617

[a]6cHS, 6 coordinate high spin; 5cHS, 5 coordinate high spin; 6cLS, 6 coordinate low spin.

the coordination and spin state of the haem. The strong band between 1350 and 1380 cm^{-1}, assigned to the ν_4 mode, is sensitive to the π-electron density of the porphyrin macrocycle and, therefore, the oxidation state of the iron. The ν_3, ν_2, and ν_{10} modes, which give rise to bands in the 1475–1510, 1550–1590, and 1600–1640 cm^{-1} regions, respectively, are sensitive to both the axial coordination and spin state of the iron.

2.2 Fe–Ligand Modes

Interactions of ligands with the protein side chains that line the haem pocket are critical for discriminating among the binding of the various ligands, and for regulating the subsequent protein conformational changes or the ligand reactivity. Steric contacts can prevent ligand binding or induce slight distortions of the porphyrin ring geometry, thus modifying the Fe–ligand electronic properties. In addition, electric fields and H-bonds associated with side chains distal to the bound ligand can polarize the Fe–ligand bond. The characterization of these interactions is fundamental in order to understand the underlying physical and chemical properties of haem protein functions. In this framework, vibrational spectroscopy offers a powerful tool since in the low-frequency region of the RR spectrum, the Fe–ligand stretching vibrations can be detected. Their frequency is strongly modulated by the interactions between the ligand and the surrounding amino acids (see below); therefore, information on the role of the haem cavity residues in ligand stabilization and the hydrogen-bonding network lining the proximal and distal sides of the haem can be obtained.

The frequencies of specific iron–ligand modes are typically located in the 200 and 800 cm^{-1} frequency region of the RR spectrum and are intensified with excitation in the Soret and/or CT bands (Droghetti et al., 2011; Kerr & Yu, 1988; Spiro, 1988). The haem protein ferric derivatives of a variety of small molecules, such as H_2O, NO, CN^-, H_2S, and F^-, have been identified. The most frequently studied ferrous haem ligands are oxygen-containing diatomic molecules, such as O_2, CO, and NO. Moreover, the RR band associated with the iron–histidine stretching mode has been observed only in the five-coordinate ferrous derivatives (Kitagawa, 1988).

2.2.1 The Proximal Iron–Histidine Stretching Mode

The Fe–His bond stretching mode, ν(Fe–His), is an optimum probe of the proximal cavity structure as it is very sensitive to the protein matrix, its frequency being strongly affected by the H-bonds between the N_δ atom of the proximal His and nearby residues (Kitagawa, 1988; Stein, Mitchell, & Spiro,

1980; Teraoka & Kitagawa, 1980). In the case of a proximal ligand containing a neutral His, the ν(Fe–His) mode has been found at around 200 cm^{-1}, while deprotonation at the N_{δ} position increases the frequency to the region between 240 and 260 cm^{-1}. Such up-shifted frequencies are typical of haem-containing peroxidases, which have an imidazolate ligand resulting from the strong H-bonding interaction between the proximal His and a nearby carboxylate group of an Asp residue (Smulevich, Feis, Howes, & Ivancich, 2010). In myoglobin (Mb), where the N_{δ} proton is H-bonded to a neutral backbone carbonyl group, the ν(Fe–His) stretch was found at 220 cm^{-1} (Kitagawa, Nagai, & Tsubaki, 1979), while in the tetrameric human Hb (HbA) the frequency shifts to 214 cm^{-1} since the iron–histidine bond is strained by the F-helix due to the T-state quaternary interactions (Hori & Kitagawa, 1980; Kitagawa, 1988; Spiro & Li, 1988).

2.2.2 The Distal Ligands

A key objective of the characterization of haem proteins is to establish well-defined correlations between functional aspects and chemical properties of the active site. In general, the reactivity of the active site depends on the oxidation and spin state of the iron ion, the status of the proximal iron ligand, and the nature of the distal cavity residues. In particular, the distal protein amino acids can stabilize (or destabilize) the iron-bound ligands through a manifold of interactions. H-bonds play a significant role in determining ligand stabilization, and different interactions can be envisaged, e.g., single residue-to-ligand H-bonds (Streit, Blanc, Lukat-Rodgers, Rodgers, & DuBois, 2010), water-mediated H-bonds (Edwards & Poulos, 1990), or even ligand inclusive H-bond networks (Egawa & Yeh, 2005; Feis et al., 2008; Ouellet et al., 2003; Rodriguez, Zeng, Wilks, & Rivera, 2007).

In general, the metal–ligand stretching modes can be selectively enhanced by tuning the laser frequency within the Soret or charge transfer transition bands (CT1 or CT2) (Kerr & Yu, 1988; Spiro, 1982). The nature and sensitivity of the metal–ligand bond to the haem structure and haem environment make the frequency of these stretching modes an ideal probe able to provide detailed information on the haem–ligand–protein interactions.

2.2.2.1 H_2O/OH^-

Water is the most common ligand for ferric haem proteins, giving rise to a 6cHS aquo-met derivative with an axially coordinated water ligand in the distal position. The ν(Fe–H_2O) stretching mode, however, has never been observed in the RR spectrum of any haem protein due to its weak intensity. At alkaline pH, the majority of haem proteins undergo an acid–alkaline transition, characterized by a hydroxide ligand bound to the haem iron. The ν(Fe–OH) stretching frequency is very sensitive to the distal environment. An increase of the H-bond strength between the hydroxo group and nearby proton donor/acceptor residues determines a decrease of the force constant of the Fe–OH^- bond with the concomitant decrease of the ν(Fe–OH) stretching frequency.

In Mb and HbA, the haem-bound hydroxide group is stabilized by a weak H-bond with the distal His residue, and two ν(Fe–OH) modes have been observed at about 490 and 550 cm^{-1}. They correspond to the six coordinate HS and low-spin (LS) species, respectively (Feis, Marzocchi, & Smulevich, 1994). In contrast, in haem-containing peroxidases, where both the distal Arg and His residues are involved in a concerted manner via H-bonds in the stabilization of the ligand, the haem-bound hydroxide gives rise to a 6cLS haem, with a ν(Fe–OH) frequency around 500 cm^{-1} (Feis et al., 1994; Howes, Feis, Indiani, Marzocchi, & Smulevich, 2000; Howes, Rodriguez-Lopez, Smith, & Smulevich, 1997; Nissum, Feis, & Smulevich, 1998; Sitter, Shifflett, & Terner, 1988). This frequency is about 50 cm^{-1} wavenumbers lower than that observed in the globins, confirming the presence of strong H-bonds. These results clearly show the high sensitivity of the ν(Fe–OH) frequency to the distal environment.

2.2.2.2 Fluoride

Another ligand that recently has attracted attention as a probe of the H-bonding interaction between the ligand and the distal cavity is fluoride. Fluoride is an anionic weak-field ligand that coordinates ferric haems at pH values below 7.0, with the formation of a 6cHS complex. These complexes are characterized by a strong Soret band with maxima between 404 and 406 nm and a weak Q band at 490 nm. In addition, the electronic absorption spectrum displays an interesting feature in the 600–620 nm range, where a weak band due to a charge transfer transition (CT1) appears. Its wavelength is a sensitive probe of axial ligand polarity and of its interaction with the distal protein residues. It has been found that it blue-shifts when the p- and/or π-donor capability of the axial ligand increases or when the ligand acts as a H-bond donor (e.g. H_2O or H_3O^+), since the interaction between the p/π orbitals of the ligand and the iron dπ orbitals raises the energy of the latter (Smulevich, Neri, Marzocchi, & Welinder, 1996; Smulevich et al., 1997). The opposite effect is observed when the ligand acts as an H-bond acceptor, as in the case of the fluoride ligand (Neri, Kok, Miller, & Smulevich, 1997). In particular, an extensive study on haem peroxidases from various organisms suggested that the position of the CT1 maximum could be related to structural properties of the protein cavity surrounding the bound fluoride anion, demonstrating the sensitivity of the fluoride ligand to the distal cavity environment. For example, distal mutations markedly affected the CT1 position of the fluoride complex of cytochrome *c* peroxidase (CCP). In particular, the CT1 maximum observed at 617 nm as a result of three H-bonds donated to the fluoride by Arg48, Trp51, and a water molecule, blue-shifted in the W51F mutant where replacement of the distal Trp by a non-H-bonding Phe residue reduces the number of H-bonds of the ligand with the protein matrix from three to two. This confirmed that when a ligand acts as a hydrogen-bond donor, the stronger the hydrogen bond and the charge donation to the iron atom, the lower is the wavelength of the CT1 band.

In addition to the UV–vis spectra, the RR frequency of the ν(Fe–F) stretching mode in the low-frequency region provides direct evidence for H-bonding of the fluoride ligand to distal residues via information on the Fe–ligand strength. In fact, the ν(Fe–F) stretching frequency is directly affected by H-bonding to the fluoride ion. This is mainly due to the reduction of the Fe–F bond strength in H-bonded complexes, as a consequence of a decreased electron density on the fluoride ion. The ν(Fe–F) stretching modes appear in the region between 370 and 470 cm^{-1}. In Mb, at neutral pH, a band at 462 cm^{-1} has been assigned to a Fe–F stretching mode on the basis of the 2 cm^{-1} up-shift observed upon $^{56}Fe/^{54}Fe$ isotopic substitution (Desbois, Lutz, & Banerjee, 1979). A clear correlation between the absorption maxima of the CT1 and the H-bonding to the bound anion has been found, thus indicating that fluoride is a sensitive probe of the distal environment in peroxidases (Neri et al., 1997). Moreover, a low ν(Fe–F) stretching frequency has been shown to correlate with a red-shifted CT1 band in the complexes where fluoride is strongly H-bonded to distal pocket residues.

2.2.2.3 H_2S

Hydrogen sulphide, a well-known poisonous gas, is emerging as the fourth (together with O_2, NO, and CO) gas molecule which can act as a signal for intermolecular transduction. In fact, the discovery that H_2S is produced naturally in the human body has changed the perception of this molecule from a toxic pollutant to a biologically relevant molecule (Pietri, Roman-Morales, & Lopez-Garriga, 2011).

Interaction of H_2S with the haem iron is very complex due to the high reactivity of the molecule and many diverse reactions have been observed. It has been known for a long time that the reaction of excess H_2S with metMb or metHb in the presence of O_2 forms a green derivative, as a consequence of the formation of a sulphhaem complex in which the sulphur atom is inserted within the macrocycle ring (Keilin, 1933; Nicholls, 1961; Ríos-Gonzalez, Roman-Morales, Pietri, & Lopez-Garriga, 2014). In addition, in haem proteins, residues from the proximal (Bieza et al., 2015) and distal sides (Ríos-Gonzalez et al., 2014) of the haem cavity appear to modulate the reactivity of the haem group, since some haem proteins have been found to be reduced by sulphides (Hill & Nicholls, 1980; Nicoletti, Thompson, Franzen, & Smulevich, 2011; Pietri et al., 2009), whereas others form remarkably stable FeIII–S(sulphide) adducts (Boffi, Rizzi, Monacelli, & Ascenzi, 2000; Nicoletti et al., 2010).

Interestingly, the HS^- anion has recently been suggested to be a potential physiological ligand for haem proteins. A wide range of organisms have been suggested to be involved in such a mechanism from bacteria, e.g., the trHbs from *B. subtilis* and *T. fusca* (Nicoletti et al., 2010), to marine invertebrates living in sulphide-rich environments, such as the clam Hb of *Lucina pectinata* (Bailly & Vinogradov, 2005; Kraus & Wittenberg, 1990), and dehaloperoxidase-haemoglobin (DHP), an enzyme isolated from the marine worm *Amphitrite ornata*, that serves as the respiratory Hb and has a detoxification function for the organism (Nicoletti, Thompson, et al., 2011). Sulphide binding to these ferric proteins resulted in the formation of a 6cLS complex with a typical electronic absorption spectrum characterized by a Soret maximum at about 424–427 nm and a RR ν(Fe–S) stretching mode around 375 cm^{-1} (Cerda, Echevarria, Morales, & Lopez-Garriga, 1999; Nicoletti et al., 2010; Nicoletti, Thompson, et al., 2011).

2.2.2.4 O_2

Among the various ligands which bind the haem iron in its ferrous state, the most extensively studied have been oxygen-containing diatomic molecules (X–O), such as O_2, CO, and NO (Spiro, Soldatova, & Balakrishnan, 2013). In fact, the function of many haem globin proteins is the transport of O_2 (Jensen, 2009). The first identification of the RR iron–dioxygen ν(Fe–O_2) stretching mode was in the RR spectrum of HbA in erythrocytes by Brunner in 1974 (Brunner, 1974). The ν(Fe–O_2) stretching mode has now been identified for several O_2-bound haem proteins by RR spectroscopy. In oxyMb and oxyHb, the ν(Fe–O_2) vibration has been identified, on the basis of the $^{16}O_2$ and $^{18}O_2$ isotopic substitution experiments, around 570 cm^{-1} (Duff, Appelman, Shriver, & Klots, 1979; Hirota et al., 1996; Nagai, Kitagawa, & Morimoto, 1980). The X-ray structure of MbO_2 revealed that the haem-bound dioxygen is stabilized by the E7 His through H-bonding interactions to both oxygen atoms of the dioxygen. Unlike globins, trHbs are not oxygen transporters as they are not capable of fast oxygen delivery. Their very low rates of oxygen release are in fact a consequence of strong H-bonding interactions between the haem-bound O_2 and the amino acid residues lining the distal haem pocket; hence, the oxygen atom is tightly bound to the haem (Arroyo Mañez et al., 2011; Giangiacomo, Ilari, Boffi, Morea, & Chiancone, 2005; Ouellet et al., 2003). Therefore, in general, a very slow O_2 off-rate has been measured together with a very low ν(Fe–O_2) stretching frequency (Egawa & Yeh, 2005). For example, the ν(Fe–O_2) frequency of trHb-I from the green algae *C. eugametos* has been observed at 554 cm^{-1} (Das, Couture, Ouellet, Guertin, & Rousseau, 2001), in trHb-I from the eubacterium *M. tuberculosis* at 560 cm^{-1} (Couture et al., 1999), in Mt-trHbO(trHb-III) at 559 cm^{-1} (Mukai, Savard, Ouellet, Guertin, & Yeh, 2002), and in Cj-trHbP(trHb-III) at 542 cm^{-1} (Arroyo Mañez et al., 2011).

2.2.2.5 CO

While the function of many haem proteins is the transport of O_2 and NO, including globins (Jensen, 2009), the CO adduct has been used extensively as a probe of the nature and architecture of the haem-binding pocket, which can determine the reactivity of the haem towards other ligands, particularly O_2 and NO. Since the FeCO back-bonding is modulated by polar interactions and, in particular, by the formation of H-bonds between the bound CO and the distal protein residues (Spiro & Wasbotten, 2005), the electrostatic interaction between the FeCO unit and the protein alters the electron distribution in the FeCO unit with the consequent change of the order of the C–O bond. A positively charged electrostatic field favours back-donation, which strengthens the Fe–C bond and correspondingly weakens the C–O bond, thereby increasing the ν(FeC) vibrational frequencies and decreasing the ν(CO) frequencies (Phillips, Teodoro, Li, Smith, & Olson, 1999). On this basis, an inverse linear correlation has been established between the frequencies of the ν(FeC) and ν(CO) stretching modes for a large class of haem protein CO complexes containing imidazole as the fifth iron ligand, the most common proximal ligand in haem proteins (Kerr & Yu, 1988) (see Fig. 2). The ν(FeC)/ν(CO) position along the correlation

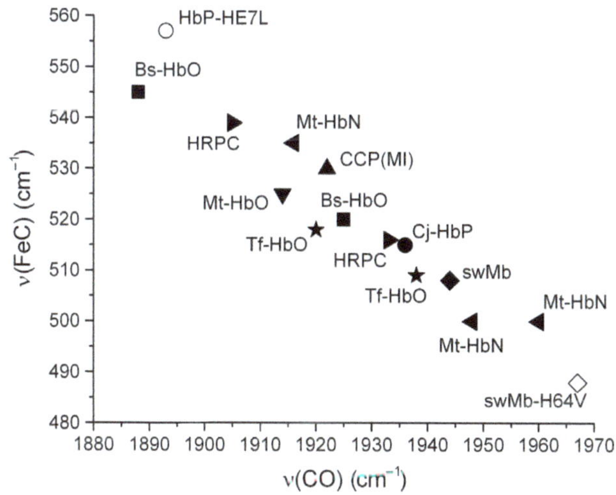

Figure 2 Back-bonding correlation plot for the ν(Fe–C) and ν(C–O) stretching frequencies of various truncated Hbs, peroxidases, and Mb: Bs-HbO (■), Mt-HbO (▼), Tf-HbO (★), Mt-HbN (◄), Cj-HbP (●) and its HE7L mutant (○), HRPC (►), CCP(MI) (▲), and sperm whale (sw) Mb (◆) and its H64V mutant (◇). The data points and corresponding references are reported in Table 2.

line reflects the type and strength of distal polar interactions (Spiro & Wasbotten, 2005), as clearly shown by the CO adducts of a series of Mb distal variants (Phillips et al., 1999). The calculated electrostatic potential field distributions of the bound CO in the distal pockets correlate inversely with ν(C–O) and directly with ν(Fe–CO). The distal histidine residue (H64) of wild-type (WT) Mb forms a weak H-bond interaction with the bound CO, resulting in a point located in the central region of the back-bonding line. When the distal histidine is replaced by hydrophobic residues (e.g. H64V), the ν(FeC)/ν(CO) point slides down the line reflecting the expected decreased back-bonding. On the contrary, proteins with distal residues that strongly interact with the bound CO via H-bonds lie higher on the line, as shown by horseradish peroxidase (HRPC) and recombinant CCP [CCP (MI)] and a series of Hbs from all three classes of trHbs (Fig. 2, Table 2). In fact, peroxidases and trHbs, which are both characterized by polar distal cavities, display similar CO-binding properties. A comparison between the CO binding in globins and peroxidases clearly indicates that back-donation is stronger in peroxidases than in globins. Studies involving site-directed mutagenesis to trace the distal interactions with CO have demonstrated that in peroxidases, the distal arginine and histidine residues that play critical roles in peroxide activation form strong H-bonds with the bound CO. As a

Table 2 Vibrational Frequencies (cm^{-1}) of the CO Complexes of Various Truncated Hbs, Peroxidases, and Mb

Protein	ν(Fe–C)	ν(C–O)	Reference
Truncated Hb Class I			
M. tuberculosis HbN	535	1916	Samuni, Ouellet, Guertin, Friedman, and Yeh (2004)
	500	1948	
	500	1960	
Truncated Hbs Class II			
B. subtilis HbO	545 (form 1)	1888	Feis et al. (2008)
	520 (form 2)	1925	
M. tuberculosis HbO	525	1914	Mukai et al. (2002)
T. fusca HbO	518 (form 1)	1920	Droghetti et al. (2010)
	509 (form 2)	1938	
Truncated Hbs Class III			
C. jejuni HbP	515	1936	Wainwright, Wang, Park, Yeh, and Poole (2006) and Lu, Egawa, Wainwright, Poole, and Yeh (2007)
C. jejuni HbP-HE7L	557	1893	
Peroxidases			
HRPC	539 (form 1)	1905	Feis, Rodriguez-Lopez, Thorneley, and Smulevich (1998), Uno et al. (1987), and Evangelista-Kirkup, Smulevich, and Spiro (1986)
	516 (form 2)	1933	
CCP(MI)	530	1922	Smulevich, Evangelista-Kirkup, English, and Spiro (1986)
Mb			
swMb	508	1944	Li, Quillin, Philips, and Olson (1994), Tsubaki, Srivastava, and Yu (1982), and Howes, Helbo, Fago, and Smulevich (2012)
swMb H64V	488	1967	Li et al. (1994) and Ling, Li, Olson, and Bocian (1994)

consequence, their ν(FeC)/ν(CO) points lie in the upper left corner of the correlation line (Smulevich, Feis, & Howes, 2005).

3. COMPUTER SIMULATION TECHNIQUES

Computer simulation techniques to model biomolecular systems have emerged during the last few decades as an important tool to complement experimental studies (Karplus & McCammon, 2002). The *in silico* generated models and the information that can be obtained through their study have proven to be essential for analysing the structural, spectroscopic, and kinetic data provided by experimental methods (Elber, 2010).

In particular, computer simulations allow a systematic and economical tool to analyse the dependence of a property of interest on static (e.g. amino acid sequence) and dynamic factors. In recent years, the increase in computing power and the accuracy of the models have made it possible to draw biologically relevant conclusions and propose new hypotheses based mainly on computer-generated data (Guallar & Wallrapp, 2011; Norberg & Nilsson, 2002; Roux & Schulten, 2004; Zhao & Caflisch, 2015). The hydrogen-bond network of trHbs is dynamic and several conformations should be considered; consequently, molecular dynamics (MD) simulations emerge as a very important tool for studying the active site of haem proteins. It is important to note that the dynamical features of the active sites of trHbs do not involve the formation or breaking of chemical bonds, and thus, they can be modelled by using classical force fields. Such force fields require parameters that should be obtained by careful quantum-mechanical calculations.

During the last decade, our group has studied several haem proteins using MD simulations (Arroyo Mañez et al., 2011; Bidon-Chanal et al., 2006; Boechi et al., 2008; Forti et al., 2011). Herein, we focus our attention on the interactions between distal site residues and the ligand (O_2, CO, F^-, SH^-, OH^-) coordinated to the haem iron of trHbs. The hydrogen-bond network formed by all those actors, and especially their dynamical properties, allow us to address the molecular basis of protein behaviour towards these ligands, and thus to propose possible protein functions. Furthermore, comparison of single point mutants performed *in silico* also allows the determination of individual residue contributions to ligand affinity. Moreover, the combination of this kind of computer simulation with the previously described spectroscopic techniques constitutes a powerful approach to achieve a better understanding of the active sites of haem proteins.

The natural ligand of many haem proteins is the oxygen molecule, which adopts a negative charge density when coordinated to the iron of the haem group due to the significant charge transfer (π-back-bonding) from the haem group to the ligand. Thus, the coordinated oxygen is stabilized by polar distal site residues able to act as H-bond donors (Capece et al., 2013). Other ligands interact to a greater or lesser extent with the distal site residues depending on the charge density of the iron coordinated species.

4. *THERMOBIFIDA FUSCA* HB

The trHb from *T. fusca* (Tf-trHb) is an archetypal trHb in that it exemplifies the structural properties of group II trHbs (Bonamore et al., 2005). Interestingly, the presence of a gene coding for the Tf-trHb coexists with at least three different globin genes, including a gene encoding for a Pgb. Tf-trHb, similar to the highly homologous mycobacterial Mt-trHbO and *Bacillus* Bs-trHbO, is characterized by a very high affinity for oxygen, a significant NO dioxygenase activity, and a surprisingly high peroxidase-like activity. Besides these notable enzymatic properties, its role within the physiology of the microorganism has not been clarified.

Structure-based sequence alignments show unambiguously that Tf-trHb is closely related to Mt-trHbO and Bs-trHbO (Bonamore et al., 2005). The overall fold of Tf-trHb conforms to that of the other two group II Hbs, for which the X-ray structures are available: Mt-trHbO (PDB code: 1NGK) (Milani et al., 2003) and Bs-trHb (PDB code: 1UX8) (Giangiacomo et al., 2005). Comparison of the crystal structure of Tf-trHb with Bs-trHbO and Mt-trHbO indicates that there is a notable correspondence regarding the amino acids that constitute the haem pocket as discussed below.

4.1 Haem Cavity Structure

Distal haem pocket residues in the key topological positions that govern ligand-binding properties (PheB9, TyrB10, TyrCD1, AlaE7, LeuE11, and TrpG8) are all conserved in Tf-trHb and Mt-trHbO, whereas the key residue TyrB10 is substituted by a Phe residue in Bs-trHbO.

The potentially important role of these polar residues in ligand stabilization and the unique architecture of the active site of Tf-trHb, and other globins belonging to the same family, has stimulated extensive studies aimed at understanding the interplay between iron-bound ligands and distal amino acids. X-Ray crystallography has provided fundamental snapshots of the active site structural framework and highlighted the network of interactions that contributes to modulate the functional properties of the macromolecule around the metal-binding site (Bonamore et al., 2005; Giangiacomo et al., 2005; Milani et al., 2003). On the basis of X-ray diffraction data and static or dynamic spectroscopic observations, it is now clear that iron-bound ligands in the haem-binding pocket experience a network of electrostatic fields and hydrogen-bonding interactions exerted by nearby amino acid side chains. In fact, specific short-range amino acid side chain interactions with ligands appear to contribute to the diversity of haem protein functional behaviour at a level of structural detail that is not always achievable even with high-resolution crystallographic data. Moreover, even if the crystal structure of a given protein is available, it is typically limited to a specific oxidation and coordination state, usually the resting state with no ligand in the distal cavity.

Therefore, a spectroscopic approach, using UV–vis, RR, FTIR, EPR, and/or NMR, complements the crystallographic studies. These techniques are helpful in revealing the dynamic nature of Hbs under various physical and chemical conditions. Among these spectroscopic tools, RR spectroscopy is especially versatile and informative. Combined with mutagenesis studies and MD simulations, RR has uncovered important structural features of several Hbs from unicellular organisms (Egawa & Yeh, 2005; Lu et al., 2008).

5. SPECTROSCOPY AND COMPUTER SIMULATION OF TF-trHb

5.1 Ferric

The UV–vis and RR spectra of the ferric derivative at neutral pH shows a mixture of 6cLS and 6cHS configurations, characteristic of a Fe–OH^- haem (UV–vis maxima at 541 and 577 nm) and a 6cHS aquo form (CT1 bands around 498 and 634 nm, respectively). Classical MD simulations reveal that the coordinated water is stabilized by strong H-bonds with the indole N proton of WG8 (H_2O–N_{WG8} = 1.9 Å) and the hydroxylic hydrogen of YCD1 (OH_2O–O_{YCD1} = 1.5 Å), confirming the spectroscopic findings that suggest a Fe–OH^- character of the ligand even at pH 6.1 (Nicoletti et al., 2014, 2013) (Fig. 3A).

The contribution from the LS component increases as the pH is raised, reflecting the complete deprotonation of the haem-bound water and the formation of a hydroxide derivative. In the RR spectrum at pH 9.8,

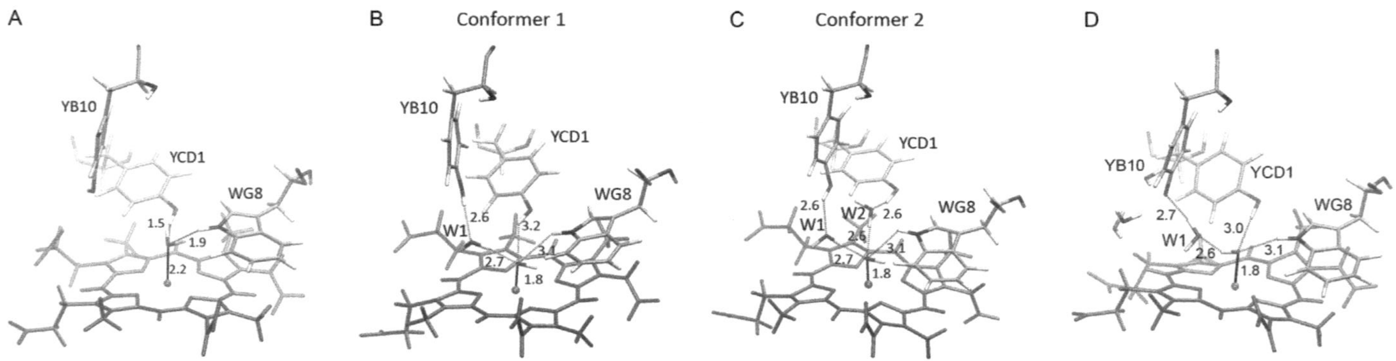

Figure 3 Schematic representation of the distal side of wild-type (WT) Tf-trHb at neutral (A) and alkaline pH (B–D). The dotted lines represent the hydrogen-bond network stabilizing the iron-bound H_2O (A) or the iron-bound OH^- in the two conformations of WT Tf-trHb at alkaline pH when both YB10 and YCD1 are protonated (B and C), or when only YB10 is deprotonated (D), on the basis of MD simulations. Distances are in angstroms. (See the colour plate.)

a ν(Fe–OH) stretching mode has been observed at 485 cm^{-1} (Nicoletti et al., 2014, 2013). The remarkably low frequency, about 65 cm^{-1} lower than that observed for metMb and metHb (Feis et al., 1994), indicates the presence of strong multiple hydrogen bonds between the OH$^-$ ligand and distal residues. In fact, with an increase of the H-bond strength, a decrease of the force constant of the Fe–OH$^-$ bond with the concomitant decrease of the ν(Fe–OH) stretching frequency is expected. At alkaline pH, MD simulations show slightly different results depending on the protonation/deprotonation status of the distal YB10. When both tyrosines (YB10 and YCD1) were considered to be protonated, two conformations in which there is a strong hydrogen-bond network involving all three distal site residues and water molecules are obtained. In the first conformation, the coordinated anion is stabilized by direct interaction with TrpG8 and TyrCD1, and an indirect water-mediated interaction with TyrB10. In the second conformation, TyrCD1 and TyrB10 stabilize the coordinated anion by a water-mediated bridge (Nicoletti et al., 2013) (Fig. 3B and C). However, at extreme alkaline pH the tyrosines in the active site can be deprotonated, and thus, tyrosine becomes an hydrogen-bond acceptor modifying the hydrogen-bond network that stabilizes the coordinated anion (Nicoletti et al., 2014). Therefore, the MD simulations of the WT form containing a deprotonated YB10 residue show only one conformation. The OH$^-$ ligand is acting as acceptor of direct H-bonds with TrpG8 and YCD1, whereas YB10 is H-bonded to one or two water molecules, one of which is H-bonded to the OH$^-$ ligand as donor. This conformation is characterized by H-bonds that stabilize the OH$^-$ ligand (Fig. 3D) more favourably than the case when both tyrosines are protonated (Fig. 3B and C).

Mutation of the distal residues with a Phe residue markedly affects the RR spectrum, resulting in an increased ν(Fe–OH) stretching frequency as a consequence of the change in the H-bond strength. The combined analysis of the RR and MD simulations clearly indicates (Fig. 4) that the singly mutated proteins show a similar, remarkable up-shift (between 20 and 28 cm^{-1}) of the ν(Fe–OH) frequency resulting from the rupture of one H-bond interaction. However, the strength of the H-bonding between the various distal residues and the hydroxo ligand is quite different, as highlighted by the different shift observed in the doubly mutated variants. The YB10F–YCD1F mutant, in which both distal Tyr residues are missing, behaves similar to the single mutants, whereas the YB10F–WG8F mutant, where the YCD1 is the only putative H-bond donor, shows a ν(Fe–OH) frequency comparable to that of WT, indicating a rearrangement of the

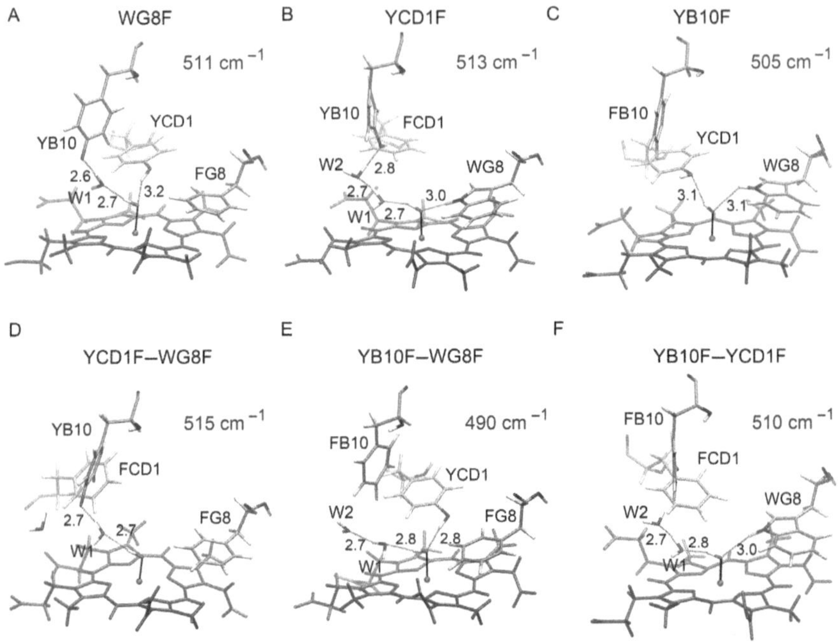

Figure 4 Schematic representation of the distal haem cavity of the mutated proteins at alkaline pH showing, on the basis of MD simulations, the hydrogen-bond network (dotted lines) stabilizing the iron-bound OH^- for single (A–C) and double (D–F) mutant proteins. As TyrB10 is deprotonated it acts as a H-bond acceptor, whereas WG8 and YCD1 are H-bond donors. Distances are in angstroms. The RR ν(Fe–OH) stretching mode frequencies for each mutant are shown in red. *Reprinted with permission from Nicoletti et al. (2014). Copyright (2014) American Chemical Society.* (See the colour plate.)

residue and a strong H-bond interaction. In contrast, for the YCD1–WG8F mutant, where only YB10 can act as a potential H-bond donor, the highest frequency up-shift among all the variants is observed, thus indicating the weakest H-bond interaction to the hydroxo ligand among the three distal polar residues.

The behaviour described above for Tf-trHb is quite different from that observed for the group II Hb Mt-trHbO, for which a covalent bond between the phenyl oxygen of the TyrB10 and the Ce2 of the TyrCD1 is present in six subunits of the dodecamer. In the other six subunits, although the aromatic side-chain groups of the TyrB10 and TyrCD1 residues maintain a similar orthogonal orientation, they are only in very close contact but without the covalent linkage (Milani et al., 2003). This unique CD1/B10Tyr pair creates a tight and rigid distal environment for the incoming ligands. Consequently, while at neutral pH Mt-trHbO is a mixture of HS and LS species, at pH 10.5 two ν(Fe–OH) modes were found at 446 and 533 cm^{-1} (Mukai et al., 2002). The 446 cm^{-1} mode disappeared when the TyrCD1 was mutated to Phe, while it was unchanged when the TyrB10 was mutated to Phe. Therefore, the 446 cm^{-1} mode has been associated with a conformation in which the haem-bound hydroxide is stabilized by an H-bond provided by the TyrCD1, but not the TyrB10. This scenario is consistent with the crystal structure of the ferric cyanide complex, in which the Fe-bound cyanide accepts H-bonds from the TyrCD1 and the TrpG8, but not from the TyrB10. The 533 cm^{-1} mode has been assigned to an open conformation in which the H-bonding interactions are absent (Mukai et al., 2002). In the TyrB10Phe or TyrCD1Phe mutant, the 533 cm^{-1} mode shifts to 510 cm^{-1}, presumably due to a reorganization of the distal pocket induced by the deletion of the covalent bond between the TyrB10 and the TyrCD1.

5.2 Ferrous

WT Tf-trHb and mutated proteins at pH 7.0 display electronic absorption Soret and Q bands (428–431 and 558 nm, respectively) and RR spectra typical of a 5cHS species (Droghetti et al., 2010). The ν(Fe–His) mode is at 223 cm^{-1}, intermediate between that of swMb (220 cm^{-1}) (Kitagawa et al., 1979) and that of other trHbs, like Mt-trHbO (226 cm^{-1}) (Samuni et al., 2004), and *Pseudoalteromonas haloplanktis* TAC125 trHbO (222 cm^{-1}) (Giordano et al., 2015, 2011). As for the latter proteins, the relatively short Fe–His bond can be ascribed to the staggered orientation of the imidazole plane of the proximal His with respect to the haem nitrogen atoms (see Fig. 5), in contrast to the eclipsed orientation in Mb and HbA (Droghetti et al., 2010; Samuni et al., 2004). Hence, the higher frequency

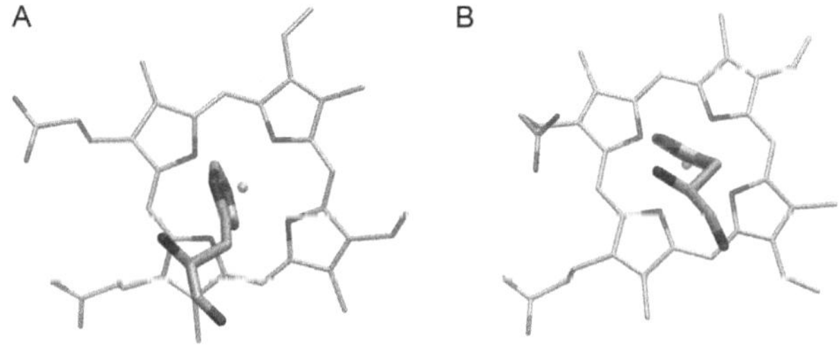

Figure 5 View of the proximal side of Tf-trHb (A) and Mb (B). The proximal histidine plane (His106) of Tf-trHb is shown to be staggered with respect to the haem nitrogen atoms (A), whereas that of Mb is eclipsed (B). Data from PDB entries 2BMM and 1VXG, respectively.

of the Fe–His stretching modes could, in part, be a consequence of the weakening of the repulsive interactions between the haem and the proximal His, which would allow a stronger Fe coordination.

Moreover, the RR spectra of ferrous WT Tf-trHb share other characteristics with those of other group II bacterial trHbs, such as a haem distortion and the weak intensity of the propionyl bending bands. The haem distortion, revealed by the activation of many out-of-plane modes in the RR low-frequency region, is a consequence of the strain induced on the relatively rigid haem tetrapyrrole ring by the interaction with the protein matrix. Moreover, the weak intensity of the propionyl bending bands is a spectral feature shared by both Tf-trHb and Mt-trHbO, which have an opposite orientation of the haem group with respect to the α,γ-meso axis. It has been shown for Mb, where the haem group is 'reversed', i.e., rotated by 180° around the α,γ-meso axis, that the intensity of the propionyl bending bands is weaker when the haem group has a reversed orientation (Rwere, Mak, & Kincaid, 2008). This finding can be assumed to be a fingerprint of the haem orientation, at least for structurally related proteins.

In conclusion, the RR spectra of the ferrous forms are typical of Class II bacterial Hb and mutations on the distal side of the cavity do not influence the haem distortion or the strength of the proximal His–Fe coordination bond.

5.3 Exogenous Ligands

5.3.1 O_2

At present, although no $\nu(Fe–O_2)$ stretching band has been detected by Raman for Tf-trHb, important information on the FeO_2 moiety of Tf-trHb has been obtained by means of MD simulations. After the standard equilibration process to avoid any crash in the starting structure (PDB entry: 2BMM), we performed MD simulations (50 ns) of the Tf-trHb FeO_2 moiety. The results show that two distal site residues (TrpG8 and TyrCD1) interact with the coordinated oxygen molecule (Fig. 6). Both residues act as hydrogen-bond donors to the oxygen molecule, since the coordinated oxygen bears a significant negative charge density due to metal–ligand π-back donation (Marti et al., 2006). The two hydrogen-bond donors provide strong stabilization of the coordinated ligand, which is consistent with a very low dissociation rate constant (k_{off}=0.001 s^{-1}). These results agree well with those proposed for Mt-tHbO, in which the dioxygen is stabilized by H-bonds provided by the TyrCD1 and the TrpG8 residues, resulting in the extremely slow O_2 off-rate (0.0014 s^{-1}) that is about 10,000-fold slower than for Mb (Ouellet et al., 2003). Moreover, the interplay of

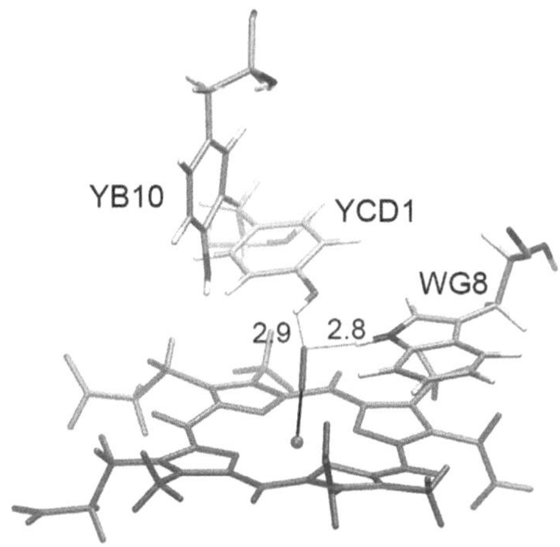

Figure 6 Schematic representation on the basis of MD simulations of a typical snapshot of the distal environment of oxy Tf-trHb showing the H-bonds involving the iron-bound O_2. Distances are in angstroms.

hydrogen-bonding interactions between the haem-bound ligand and the surrounding TyrB10, TrpG8, and HisE7 distal residues in Cj-trHbP has been found to lead to an extremely high oxygen affinity due to a slow dissociation rate (0.0041 s^{-1}) (Arroyo Mañez et al., 2011; Egawa, Wainwright, Poole, & Yeh, 2007).

5.3.2 CO

Two isotope-sensitive peaks (^{12}CO vs. ^{13}CO), assigned to ν(FeC) stretching modes, have been identified in the low-frequency region of the RR spectra of the WT Tf-trHb-CO complex at 509 and 518 cm^{-1}. Accordingly, two ν(CO) stretching modes have been identified at 1940 and 1920 cm^{-1} (Fig. 7). The ν(FeC) and ν(CO) stretching frequencies of WT Tf-trHb and a series of its mutants together with their correlation plot are reported in Fig. 8 and Table 3. The two conformers, identified as form 1 [ν(FeC) and ν(CO) at 509 and 1938 cm^{-1}] and form 2 [ν(FeC) and ν(CO) at 518 and 1920 cm^{-1}], reside in the central and higher parts of the plot, indicating that the bound CO is stabilized by polar interactions. MD simulations have shown that, during the time scale of the simulation, TrpG8 is H-bonded to the coordinated CO. In addition, TyrCD1 is very flexible and can interact with CO, though more weakly than TrpG8 (Fig. 9).

Taken together, these results suggest that the less polar form 1 corresponds to the TrpG8 residue H-bonded to CO, and the more polar form

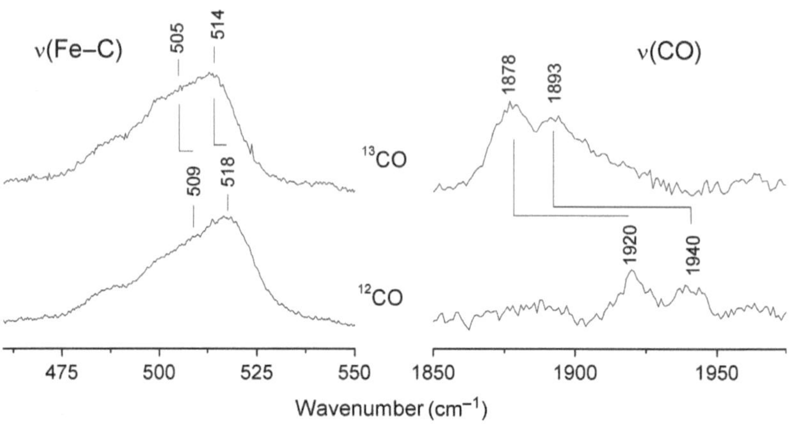

Figure 7 RR spectra of the ^{12}CO and ^{13}CO complexes of WT Tf-trHb. Low-frequency (left) and high-frequency (right) regions showing the ν(Fe–CO) and ν(C–O) stretching modes, respectively.

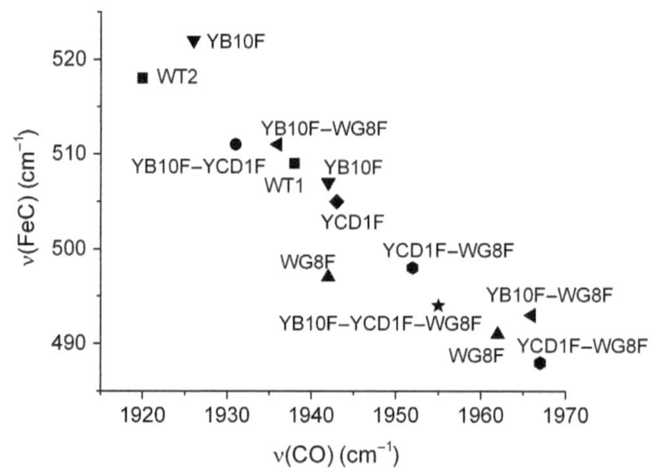

Figure 8 Back-bonding correlation plot for the ν(Fe–C) and ν(C–O) stretching frequencies of WT Tf-trHb and its WG8F, YB10F, YCD1F, YB10F–WG8F, YCD1F–WG8F, YB10F–YCD1F, and YB10F–YCD1F–WG8F mutants. The ν(Fe–C) and ν(C–O) frequencies are reported in Table 3.

2 corresponds to TrpG8 and TyrCD1 that are both H-bonded to the coordinated CO (Fig. 10).

The important role of H-bonding between CO and TrpG8 in the stabilization of both forms 1 and 2 has been highlighted by the spectra of the WG8F mutant (Fig. 11). In fact, for this single mutation the FeCO frequencies are consistent with conformers where polar interactions with the

Table 3 Vibrational Frequencies (cm^{-1}) of the CO Complexes of WT Tf-trHb and Its Mutants (Droghetti et al., 2010)

Proteins	ν(Fe–C) RR	ν(C–O) IR
WT1 (form 1)	509	1938
WT2 (form 2)	518	1920
WG8F	491	1962
	497	1942
YB10F	507	1942
	522	1926
YCD1F	505	1943
YB10F–WG8F	493	1966
	511	1936
YCD1F–WG8F	488	1967
	498	1952
YB10F–YCD1F	511	1931
YB10F–YCD1F–WG8F	494	1955

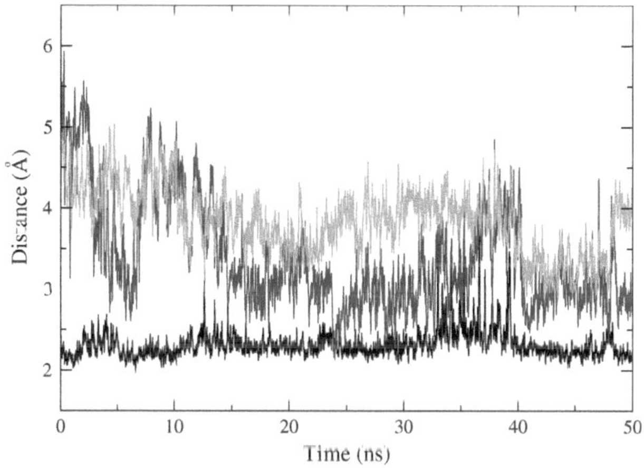

Figure 9 Time evolution of selected hydrogen-bond distances between the H atom of TrpG8Nε (black), the hydroxylic H of TyrCD1 (red (dark grey in the print version)), and the hydroxylic H of TyrB10 (green (grey in the print version)) with the O atom of the coordinated CO. *Adapted with permission from Droghetti et al. (2010). Copyright (2010) American Chemical Society.*

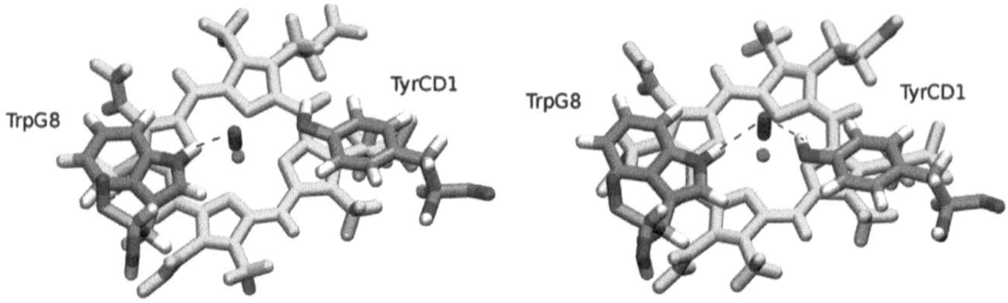

Figure 10 Schematic representation of the distal side of Tf-trHb, showing the H-bonds involving the iron-bound CO in the two forms on the basis of MD simulations. *Adapted with permission from Droghetti et al. (2010). Copyright (2010) American Chemical Society.*

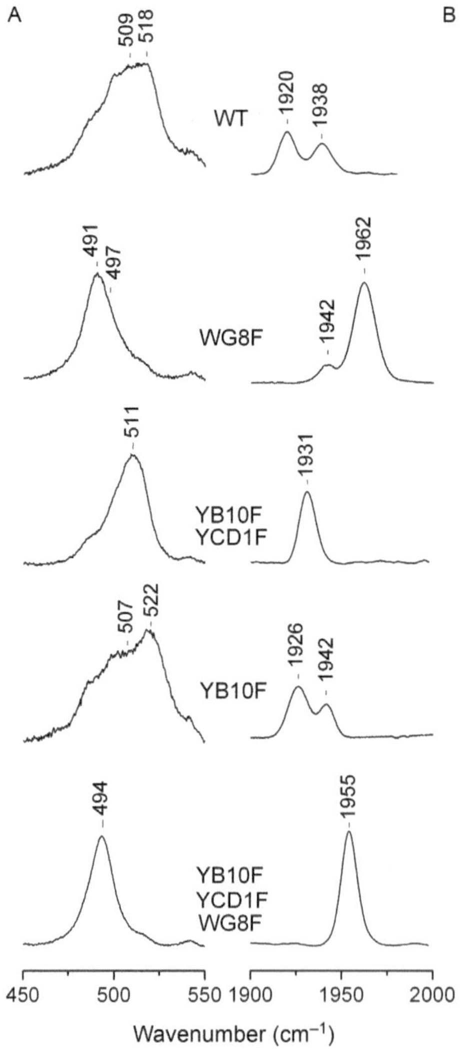

Figure 11 RR (A) and IR (B) spectra of the CO complexes of WT Tf-trHb and selected mutants showing the ν(Fe–CO) and ν(C–O) stretching modes, respectively. *Adapted with permission from Droghetti et al. (2010). Copyright (2010) American Chemical Society.*

surrounding amino acids are absent [ν(FeC) and ν(CO) at 491 and 1962 cm^{-1}, respectively] or weak [ν(FeC) and ν(CO) at 497 and 1942 cm^{-1}, respectively]. Moreover, the spectrum of the double mutant YB10F–YCD1F displays a fairly strong CO-Trp polar interaction [ν(FeC) and ν(CO) at 511 and 1931 cm^{-1}, respectively]. No effect on the characteristics of the CO complex is observed upon mutation of TyrB10 to Phe (YB10F), indicating that this residue is not involved in the stabilization of the CO ligand. Interestingly, as expected, only one apolar form [ν(FeC) at 494 cm^{-1} and ν(CO) at 1955 cm^{-1}] is observed in the case of the triple Phe mutant.

In spite of the similarity of the polar distal cavity with that of Tf-trHb, the CO complex of Mt-trHbO displays only one conformer [ν(FeC) and ν(CO) at 525 and 1914 cm^{-1}, respectively] (see Fig. 2). The mutation of the TyrB10 to Phe does not affect the ν(Fe–CO) frequency; in contrast, the mutation of the TyrCD1 to Phe shifts the ν(Fe–C) frequency to 515 cm^{-1} (Mukai et al., 2002). This suggests that the haem-bound CO in the WT protein is stabilized by an H-bond donated by the phenolic side-chain group of the TyrCD1, but not the TyrB10, which is analogous to the H-bonding network concluded for the hydroxide-bound ferric derivative. The combination of data obtained from RR spectroscopy, mutagenesis, and MD indicates that the haem-bound CO mainly interacts with TrpG8 and TyrCD1 (Guallar, Lu, Borrelli, Egawa, & Yeh, 2009; Ouellet, Milani, et al., 2007; Ouellet, Ranguelova, et al., 2007). These results clearly demonstrate that the distal interactions in Mt-trHbO and Tf-trHb are different, in spite of the occurrence of the Trp–Tyr–Tyr triad in both proteins. The structural difference between the two proteins can possibly be identified with an increased mobility of the distal Tyr residues in Tf-trHb, leading to the appearance of form 1, which is singly H-bonded with TrpG8.

The comparison with Bs-trHbO shows even more striking differences, which can be expected on the basis of the distal cavity structure that is characterized by the polar amino acids TrpG8, TyrB10, and GlnE11 (Feis et al., 2008). The case of the CO complexes of Bs-trHbO can be considered particularly extreme, because the effect of back-bonding on the FeCO vibrational frequencies is one of the strongest observed. In fact, two conformers are observed, the more polar of which displays ν(FeC) and ν(CO) at 545 and 1888 cm^{-1}, respectively (see Fig. 2). Moreover, classical MD simulations show that the CO coordinated to Bs-trHbO interacts strongly with TrpG8 and weakly with GlnE11, which pivots forming and breaking the bond with the coordinated CO (Boechi, Manez, Luque, Marti, & Estrin, 2010).

5.3.3 Fluoride

The fluoride anion binds ferric haem proteins. Although this ligand does not have any physiological role, the combination of RR and UV–vis spectroscopies of fluoride complexes can provide a simple and direct method to monitor the interaction between the distal haem pocket and the iron-bound ligand (Droghetti et al., 2011; Nicoletti, Droghetti, et al., 2011). In fact, as hydrogen bonding is the only effective stabilization mechanism for the haem-bound fluoride, the high sensitivity of fluoride complexes can be used to probe hydrogen bonding in the distal cavity of haem proteins. Recently, the combination of UV–vis, RR, and MD simulations has been applied to study the fluoride complexes of native and distal single, double, and triple mutants of the key WG8, YCD1, and YB10 residues of Tf-trHb (Nicoletti, Droghetti, et al., 2011).

From the comparison of the electronic absorption spectra of the fluoride complexes of WT Tf-trHb, Mb, and HRPC (Fig. 12, left panel, Table 4), it can be seen that unlike the CT2 band maxima (between 451 and 454 nm),

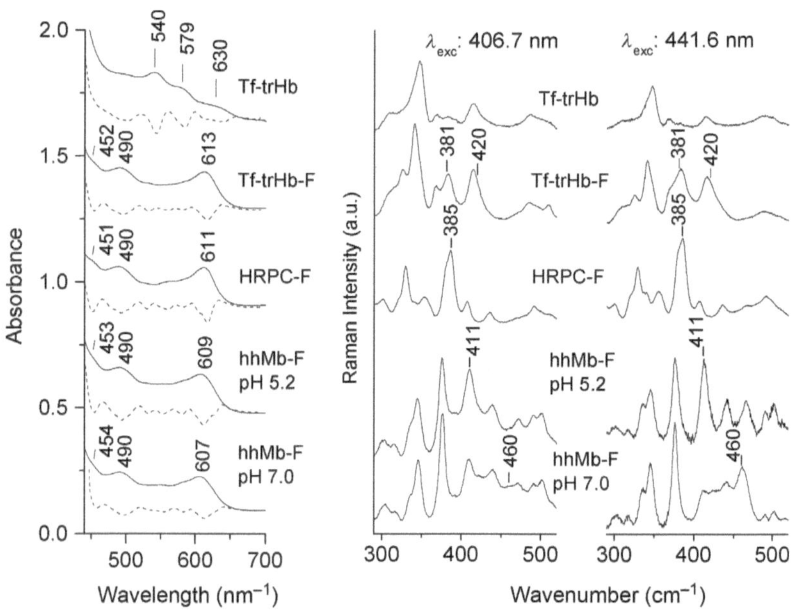

Figure 12 Comparison of the electronic absorption (continuous line), second derivative (dotted line) spectra in the visible region (left), and RR spectra in the low-frequency region taken with excitation in resonance with the Soret (406.7 nm) and with the CT2 band (441.6 nm) (right) of ferric Tf-trHb, together with the fluoride complexes of Tf-trHb, HRPC, and hhMb at pH 7.0 and 5.2. *Adapted from Droghetti et al. (2011). Copyright (2011), with permission from Elsevier.*

Table 4 CT1 Band Maxima (Expressed in nm and cm^{-1}) and ν(Fe–F) Frequencies of WT Tf-trHb and Its Mutants Compared to Those of Mb and HRPC

	CT1 (nm)	CT1 (cm^{-1})	ν(Fe–F) (cm^{-1})	References
YB10F–YCD1F–WG8F	602	16,611	471	Nicoletti, Droghetti, et al. (2011)
YCD1F–WG8F	605	16,529	432	Nicoletti, Droghetti, et al. (2011)
WG8F	609	16,420	421	Nicoletti, Droghetti, et al. (2011)
YCD1F	609	16,420	421	Nicoletti, Droghetti, et al. (2011)
YB10F–YCD1F	609	16,420	419	Nicoletti, Droghetti, et al. (2011)
YB10F–WG8F	610	16,393	415	Nicoletti, Droghetti, et al. (2011)
WT	612	16,340	381 (I) 420 (II)	Nicoletti, Droghetti, et al. (2011)
YB10F	613	16,313	381 (I) 420 (II)	Nicoletti, Droghetti, et al. (2011)
hhMb (pH 7.0)	607	16,474	460	Droghetti et al. (2011)
hhMb (pH 5.2)	609	16,420	411	Droghetti et al. (2011)
HRPC	611	16,367	385	Neri et al. (1997)

which are very similar among the various haem proteins, small but significant differences in the wavelength of the CT1 band are observed. In fact, as expected (Neri et al., 1997), the wavelength of the CT1 band is a sensitive probe of axial ligand polarity and of its interaction with the distal protein residues. In the case of Mb-F complex at neutral pH, the X-ray structure shows that the ligand is doubly H-bonded, with a water molecule and the distal His (Aime et al., 1996). The observed CT1 band occurs at 607 nm, but up-shifts to 609 nm at acidic pH, where a protonated histidine interacts with the ligand. A further up-shift is observed for the HRPC-F complex (CT1 at 611 nm), consistent with the fact that in HRPC the positively charged guanidinium group of the distal Arg together with the distal His stabilizes the anion via H-bonds (Neri et al., 1997). On the basis of these findings, the CT1 maximum at 613 nm observed for Tf-trHb-F is consistent with the presence of multiple H-bonds donated to the fluoride by the distal residues. The RR spectra revealed that Tf-trHb-F is characterized by two conformers with ν(Fe–F) at 420 (conformer 2) and 381 cm^{-1} (conformer 1) (Nicoletti, Droghetti, et al., 2011). This latter is similar to the spectroscopic signature of haem-containing peroxidases, in agreement with the presence of strong H-bond interactions (Fig. 12, right), while the RR spectrum of horse heart (hh) Mb shows a band at 460 cm^{-1} band at neutral pH which downshifts to 411 cm^{-1} at acid pH (Droghetti et al., 2011).

The presence of multiple strong H-bond interactions in Tf-trHb-F is also demonstrated by the single TrpG8 and TyrCD1 mutants; in the absence of either TrpG8 or TyrCD1, the ν(Fe–F) band shifts to higher frequency, with a concomitant blue-shift of the CT1 band (see Table 4). This observation clearly indicates that the bound fluoride is stabilized by H-bond interactions donated by both TrpG8 and TyrCD1. Conversely, the absence of TyrB10 leads to a strengthening of the H-bond, as observed in the single YB10F mutant and the double YB10F–WG8F and YB10–YCD1 mutants. Based on the Raman measurements, we concluded that in the presence of TrpG8 and TyrCD1, TyrB10 does not interact with the ligand. In fact, TyrB10 is able to weakly interact with fluoride only when H-bond interactions with both TrpG8 and TyrCD1 are missing (Nicoletti, Droghetti, et al., 2011).

As observed for the case of the oxygen ligand and in agreement with the RR results, MD simulations of the fluoride adducts of the WT protein show that two distal site residues interact with the coordinated ligand: TrpG8 and TyrCD1 act as hydrogen-bond donors for the polar coordinated fluoride (Nicoletti, Droghetti, et al., 2011). However, an MD simulation (50 ns) performed recently shows that TyrB10 can interact weakly with the fluoride via a water molecule in one of the protein conformations (conformer 1), while no interaction is present in conformer 2 (Fig. 13A and B). The MD simulations of the single and double mutants well agree with the RR results (Fig. 13C–H), showing that when TrpG8 and/or TyrCD1 are absent, TyrB10 interacts with the fluoride and the interaction depends on the presence of water molecules that occupy the distal site dynamically (Fig. 13C, E, and F).

A linear correlation has been found between the frequencies of the ν(Fe–F) stretching mode and the CT1 band for the Tf-trHb fluoride complexes of the WT and mutant proteins. The validity of the correlation has been confirmed by its application to the literature data of Mb, HRPC, and DHP (Nicoletti, Droghetti, et al., 2011). As can be seen from Fig. 14, the spectroscopic markers of the fluoride complexes [ν(Fe–F) and CT1 wavenumbers] are fitted well by a straight line with positive slope. The (CT1)/ν(Fe–F) position along the correlation line reflects the extent of distal H-bonding interactions. Fluoride complexes in which the ligand is stabilized by multiple, strong H-bond interactions, like those of the WT Tf-trHb and

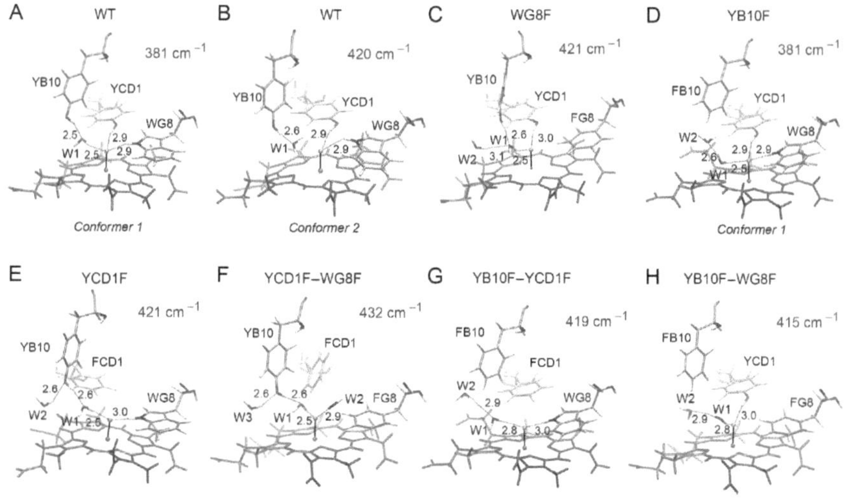

Figure 13 Schematic representation of the distal haem cavity of the mutated proteins with fluoride showing, on the basis of MD simulations, the hydrogen-bond network (dotted lines) stabilizing the fluoride for WT (A and B), single (C–E), and double (F–H) mutant proteins. TyrB10, WG8, and YCD1 are H-bond donors. Distances are in angstroms. The RR ν(Fe–F) stretching frequencies for each mutant are shown in red. For the YB10F mutant, only conformer 1 is shown. (See the colour plate.)

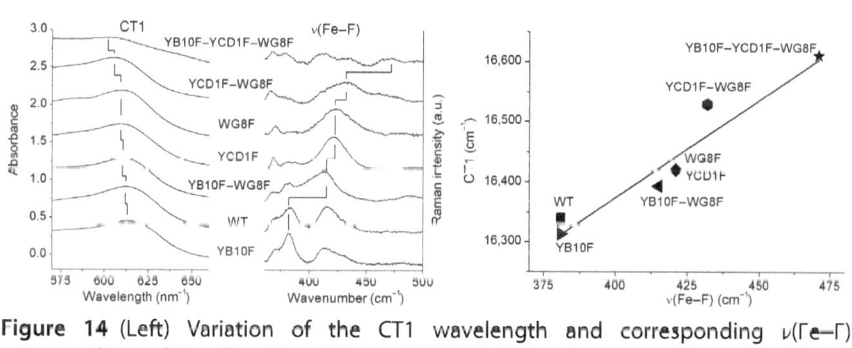

Figure 14 (Left) Variation of the CT1 wavelength and corresponding ν(Fe–F) wavenumbers of the fluoride adducts of WT Tf-trHb and its mutants YB10F, YCD1F, WG8F, YB10F–WG8F, YCD1F–WG8F, and YB10F–YCD1F–WG8F. (Right) Empirical correlation plot between the ν(Fe–F) wavenumbers and the CT1 energy of WT Tf-trHb and its mutants. The experimental data are reported in Table 4. The line represents a least-squares fit of the experimental data of WT Tf-trHb and the combinatorial set of its mutants considering only the strongly H-bonded forms of WT and YB10F with the ν(Fe–F) stretch at 381 cm^{-1}.

YB10F conformer that is characterized by a ν(Fe–F) frequency at 381 cm^{-1}, are located at the extreme left side. In fact, H-bonding decreases both the ν(Fe–F) stretching frequency and the energy of the $e_g(d\pi)$ orbitals, thereby lowering the energy of the $a_{2u}(\pi) \rightarrow e_g(d\pi)$ CT1 transition. At the other extremity of the line, fluoride complexes with very weak H-bond interactions are found, like that of the triple mutant (YB10F–YCD1F–WG8F). In the central region, the complexes characterized by one or two H-bonds can be found.

5.3.4 H_2S

The first observation of a high-affinity sulphide-binding Hb concerned haemoglobin I (HbI) from the mollusc *L. pectinata*, which binds sulphide in its ferric state and functions as a sulphide carrier (Kraus & Wittenberg, 1990). Recently, Bs-trHbO and Tf-trHbO (Fig. 15) were found to bind sulphide (Nicoletti et al., 2010). The electronic absorption and RR data indicate that the sulphide adducts are typical of a ferric iron LS derivative characterized, in analogy with other LS ferric sulphide adducts, by the strong Fe–S stretching band at 375 cm^{-1}.

Both proteins possess a high affinity for hydrogen sulphide at pH 7.0, though lower than that reported previously for *L. pectinata* HbI (Kraus & Wittenberg, 1990) (Table 5). The high sulphide affinity of these proteins must have a physiological significance as H_2S is produced in bacteria at metabolic steps involved in cysteine biosynthesis and, hence, in thiol redox homeostasis. Thiol-based switches have recently been demonstrated to be common in protein sensors of antioxidant cascades, in both prokaryotic and eukaryotic organisms. Several redox sensors, under the master control of OxyR, respond to hydrogen peroxide stress by formation of disulphide bonds. TrHbs have been shown to be part of an operon linked to OxyR, at least in Bacillus species (Larsson, Rogstam, & von Wachenfeldt, 2007). In this framework, a role of microbial globins in modulating H_2S signalling and/or thiol oxidation (the so-called disulphide stress) has been proposed (Nicoletti et al., 2010).

From a kinetic point of view, the overall high affinity resides in the slow rate of sulphide release, attributed to hydrogen-bonding stabilization of the bound ligand by the distal WG8 residue. A set of point mutants in which distal residues have been replaced with Phe indicates that the WG8 residue represents the major kinetic barrier to the escape of the bound sulphide species. Accordingly, classical MD simulations of Tf-trHbO with coordinated HS^- indicate that WG8 plays a key role in the stabilization of coordinated

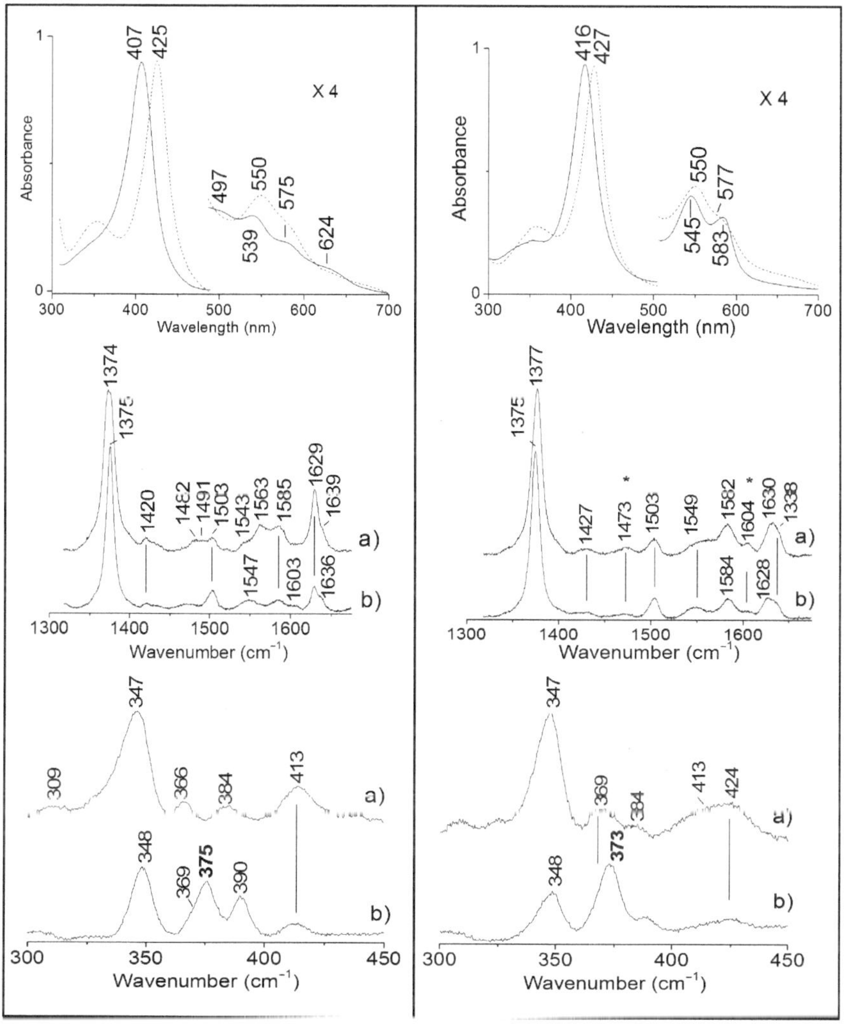

Figure 15 Comparison of the electronic absorption and resonance Raman spectra of WT ferric *T. fusca* (left) and *B. subtilis* (right), in the absence (—, a) and presence of Na_2S (---, b). (Top) Absorption spectra, (middle) high-frequency, and (bottom) low-frequency RR spectra. The ν(Fe–S) is indicated in bold. The asterisk indicates spurious bands of the deoxy form deriving from the photolysis of the ferrous-oxy species present in Bs-trHb in small quantities. *Adapted with permission from Nicoletti et al. (2010). Copyright (2010) American Chemical Society.*

SH^-, whereas the YCD1 and YB10 contributions are negligible (Fig. 16). Interestingly, when TrpG8 is mutated to a hydrophobic residue, such as Phe (both in the single mutant WG8F and in the triple mutant WG8F–YB10F–YCD1F), the dissociation constant increases by about two orders of

Table 5 Thermodynamic and Kinetic Constants for Sulphide Binding to Various Haem Proteins (Nicoletti et al., 2010)

Protein	$K_{(HS^-)}(M^{-1})$	$k_{(HS^-)}(M^{-1}s^{-1})$	$k'_{(HS^-)}(s^{-1})$	References
Bs-trHb	5.0×10^6	1.3×10^4	0.0026	Nicoletti et al. (2010)
Tf-trHb	2.8×10^6	5×10^3	0.0018	Nicoletti et al. (2010)
WG8F–YB10F–YCD1F *Tf-trHb*	2.0×10^7	4.4×10^6	0.22	Nicoletti et al. (2010)
swMb	5×10^4			Jorgensen, Chandrasekar, Madura, Impey, and Klein (1983)
L29F–H64Q–V68F swMb	3.7×10^7			Jorgensen et al. (1983)
L. pectinata HbI	1.1×10^7	2.3×10^5	0.00022	Kraus and Wittenberg (1990)
L. pectinata HbII	1.7×10^4	1.13×10^4	0.017	Kraus and Wittenberg (1990)
L. pectinata HbIII	2.5×10^4	$1–4 \times 10^4$	0.016	Kraus and Wittenberg (1990)

magnitude, implying a weakened stabilization of the iron-bound sulphide (Table 5). This is in agreement with the notion that hydrogen bonding is the main kinetic control of ligand dissociation in haem proteins and confirms the role of TrpG8 in sulphide stabilization in trHbs.

6. CONCLUDING REMARKS

The body of experimental observations and MD simulations here reviewed indicates that a multiplicity of interactions modulate the properties of Fe–ligand moieties in trHbs and, hence, dominate the dynamics of ligand entry and escape and possibly also modulate ligand stabilization. The subtle features of each Fe–ligand complex must be considered in order to account for the observed differences in ligand-binding behaviour. For example, both the experimental and MD simulations data indicate that, as in the case of the CO adducts (Droghetti et al., 2010), TrpG8 and TyrCD1 can form strong hydrogen bonds with fluoride, whereas TyrB10 plays only a minor role in

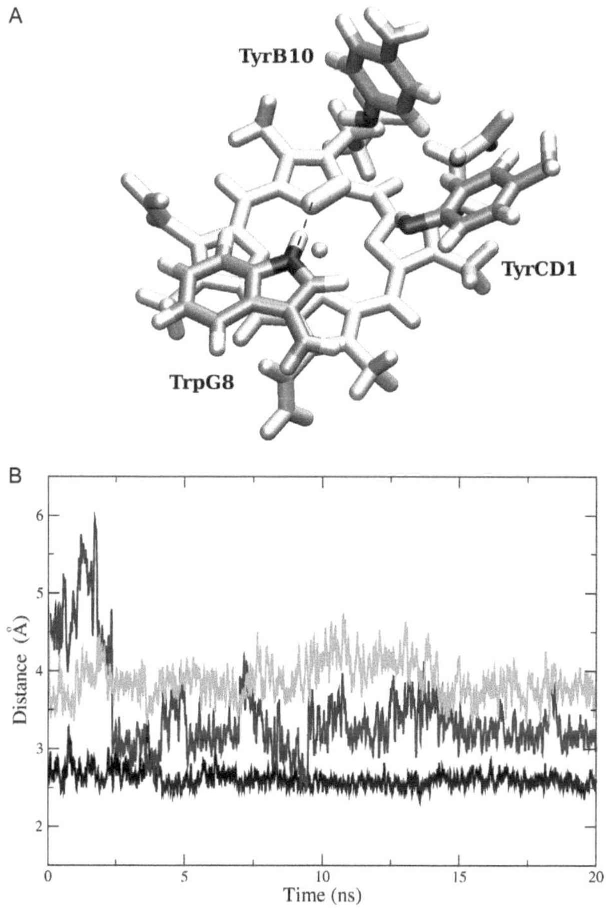

Figure 16 (A) Schematic representation of the WT Tf-trHb distal site with coordinated SH^-, showing the residues of the distal site (TrpG8, TyrCD1, and TyrB10). (B) Time evolution of selected distances among distal residues and haem-bound SH^- of WT Tf-trHb. The distances correspond to those between the S atom of the coordinated SH^- and the indole N proton of TrpG8 (black), the hydroxyl hydrogen of TyrCD1 (red (dark grey in the print version)), and the hydroxyl hydrogen of TyrB10 (green (grey in the print version)). *Adapted with permission from Nicoletti et al. (2010). Copyright (2010) American Chemical Society.*

the stabilization of the ligand. However, when the effects of the mutations on the RR spectra are compared, the corresponding complexes exhibit different behaviour, since back-bonding in the CO complexes depends on many kinds of polar interactions with the neighbouring amino acids, while hydrogen bonding is the only effective stabilization mechanism for the haem-bound fluoride. Moreover, the two types of complex have different chemical properties since carbon monoxide bears less charge density than fluoride. On the other hand, the HS^- ligand establishes strong H-bonding interactions between the S-atom of the coordinated sulphide, which bears a significant negative charge density, and H-bond donating amino acids. In turn, molecular oxygen offers the possibility of concomitant interactions with all three main H-bond donors. As a consequence, iron-bound oxygen remains confined within the H-bonding network with little possibility to escape from within the haem pocket.

Taken together, these results highlight the considerable potential afforded by the combination of RR spectroscopy and the microscopic complementary insight provided by MD simulations to widen our understanding on the complex interactions that modulate the relationship between structure and reactivity in this important group of proteins.

ACKNOWLEDGEMENTS

We thank members of our research groups and co-workers who are responsible for some of the work reviewed in this chapter; their names are listed in the appropriate references. We would particularly like to thank for their contribution Dr. Alessandra Bonamore, Mr. Juan P. Bustamante, Dr. Enrica Droghetti, Dr. Alessandro Feis, and Dr. Francesco P. Nicoletti.

REFERENCES

Adar, F. (1978). Electronic absorption spectra of hemes and hemoproteins. In D. Dolphin (Ed.), *Physical chemistry, part A*: *Vol. 3. The porphyrins* (pp. 167–209). New York: Academic Press.

Aime, S., Fasano, M., Paoletti, S., Cutruzzola, F., Desideri, A., Bolognesi, M., et al. (1996). Structural determinants of fluoride and formate binding to hemoglobin and myoglobin: Crystallographic and 1H-NMR relaxometric study. *Biophysical Journal*, *70*, 482–488.

Arroyo Mañez, P., Lu, C., Boechi, L., Marti, M. A., Shepherd, M., Wilson, J. L., et al. (2011). Role of the distal hydrogen-bonding network in regulating oxygen affinity in the truncated hemoglobin III from *Campylobacter jejuni*. *Biochemistry*, *50*, 3946–3956.

Arya, S., Sethi, D., Singh, S., Hade, M. D., Singh, V., Raju, P., et al. (2013). Truncated hemoglobin, HbN, is post-translationally modified in *Mycobacterium tuberculosis* and modulates host-pathogen interactions during intracellular infection. *The Journal of Biological Chemistry*, *288*, 29987–29999.

Bailly, X., & Vinogradov, S. (2005). The sulfide binding function of annelid hemoglobins: Relic of an old biosystem? *Journal of Inorganic Biochemistry*, *99*, 142–150.

Bidon-Chanal, A., Martí, M. A., Crespo, A., Milani, M., Orozco, M., Luque, F. J., et al. (2006). Ligand-induced dynamical regulation of NO conversion in *Mycobacterium tuberculosis* truncated hemoglobin-N. *Proteins*, *64*, 457–464.

Bieza, S. A., Boubeta, F., Feis, A., Smulevich, G., Estrin, D. A., Boechi, L., et al. (2015). Reactivity of inorganic sulfide species toward a heme protein model. *Inorganic Chemistry*, *54*, 527–533.

Boechi, L., Manez, P. A., Luque, F. J., Marti, M. A., & Estrin, D. A. (2010). Unraveling the molecular basis for ligand binding in truncated hemoglobins: The trHbO *Bacillus subtilis* case. *Proteins*, *78*, 962–970.

Boechi, L., Martí, M. A., Milani, M., Bolognesi, M., Luque, F. J., & Estrin, D. A. (2008). Structural determinants of ligand migration in *Mycobacterium tuberculosis* truncated hemoglobin O. *Proteins*, *73*, 372–379.

Boffi, A., Rizzi, M., Monacelli, F., & Ascenzi, P. (2000). Determination of H2S solubility via the reaction with ferric hemoglobin I from the bivalve mollusc *Lucina pectinata*. *Biochimica et Biophysica Acta*, *1523*, 206–208.

Bonamore, A., Ilari, A., Giangiacomo, L., Bellelli, A., Morea, V., & Boffi, A. (2005). A novel thermostable hemoglobin from the actinobacterium *Thermobifida fusca*. *FEBS Journal*, *272*, 4189–4201.

Brunner, H. (1974). Identification of the iron-ligand vibration of oxyhemoglobin. *Naturwissenschaften*, *61*, 129.

Capece, L., Boechi, L., Perissinotti, L. L., Arroyo-Mañez, P., Bikiel, D. E., Smulevich, G., et al. (2013). Small ligand-globin interactions: Reviewing lessons derived from computer simulation. *Biochimica et Biophysica Acta*, *1834*, 1722–1738.

Cerda, J., Echevarria, Y., Morales, E., & Lopez-Garriga, J. (1999). Resonance Raman studies of the heme-ligand active site of haemoglobin I from *Lucina pectinata*. *Biospectroscopy*, *5*, 289–301.

Choi, S., Spiro, T. G., Langry, K. C., Smith, K. M., Budd, D. L., & La Mar, G. N. (1982). Structural correlations and vinyl influences in resonance Raman spectra of protoheme complexes and proteins. *Journal of the American Chemical Society*, *104*, 4345–4351.

Couture, M., Yeh, S.-R., Wittenberg, B. A., Wittenberg, J. B., Ouellet, Y., Rousseau, D. L., et al. (1999). A cooperative oxygen-binding hemoglobin from *Mycobacterium tuberculosis*. *Proceedings of the National Academy of Sciences of the United States of America*, *96*, 11223–11228.

Das, T. K., Couture, M., Ouellet, Y., Guertin, M., & Rousseau, D. L. (2001). Simultaneous observation of the O—O and Fe—O_2 stretching modes in oxyhemoglobins. *Proceedings of the National Academy of Sciences of the United States of America*, *98*, 479–484.

Desbois, A., Lutz, M., & Banerjee, R. (1979). Low-frequency vibrations in resonance Raman spectra of horse heart myoglobin. Iron-ligand and iron-nitrogen vibrational modes. *Biochemistry*, *18*, 1510–1518

Droghetti, E., Nicoletti, F. P., Bonamore, A., Boechi, L., Arroyo-Mañez, P., Estrin, D. A., et al. (2010). Hemepocket structural properties of a bacterial truncated hemoglobin from *Thermobifida fusca*. *Biochemistry*, *49*, 10394–10402.

Droghetti, E., Nicoletti, F. P., Bonamore, A., Sciamanna, N., Boffi, A., Feis, A., et al. (2011). The optical spectra of fluoride complexes can effectively probe H-bonding interactions in the distal cavity of heme proteins. *Journal of Inorganic Biochemistry*, *105*, 1338–1343.

Duff, L. L., Appelman, E. H., Shriver, D. F., & Klots, I. M. (1979). Steric disposition of O_2 in oxyhemoglobin as revealed by its resonance Raman spectrum. *Biochemical and Biophysical Research Communications*, *90*, 1098–1103.

Edwards, S. L., & Poulos, T. L. (1990). Ligand binding and structural perturbations in cytochrome c peroxidase. *The Journal of Biological Chemistry*, *265*, 2588–2595.

Egawa, T., Wainwright, L. M., Poole, R. K., & Yeh, S.-R. (2007). Structural and functional properties of a truncated hemoglobin from a food-borne pathogen *Campylobacter jejuni*. *The Journal of Biological Chemistry*, *282*, 13627–13636.

Egawa, T., & Yeh, S.-R. (2005). Structural and functional properties of hemoglobins from unicellular organisms as revealed by resonance Raman spectroscopy. *Journal of Inorganic Biochemistry*, *99*, 72–96.

Elber, R. (2010). Ligand diffusion in globins: Simulations versus experiment. *Current Opinion in Structural Biology*, *20*, 162–167.

Evangelista-Kirkup, R., Smulevich, G., & Spiro, T. G. (1986). Alternative carbon monoxide binding modes for horseradish peroxidase studied by resonance Raman spectroscopy. *Biochemistry*, *25*, 4420–4425.

Feis, A., Lapini, A., Catacchio, B., Brogioni, S., Foggi, P., Chiancone, E., et al. (2008). Unusually strong H-bonding to the heme ligand and fast geminate recombination dynamics of the carbon monoxide complex of *Bacillus subtilis* truncated hemoglobin. *Biochemistry, 47*, 902–910.

Feis, A., Marzocchi, M. P., & Smulevich, G. (1994). Spin-state and axial ligand bonding in the hydroxy complexes of metmyoglobin, methemoglobin, and horseradish peroxidase at room and low temperatures. *Biochemistry, 33*, 4577–4583.

Feis, A., Rodriguez-Lopez, J. N., Thorneley, R. N. F., & Smulevich, G. (1998). The distal cavity structure of carbonyl horseradish peroxidase as probed by the resonance Raman spectra of His42Leu and Arg38Leu mutants. *Biochemistry, 37*, 13575–13581.

Forti, F., Boechi, L., Bikiel, D., Martí, M. A., Nardini, M., Bolognesi, M., et al. (2011). Ligand migration in *Methanosarcina acetivorans* protoglobin: Effects of ligand binding and dimeric assembly. *The Journal of Physical Chemistry B, 115*, 13771–13780.

Freitas, T. A., Hou, S., Dioum, E. M., Saito, J. A., Newhouse, J., Gonzalez, G., et al. (2004). Ancestral hemoglobins in Archaea. *Proceedings of the National Academy of Sciences of the United States of America, 101*, 6675–6680.

Frey, A. D., & Kallio, P. T. (2003). Bacterial hemoglobins and flavohemoglobins: Versatile proteins and their impact on microbiology and biotechnology. *FEMS Microbiology Reviews, 27*, 525–545.

Giangiacomo, L., Ilari, A., Boffi, A., Morea, V., & Chiancone, E. (2005). The truncated oxygen-avid hemoglobin from *Bacillus subtilis:* X-ray structure and ligand binding properties. *The Journal of Biological Chemistry, 280*, 192–202.

Giordano, D., Pesce, A., Boechi, L., Bustamante, J. P., Caldelli, E., Howes, B. D., et al. (2015). Structural flexibility of the heme cavity in the cold-adapted truncated hemoglobin from the Antarctic marine bacterium *Pseudoalteromonas haloplanktis* TAC125. *The FEBS Journal, 282*, 2948–2965.

Giordano, D., Russo, R., Ciaccio, C., Howes, B. D., di Prisco, G., Marden, M. C., et al. (2011). Ligand- and proton-linked conformational changes of the ferrous 2/2 hemoglobin of *Pseudoalteromonas haloplanktis* TAC125. *IUBMB Life, 63*, 566–573.

Gouterman, M. (1978). Optical spectra and electronic structure of porphyrins and related rings. In D. Dolphin (Ed.), *The porphyrins: Vol. 3*. (pp. 1–165). New York: Academic Press.

Guallar, V., Lu, C., Borrelli, K., Egawa, T., & Yeh, S. R. (2009). Ligand migration in the truncated hemoglobin-II from *Mycobacterium tuberculosis*: The role of G8 tryptophan. *The Journal of Biological Chemistry, 284*, 3106–3116.

Guallar, V., & Wallrapp, F. H. (2011). QM/MM methods: Looking inside heme proteins biochemistry. *Biophysical Chemistry, 149*, 1–11.

Hill, B. C., & Nicholls, P. (1980). Cytochrome c reduction by cysteine plus copper: A pseudo substrate system for cytochrome c oxidase. *Canadian Journal of Biochemistry, 58*, 499–503.

Hirota, S., Li, T., Phillips, G. N., Jr., Olson, J. S., Mukai, M., & Kitagawa, T. (1996). Perturbation of the Fe-O_2 bond by nearby residues in heme pocket: Observation of Fe-O_2 Raman bands for oxymyoglobin mutants. *Journal of the American Chemical Society, 118*, 7845–7846.

Hori, H., & Kitagawa, T. (1980). Iron-ligand stretching band in the resonance Raman spectra of ferrous iron porphyrin derivatives. Importance as a probe band for quaternary structure of hemoglobin. *Journal of the American Chemical Society, 102*, 3608–3613.

Howes, B. D., Feis, A., Indiani, C., Marzocchi, M. P., & Smulevich, G. (2000). Formation of two types of low-spin heme in horseradish peroxidase isoenzyme A2 at low temperature. *Journal of Biological Inorganic Chemistry, 5*, 227–235.

Howes, B. D., Helbo, S., Fago, A., & Smulevich, G. (2012). Insights into the anomalous heme pocket of rainbow trout myoglobin. *Journal of Inorganic Biochemistry, 109*, 1–8.

Howes, B. D., Rodriguez-Lopez, J. N., Smith, A. T., & Smulevich, G. (1997). Mutation of distal residues of horseradish peroxidase: Influence on substrate binding and cavity properties. *Biochemistry*, *36*, 1532–1543.

Jensen, F. B. (2009). The dual roles of red blood cells in tissue oxygen delivery: Oxygen carriers and regulators of local blood flow. *Journal of Experimental Biology*, *212*, 3387–3393.

Jorgensen, W. L., Chandrasekar, J., Madura, J., Impey, R. W., & Klein, M. L. (1983). Comparison of simple potential functions for simulating liquid water. *Journal of Chemical Physics*, *79*, 926–935.

Karplus, M. L., & McCammon, J. A. (2002). Molecular dynamics simulations of biomolecules. *Nature Structural Biology*, *9*, 646–652.

Keilin, D. (1933). On the combination of methaemoglobin with H2S. *Proceedings of the Royal Society B*, *133*, 393–404.

Keilin, D. (1953). Haemoglobin in fungi. Occurrence of haemoglobin in yeast and the supposed stabilization of the oxygenated cytochrome oxidase. *Nature*, *172*, 390–393.

Kerr, E. A., & Yu, N.-T. (1988). Vibrational modes of coordinated CO, CN^- and NO. In T. G. Spiro (Ed.), *Biological applications of Raman spectroscopy: Vol. 3*. (pp. 39–95). New York: John Wiley and Sons, Inc.

Kitagawa, T. (1988). The heme protein structure and the iron histidine stretching mode. In T. G. Spiro (Ed.), *Biological applications of Raman spectroscopy: Resonance Raman spectra of hemes and metalloproteins: Vol. 3*. (pp. 97–131). New York: Wiley.

Kitagawa, T., Nagai, K., & Tsubaki, M. (1979). Assignment of the Fe-N epsilon (His F8) stretching band in the resonance Raman spectra of deoxy myoglobin. *FEBS Letters*, *104*, 376–378.

Kraus, D. W., & Wittenberg, J. B. (1990). Hemoglobins of the *Lucina pectinata*/bacteria symbiosis. I. Molecular properties, kinetics and equilibria of reactions with ligands. *The Journal of Biological Chemistry*, *265*, 16043–16053.

Larsson, J. T., Rogstam, A., & von Wachenfeldt, C. (2007). YjbH is a novel negative effector of the disulphide stress regulator, Spx, in Bacillus subtilis. *Molecular Microbiology*, *66*, 669–684.

Li, T. S., Quillin, M. L., Philips, G. N., & Olson, J. S. (1994). Structural determinants of the stretching frequency of CO bound to myoglobin. *Biochemistry*, *33*, 1433–1446.

Ling, J. H., Li, T. S., Olson, J. S., & Bocian, D. F. (1994). Identification of the iron-carbonyl stretch in distal histidine mutants of carbonmonoxymyoglobin. *Biochimica et Biophysica Acta, Bioenergetics*, *1188*, 417–421.

Lu, C., Egawa, T., Mukai, M., Poole, R. K., & Yeh, S.-R. (2008). Hemoglobins from *Mycobacterium tuberculosis* and *Campylobacter jejuni*: A comparative study with resonance Raman spectroscopy. In R. K. Poole (Ed.), *Methods in enzymology: Vol. 437*. (pp. 256–286). New York: Elsevier.

Lu, C., Egawa, T., Wainwright, L. M., Poole, R. K., & Yeh, S. R. (2007). Structural and functional properties of a truncated hemoglobin from a food-borne pathogen *Campylobacter jejuni*. *The Journal of Biological Chemistry*, *282*, 13627–13636.

Marti, M. A., Crespo, A., Capece, L., Boechi, L., Bikiel, D. E., Scherlis, D. A., et al. (2006). Dioxygen affinity in heme proteins investigated by computer simulation. *Journal of Inorganic Biochemistry*, *100*, 761–770.

Martínková, M., Kitanishi, K., & Shimizu, T. (2013). Heme-based globin-coupled oxygen sensors: Linking oxygen binding to functional regulation of diguanylate cyclase, histidine kinase, and methyl-accepting chemotaxis. *The Journal of Biological Chemistry*, *288*, 27702–27711.

Milani, M., Savard, P. Y., Ouellet, H., Ascenzi, P., Guertin, M., & Bolognesi, M. (2003). A TyrCD1/TrpG8 hydrogen bond network and a TyrB10-TyrCD1 covalent link shape the heme distal site of *Mycobacterium tuberculosis* hemoglobin O. *Proceedings of the National Academy of Sciences of the United States of America*, *100*, 5766–5771.

Mukai, M., Savard, P. Y., Ouellet, H., Guertin, M., & Yeh, S.-R. (2002). Unique ligand protein interactions in a new truncated hemoglobin from *Mycobacterium tuberculosis*. *Biochemistry*, *41*, 3897–3905.

Nagai, K., Kitagawa, T., & Morimoto, H. (1980). Quaternary structures and low frequency molecular vibrations of haems of deoxy and oxyhaemoglobin studied by resonance Raman scattering. *Journal of Molecular Biology*, *136*, 271–289.

Neri, F., Kok, D., Miller, M. A., & Smulevich, G. (1997). Fluoride binding in hemoproteins: The importance of the distal cavity structure. *Biochemistry*, *36*, 8947–8953.

Nicholls, P. (1961). The formation and properties of sulphmyoglobin and sulphcatalase. *Biochemical Journal*, *81*, 374–383.

Nicoletti, F. P., Bustamante, J. P., Droghetti, E., Howes, B. D., Fittipaldi, M., Bonamore, A., et al. (2014). Interplay of the H-bond donor-acceptor role of the distal residues in hydroxyl ligand stabilization of *Thermobifida fusca* truncated hemoglobin. *Biochemistry*, *53*, 8021–8030.

Nicoletti, F. P., Comandini, A., Bonamore, A., Boechi, L., Boubeta, F. M., Feis, A., et al. (2010). Sulfide binding properties of truncated hemoglobins. *Biochemistry*, *49*, 2269–2278.

Nicoletti, F. P., Droghetti, E., Boechi, L., Bonamore, A., Sciamanna, N., Estrin, D. A., et al. (2011). Fluoride as a probe for H-bonding interactions in the active site of heme proteins: The case of *Thermobifida fusca* hemoglobin. *Journal of the American Chemical Society*, *133*, 20970–20980.

Nicoletti, F. P., Droghetti, E., Howes, B. D., Bustamante, J. P., Bonamore, A., Sciamanna, N., et al. (2013). H-bonding networks of the distal residues and water molecules in the active site of *Thermobifida fusca* hemoglobin. *Biochimica et Biophysica Acta*, *1834*, 1901–1909.

Nicoletti, F. P., Thompson, M. K., Franzen, S., & Smulevich, G. (2011). Degradation of sulfide by dehaloperoxidase-hemoglobin from *Amphitrite ornata*. *Journal of Biological Inorganic Chemistry*, *16*, 611–619.

Nissum, M., Feis, A., & Smulevich, G. (1998). Characterization of soybean seed coat peroxidase: Resonance Raman evidence for a structure based classification of plant peroxidases. *Biospectroscopy*, *4*, 355–364.

Norberg, J. L., & Nilsson, L. (2002). Molecular dynamics applied to nucleic acids. *Accounts of Chemical Research*, *35*, 465–472.

Ouellet, H., Juszczak, L., Dantsker, D., Samuni, U., Ouellet, Y. H., Savard, P. Y., et al. (2003). Reactions of *Mycobacterium tuberculosis* truncated hemoglobin O with ligands reveal a novel ligand-inclusive hydrogen bond network. *Biochemistry*, *42*, 5764–5774.

Ouellet, H., Milani, M., LaBarre, M., Bolognesi, M., Couture, M., & Guertin, M. (2007). The roles of Tyr(CD1) and Trp(G8) in *Mycobacterium tuberculosis* truncated hemoglobin O in ligand binding and on the heme distal site architecture. *Biochemistry*, *46*, 11440–11450.

Ouellet, H., Ranguelova, K., Labarre, M., Wittenberg, J. B., Wittenberg, B. A., Magliozzo, R. S., et al. (2007). Reaction of Mycobacterium tuberculosis truncated hemoglobin O with hydrogen peroxide: Evidence for peroxidatic activity and formation of protein-based radicals. *The Journal of Biological Chemistry*, *282*, 7491–7503.

Pesce, A., Bolognesi, M., & Nardini, M. (2013). The diversity of 2/2 (truncated) globins. *Advances in Microbial Physiology*, *63*, 49–78.

Phillips, G. N., Jr., Teodoro, M. L., Li, T., Smith, B., & Olson, J. S. (1999). Bound CO is a molecular probe of electrostatic potential in the distal pocket of myoglobin. *The Journal of Physical Chemistry B*, *103*, 8817–8829.

Pietri, R., Lewis, A., Leon, R. G., Casabona, G., Kiger, L., Yeh, S. R., et al. (2009). Factors controlling the reactivity of hydrogen sulfide with hemeproteins. *Biochemistry*, *48*, 4881–4894.

Pietri, R., Roman-Morales, E., & Lopez-Garriga, J. (2011). Hydrogen sulfide and hemeproteins: Knowledge and mysteries. *Antioxidants and Redox Signaling, 15*, 393–404.

Ríos-Gonzalez, B. B., Roman-Morales, E. M., Pietri, R., & Lopez-Garriga, J. (2014). Hydrogen sulfide activation in hemeproteins: The sulfheme scenario. *Journal of Inorganic Biochemistry, 133*, 78–86.

Rodriguez, J. C., Zeng, Y., Wilks, A., & Rivera, M. (2007). The hydrogen-bonding network in heme oxygenase also functions as a modulator of enzyme dynamics: Chaotic motions upon disrupting the H-bond network in heme oxygenase from *Pseudomonas aeruginosa*. *Journal of the American Chemical Society, 129*, 11730–11742.

Roux, B., & Schulten, K. (2004). Computational studies of membrane channels. *Structure, 12*, 1343–1351.

Rwere, F., Mak, P. J., & Kincaid, J. R. (2008). Resonance Raman interrogation of the consequences of heme rotational disorder in myoglobin and its ligated derivatives. *Biochemistry, 47*, 12869–12877.

Samuni, U., Ouellet, Y., Guertin, M., Friedman, J. M., & Yeh, S. R. (2004). The absence of proximal strain in the truncated hemoglobins from *Mycobacterium tuberculosis*. *Journal of the American Chemical Society, 126*, 2682–2683.

Sitter, A. J., Shifflett, J. R., & Terner, J. (1988). Resonance Raman spectroscopic evidence for hemeiron-hydroxide ligation in peroxidase alkaline forms. *The Journal of Biological Chemistry, 263*, 13032–13038.

Smulevich, G., Evangelista-Kirkup, R., English, A., & Spiro, T. G. (1986). Raman and infrared spectra of cytochrome *c* peroxidase-carbon monoxide adducts in alternative conformational states. *Biochemistry, 25*, 4426–4430.

Smulevich, G., Feis, A., & Howes, B. D. (2005). Fifteen years of Raman spectroscopy of engineered heme containing peroxidases: What have we learned? *Accounts of Chemical Research, 38*, 433–440.

Smulevich, G., Feis, A., Howes, B. D., & Ivancich, A. (2010). Structure-function relationships among heme peroxidases: New insights from electronic absorption, resonance Raman, and multifrequency electron paramagnetic resonance spectroscopies. In K. M. Kadish, K. M. Smith, & R. Guilard (Eds.), *Handbook of porphyrin science with applications to chemistry, physics, materials science, engineering, biology and medicine: Vol. 6* (pp. 367–455). Singapore: World Scientific.

Smulevich, G., Neri, F., Marzocchi, M. P., & Welinder, K. G. (1996). Versatility of heme coordination demonstrated in a fungal peroxidase. Absorption and resonance Raman studies of *Coprinus cinereus* peroxidase and the Asp245Asn mutant at various pH values. *Biochemistry, 35*, 10576–10585.

Smulevich, G., Paoli, M., De Sanctis, G., Mantini, A. R., Ascoli, F., & Coletta, M. (1997). Spectroscopic evidence for a conformational transition in horseradish peroxidase at very low pH. *Biochemistry, 36*, 640–649.

Spiro, T. G. (1982). The resonance Raman spectroscopy of metalloporphyrins and heme proteins. In A. B. P. Lever & H. B. Gray (Eds.), *Iron porphyrins, part II* (pp. 89–152). Reading, MA: Addison-Wesley.

Spiro, T. G. (Ed.), (1988). In *Biological applications of Raman spectroscopy: Resonance Raman spectra of hemes and metalloproteins, Vol. 3*. New York: Wiley.

Spiro, T. G., & Li, X.-Y. (1988). Resonance Raman spectra of metalloproteins. In T. G. Spiro (Ed.), *Biological applications of Raman spectroscopy: Resonance Raman spectra of hemes and metalloproteins: Vol. 3* (pp. 1–37). New York: Wiley.

Spiro, T. G., Soldatova, A. V., & Balakrishnan, G. (2013). CO, NO and O_2 as vibrational probes of heme protein interactions. *Coordination Chemistry Reviews, 257*, 511–527.

Spiro, T. G., & Wasbotten, I. H. (2005). CO as a vibrational probe of heme protein active sites. *Journal of Inorganic Biochemistry, 99*, 34–44.

Stein, P., Mitchell, M., & Spiro, T. G. (1980). Hydrogen-bond and deprotonation effects on the resonance Raman iron-imidazole mode in deoxyhemoglobin models: Implications for hemoglobin cooperativity. *Journal of the American Chemical Society*, *102*, 7795–7797.

Streit, B. R., Blanc, B., Lukat-Rodgers, G. S., Rodgers, K. R., & DuBois, J. L. (2010). How active-site protonation state influences the reactivity and ligation of the heme in chlorite dismutase. *Journal of the American Chemical Society*, *132*, 5711–5724.

Teraoka, J., & Kitagawa, T. (1980). Resonance Raman study of the heme-linked ionization in reduced horseradish peroxidase. *Biochemical and Biophysical Research Communications*, *93*, 694–700.

Tinajero-Trejo, M., & Shepherd, M. (2013). The globins of *Campylobacter jejuni*. *Advances in Microbial Physiology*, *63*, 97–145.

Torge, R., Comandini, A., Catacchio, B., Bonamore, A., Botta, B., & Boffi, A. (2009). Peroxidase-like activity of *Thermobifida fusca* hemoglobin: The oxidation of dibenzylbutanolide. *Journal of Molecular Catalysis B: Enzymatic*, *61*, 303–308.

Tsubaki, M., Srivastava, R. B., & Yu, N.-T. (1982). Resonance Raman investigation of carbon monoxide bonding in (carbon monoxy) hemoglobin and -myoglobin: Detection of iron-carbon monoxide stretching and iron-carbon-oxygen bending vibrations and influence of the quaternary structure change. *Biochemistry*, *21*, 1132–1140.

Uno, T., Nishimura, Y., Tsuboi, M., Makino, R., Iizuka, T., & Ishimura, Y. (1987). Two types of conformers with distinct Fe-C-O configuration in the ferrous CO complex of horseradish peroxidase. Resonance Raman and infrared spectroscopic studies with native and deuteroheme-substituted enzymes. *The Journal of Biological Chemistry*, *262*, 4549–4556.

Vinogradov, S. N., Bailly, X., Smith, D. R., Tinajero-Trejo, M., Poole, R. K., & Hoogewijs, D. (2013). Microbial eukaryote globins. *Advances in Microbial Physiology*, *63*, 391–446.

Vinogradov, S. N., & Moens, L. (2008). Diversity of globin function: Enzymatic, transport, storage, and sensing. *The Journal of Biological Chemistry*, *283*, 8773–8777.

Vinogradov, S. N., Tinajero-Trejo, M., Poole, R. K., & Hoogewijs, D. (2013). Bacterial and archaeal globins—A revised perspective. *Biochimica et Biophysica Acta*, *1834*, 1789–1800.

Wainwright, L. M., Wang, Y., Park, S. F., Yeh, S. R., & Poole, R. K. (2006). Purification and spectroscopic characterization of Ctb, a group III truncated hemoglobin implicated in oxygen metabolism in the food-borne pathogen *Campylobacter jejuni*. *Biochemistry*, *45*, 6003–6011.

Wajcman, H., Kiger, L., & Marden, M. C. (2009). Structure and function evolution in the superfamily of globins. *Comptes Rendus Biologies*, *332*(2–3), 273–282.

Wakabayashi, S., Matsubara, H., & Webster, D. A. (1986). Primary sequence of a dimeric bacterial haemoglobin from *Vitreoscilla*. *Nature*, *322*, 481–483.

Wittenberg, J. B., Bolognesi, M., Wittenberg, B. A., & Guertin, M. (2002). Truncated hemoglobins: A new family of hemoglobins widely distributed in bacteria, unicellular eukaryotes, and plants. *The Journal of Biological Chemistry*, *277*, 871–874.

Zhao, H., & Caflisch, A. (2015). Molecular dynamics in drug design. *European Journal of Medicinal Chemistry*, *16*(91C), 4–14.

CHAPTER THREE

Structural Biology of Bacterial Haemophores

Paolo Ascenzi*[,†,1], Alessandra di Masi[‡], Loris Leboffe[‡], Emanuela Frangipani[‡], Marco Nardini[§], Cinzia Verde[†,‡], Paolo Visca[‡]
*Laboratorio Interdipartimentale di Microscopia Elettronica, Università Roma Tre, Roma, Italy
[†]Istituto di Bioscienze e BioRisorse, Consiglio Nazionale delle Ricerche, Napoli, Italy
[‡]Dipartimento di Scienze, Università Roma Tre, Roma, Italy
[§]Dipartimento di Bioscienze, Università di Milano, Milano, Italy
[1]Corresponding author: e-mail address: ascenzi@uniroma3.it

Contents

Abstract

Iron plays a key role in a wide range of metabolic and signalling functions representing an essential nutrient for almost all forms of life. However, the ferric form is hardly soluble, whereas the ferrous form is highly toxic. Thus, in biological fluids, most of the iron is sequestered in iron- or haem-binding proteins and the level of free iron is low, making haem and iron acquisition a challenge for pathogenic bacteria during infections. Although toxic to the host, free haem is a major and readily available source of iron for several pathogenic microorganisms. Both Gram-positive and Gram-negative bacteria have developed several strategies to acquire free haem–Fe and protein-bound haem–Fe. Haemophores are a class of secreted and cell surface-exposed proteins promoting free-haem uptake, haem extraction from host haem proteins, and haem presentation to

Advances in Microbial Physiology, Volume 67
ISSN 0065-2911
http://dx.doi.org/10.1016/bs.ampbs.2015.09.002

specific outer-membrane receptors that internalize the metal-porphyrins. Here, structural biology of bacterial haemophores is reviewed focusing on haem acquisition, haem internalization, and haem-degrading systems.

ABBREVIATIONS

ABC ATP-binding cassette
ABC-t ABC transporter
BslK *Bacillus anthracis* S-layer homology protein K
haem-Hpx haem-haemopexin
haem-SA serum haem-albumin
Hal haem-acquisition leucine-rich repeat protein
HasA haem-acquisition system A
HasR haem-acquisition system receptor
Hb haemoglobin
Hb–Hp haemoglobin–haptoglobin complex
Hbp haem/Hb-binding protein
Hp haptoglobin
Hpx haemopexin
Isd iron surface determinants
IsdH-N1 IsdH-NEAT domain 1
IsdH-N2 IsdH-NEAT domain 2
IsdH-N3 IsdH-NEAT domain 3
Kgp lysine-specific gingipain
Rgp arginine-specific gingipain
Mb myoglobin
NEAT near-iron-transporter
TISS Type I Secretion System
TPS two-partner secretion system

1. INTRODUCTION

Most organisms, including bacteria, require iron-porphyrin as an essential chemical entity in a variety of metabolic and signalling pathways. More than 100 bacterial enzymes possess iron-containing cofactors. Furthermore, several transcriptional and post-transcriptional regulators interact with iron to sense its intracellular level or the current status of oxidative stress for the efficient control of the expression of genes involved mainly in iron acquisition or protection from reactive oxygen species (Braun & Hantke, 2011; Contreras, Chim, Credali, & Goulding, 2014; Farhana, Saini, Kumar, Lancaster, & Steyn, 2012; Runyen-Janecky, 2013; Taabazuing, Hangasky, & Knapp, 2014; Wandersman & Delepelaire, 2012).

To fulfil their iron needs, bacteria have evolved multiple direct and indirect acquisition systems. Direct mechanisms comprise the uptake of iron from various sources available in higher organisms, such as lactoferrin, transferrin, and ferritin. On the other hand, the indirect strategies of iron acquisition may employ: (i) the activation of the secondary metabolism for the production of small iron-chelating compounds, called siderophores, and (ii) specialized secreted and/or membrane-bound proteins, called haemophores, to acquire haem from different sources including haemoglobin (Hb, containing more than 70% of iron in humans), the haemoglobin–haptoglobin complex (Hb–Hp), serum haem-albumin (haem-SA), haem-haemopexin (haem-Hpx), and myoglobin (Mb) (Aisen, Enns, & Wessling-Resnick, 2001; Brunori, 2014; Chiabrando, Mercurio, & Tolosano, 2014; Wittenberg & Wittenberg, 1990). The cellular uptake of iron is restricted to its physiologically most relevant species, Fe(II) (ferrous iron) and Fe(III) (ferric iron). Specific systems for Fe(II) uptake are well known for being widespread in bacteria, and appear to be functional when Fe(II) is the prevalent form of iron, e.g., in anoxic/microaerobic and reducing environments. However, Fe(II) is oxidized to Fe(III) either spontaneously, by reacting with molecular oxygen, or enzymatically, during assimilation and circulation in host organisms (Ascenzi et al., 2005; Braun & Hantke, 2011; Butler, 2005; Contreras et al., 2014; Cornelis, 2010; Farhana et al., 2012; Granger & Price, 1999; Neilands, 1995; Runyen-Janecky, 2013; Schröder, Johnson, & de Vries, 2003; Taabazuing et al., 2014; Visca, Imperi, & Lamont, 2007; Vraspir & Butler, 2009; Wandersman & Delepelaire, 2012).

Both siderophores and haemophores contribute to the virulence of many pathogens (Wilks & Burkhard, 2007; Ge & Sun, 2014). Indeed, iron acquisition is pivotal for infections caused by *Bordetella* (Brickman, Vanderpool, & Armstrong, 2006), *Haemophilus* (Morton et al., 2004), *Brucella* (Paulley, Anderson, & Roop, 2007), *Vibrio* (Henderson & Paine, 1994), *Streptococcus* (Tai, Lee, & Winter, 1993), *Staphylococcus* (Mazmanian, Ton-That, Su, & Schneewind, 2002), and many other Gram-positive and Gram-negative species. This is because pathogenic bacteria need to acquire host iron to replicate during infection, although several strategies have been developed in mammals and birds to restrict the growth of invading pathogens by reducing iron availability (Ascenzi et al., 2005; Becker & Skaar, 2014; Ge & Sun, 2014). Hosts with high levels of free iron are generally more susceptible to infection (Kontoghiorghes, Kolnagou, Skiada, & Petrikkos, 2010), and mutational inactivation of iron acquisition systems was shown to attenuate infectivity and virulence in a wide range of bacterial species (Braun, 2001; Crosa, 1997; Nairz, Schroll, Sonnweber, & Weiss, 2010). The low free iron concentration in host tissues likely functions as a barrier for efficient bacterial replication (Becker & Skaar, 2014; Sawatzki, Hoffmann, & Kubanek, 1983). The pivotal role of iron acquisition during infection of mammalian hosts supports the view that inhibition of iron uptake is promising for new therapies in infections (Becker & Skaar, 2014; Foley & Simeonov, 2012; Hulkoti & Taranath, 2014).

Since their discovery in *Serratia marcescens* (Létoffé, Ghigo, & Wandersman, 1994a, 1994b), haemophores have been found in Gram-negative and Gram-positive bacteria. Haemophores (i) belong to several and distinct protein families, (ii) are either released into the supernatant or are anchored to the cell surface, (iii) can reach the cell surface and/or the extracellular medium via distinct secretion pathways, (iv) can uptake haem from haem-proteins using different mechanisms, and (v) participate to haem transfer between partners (Cescau et al., 2007; Contreras et al., 2014; Létoffé, Redeker, & Wandersman, 1998; Maresso, Garufi, & Schneewind, 2008; Rossi et al., 2001; Wandersman & Delepelaire, 2004, 2012).

Haem-acquisition genes are negatively regulated by repressors which sense the intracellular iron content, and finely tune gene expression to prevent iron overload. When iron is limiting, the cytoplasmic iron pool is low, therefore, transcriptional repressors cannot bind the DNA and expression of haem-acquisition genes can occur. An additional level of complexity of the haem uptake systems is autoregulation, i.e., the capacity of bacteria to sense haem in the extracellular milieu and induce the expression of cognate uptake genes. Under iron limiting conditions, transcription of haem uptake genes is increased either by haem (in systems lacking haemophore) or by haem-loaded haemophores via surface signalling devices (Llamas, Imperi, Visca, & Lamont, 2014; Rossi, Paquelin, Ghigo, & Wandersman, 2003; Vanderpool & Armstrong, 2003).

Here, structural biology of bacterial haemophores is reviewed focussing on haem acquisition, haem internalization, and haem degradation systems.

2. HasA-TYPE HAEMOPHORE SYSTEMS FROM *Serratia marcescens, Pseudomonas aeruginosa,* AND *Yersinia pestis*

The haem-acquisition system A (HasA) is present in several Gram-negative bacteria, including *S. marcescens*, *P. aeruginosa*, and *Y. pestis*. The haemophores form a family of highly conserved proteins without homology to any other known protein. They bind free haem and/or extract the metal-porphyrin from Hb, presenting it to specific outer-membrane receptors. Then, the haem is internalized and enzymatically degraded to release free iron. The *S. marcescens* Has haemophore was the first member of this class of transporters to be characterized in detail and is now considered as the molecular model for this class of haemophore systems.

2.1 HasA is Secreted by the Type I Secretion Pathway

HasA-type haemophores are secreted by the Type I Secretion System (TISS) (Létoffé et al., 1994a, 1994b), which represents one of the five major secretion pathways identified in Gram-negative bacteria (Wandersman & Delepelaire, 2004). Typical TISSs are invariably characterized by a hetero-trimeric complex formed by: (i) an inner-membrane ATPase belonging to the ATP-binding cassette (ABC) class of proteins (i.e. HasD) that provides the energy for the haemophore export, (ii) an inner-membrane fusion protein (i.e. HasE), and (iii) an outer-membrane protein that belongs to the TolC family (i.e. HasF), forming a tunnel between the inner and outer membranes. Under poor iron and haem conditions in the external environment, HasA is bound to the haem-acquisition system receptor (HasR). When haem is the available iron source, apo-HasA is released; after haem binding, holo-HasA is uptaken. In the Has system, HasB is a dedicated TonB analogue that, together with ExbB and ExbD, allows haem entry and apo-HasA recycling. In turn, haem is shuttled across the inner membrane by the ABC transporter (ABC-t). Signalling across the cell envelope requires the specific sigma and anti-sigma factors HasI and HasS, respectively. After internalization, the haem is enzymatically degraded (Fig. 1) (Contreras et al., 2014; Wandersman & Delepelaire, 2012).

2.2 The HasA Structure

HasA-type haemophores are grossly globular proteins of about 21 kDa, which bind haem with the 1:1 stoichiometry. The three-dimensional structures of haem-free and haem-bound HasA-type haemophores are closely similar, the main difference being the wide displacement of the loop carrying His32 in *S. marcescens* and *P. aeruginosa*. On the other hand, the geometry of the Gln32-containing loop of *Y. pestis* HasA, overlapping the conformation of the His32-containing loop of *S. marcescens* and *P. aeruginosa* holo-HasA, is substantially unaffected by haem–Fe(III) binding (Figs. 2 and 3, and Table 1)

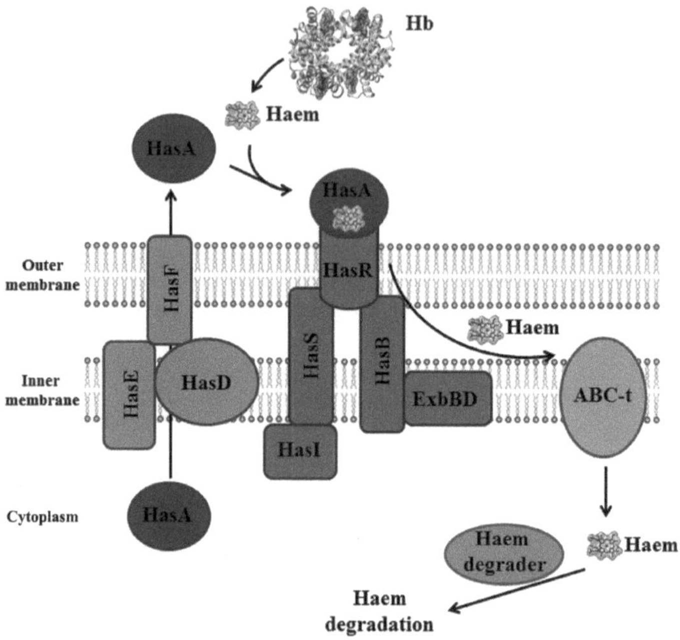

Figure 1 Haem acquisition in *S. marcescens*. When haem is available, HasA is released from the cytoplasm in the extracellular environment. HasA-type haemophores are secreted by TISS transporters that are constituted by a heterotrimeric complex formed by an inner membrane ATPase belonging to the ABC class of proteins (i.e. HasD), an inner-membrane fusion protein (i.e. HasE), and an outer-membrane protein that belongs to the TolC family (i.e. HasF). After binding of HasA to HasR, the haem is quickly transferred to HasR through an energy-independent mechanism, and apo-HasA is recycled. Haem release from HasR requires HasB, a TonB orthologue that works with two auxiliary proteins, i.e., ExbB and ExbD. Signalling across the membrane requires the presence of the sigma (i.e. HasI) and anti-sigma (i.e. HasS) factors. Once internalized, haem is degraded. For details, see the text.

(Alontaga et al., 2009; Arnoux et al., 1999; Caillet-Saguy et al., 2008; Czjzek et al., 2007; de Amorim et al., 2013; Jepkorir et al., 2010; Kumar et al., 2013, 2014; Shirataki et al., 2014; Wolff et al., 2008).

HasA displays two opposite faces, which are built up by four α-helices and seven β-strands, respectively. The His and/or Tyr residues coordinating the haem–Fe atom are positioned on two loops that separate the two faces of the HasA-type haemophores. Haem binds to *S. marcescens* and *P. aeruginosa* HasA-type haemophores by the axial ligands His32 and Tyr75. However, only Tyr75 coordinates the haem–Fe atom in *Y. pestis* HasA; the sixth coordination ligand of the haem–Fe(III) atom might be a solvent molecule and/or a loosely bound chloride ion. The strong Tyr–haem–Fe axial ligand

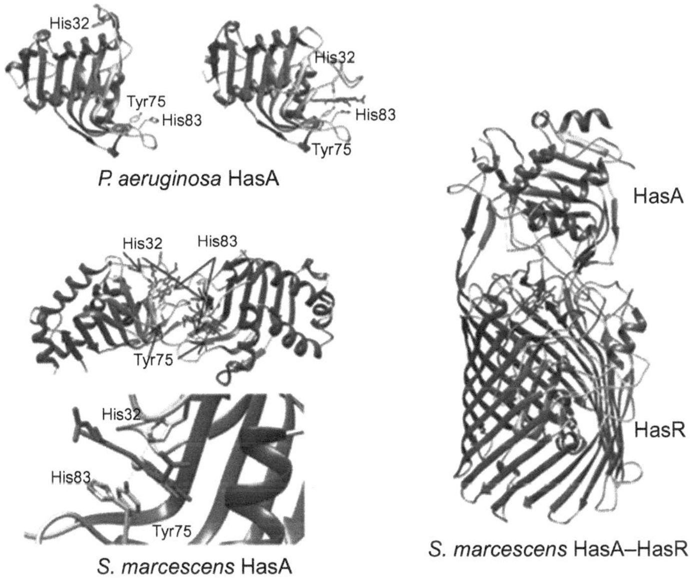

Figure 2 Crystal structures of *S. marcescens* and *P. aeruginosa* haem uptake systems. Ligand-free (left) and haem-bound (right) *P. aeruginosa* HasA (PDB ID: 3MOK and 3ELL, respectively). The haem axial ligands His32 and Tyr75, and the His83 residue stabilizing the tyrosinate form of Tyr75 are shown in stick representation. Swapped homodimeric haem-bound HasA from *S. marcescens* (PDB ID: 2CN4). The haem axial ligands His32 and Tyr75 residues of both monomers are shown in stick representation. Heterodimeric haem-bound HasA–HasR complex from *S. marcescens* (PDB ID: 3CSL). Helices are in purple, strands in forest green, coils in grey, haem in red, and the haem–Fe atom in gold. *Pictures drawn by UCSF-Chimera (Pettersen et al., 2004).* (See the colour plate.)

reflects the tyrosinate state of the Tyr coordinating residue as a consequence of hydrogen bonding to the nearby residue His83 in *S. marcescens* and *P. aeruginosa*, and His81 in *Y. pestis*. In turn, this His residue may affect haem–Fe(III) binding. Recently, it has been suggested that hydrophobic, π–π stacking, and van der Waals interactions participate to the modulation of haem binding to HasA, whereas axial coordination likely controls the slow haem release. This allows HasA to meet the challenge of capturing haem under unwelcoming conditions and releasing it selectively to HasR. In holo-HasA, haem may adopt two orientations along the α–γ meso axis with the propionates exposed to the solvent (Figs. 2 and 3) (Alontaga et al., 2009; Arnoux et al., 1999; Caillet-Saguy et al., 2008; Czjzek et al., 2007; de Amorim et al., 2013; Jepkorir et al., 2010; Kumar et al., 2013, 2014; Shirataki et al., 2014; Wolff et al., 2008).

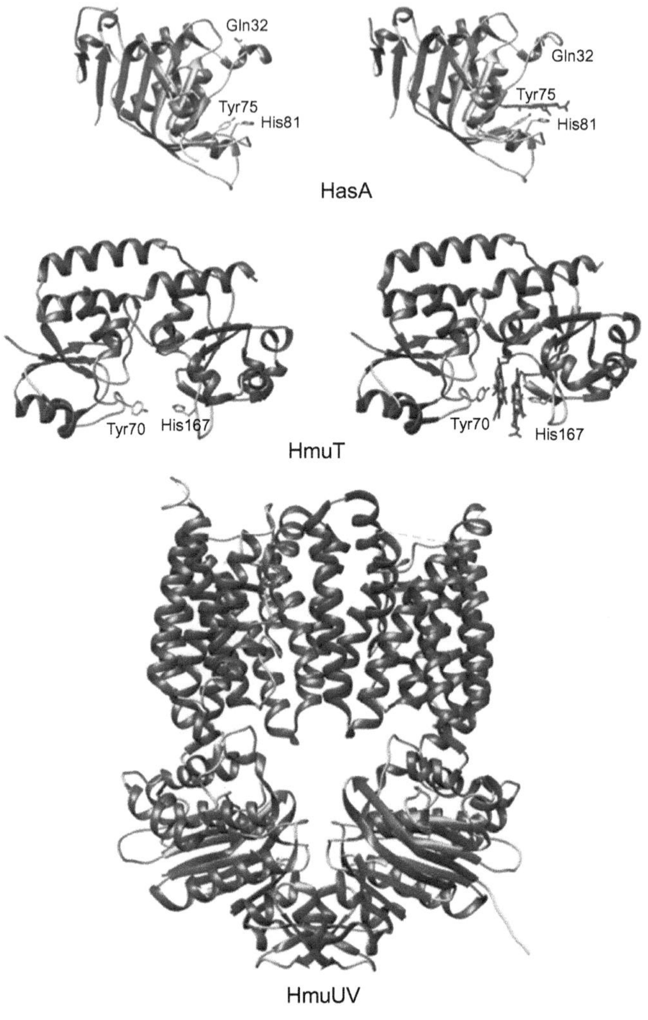

Figure 3 Crystal structures of *Y. pestis* haem uptake system. Ligand-free (left) and haem-bound (right) HasA (PDB ID: 4JER and 4JET, respectively). The haem axial ligand Tyr75 and the His81 residue stabilizing the tyrosinate form of Tyr75 are shown in stick representation. Ligand-free (left) and haem-bound (right) periplasmic haem-binding protein HmuT (PDB ID: 3MD9 and 3NU1, respectively). The haem axial ligands Tyr70 and His167 are shown in stick representation. Ligand-free haem transporter HmuUV (PDB ID: 4G1U). Helices are in purple, strands in forest green, coils in grey, haem in red, and the haem–Fe atom in gold. *Pictures drawn by UCSF-Chimera (Pettersen et al., 2004).* (See the colour plate.)

Table 1 Three-Dimensional Structures of Bacterial Haemophores

Organism	PDB ID	Protein	Ligand	Resolution (Å)	Notes [a]	References
Serratia marcescens	1B2V	HasA	Fe(III)-protoporphyrin-IX	1.90	pH 4.6	Arnoux et al. (1999)
	1DKH	HasA	Fe(III)-protoporphyrin-IX	3.20	pH 6.5	Arnoux, Haser, Izadi-Pruneyre, Lecroisey, and Czjzek (2000)
	1DK0	HasA	Fe(III)-protoporphyrin-IX	1.77	pH 8.0	Arnoux et al. (2000)
	1YBJ	HasA	Ligand-free	–[b]		Wolff et al. (2008)
	2CN4	HasA	Fe(III)-protoporphyrin-IX	2.30	Swapped homo-dimer	Czjzek et al. (2007)
	2M2K	HasB	Ligand-free	–[b]		de Amorim et al. (2013)
	2UYD	HasA	Fe(III)-protoporphyrin-IX	2.70	His83Ala	Caillet-Saguy et al. (2008)
	3CSL	HasA–HasR	Fe(III)-protoporphyrin-IX	2.70	Hetero-dimer	Krieg et al. (2009)
	3CSN	HasA–HasR	Ligand-free	3.00	Hetero-dimer	Krieg et al. (2009)
	3DDR	HasA–HasR	Fe(III)-protoporphyrin-IX	2.80	HasR Ile671Gly, hetero-dimer	Krieg et al. (2009)
Pseudomonas aeruginosa	3ELL	HasA	Fe(III)-protoporphyrin-IX	1.70		Alontaga et al. (2009)
	3MCK	HasA	Ligand-free	1.55		Jepkorir et al. (2010)
	3MOL	HasA	Fe(III)-protoporphyrin-IX	1.20	His32Ala, homo-dimer	Jepkorir et al. (2010)

Continued

Table 1 Three-Dimensional Structures of Bacterial Haemophores—cont'd

Organism	PDB ID	Protein	Ligand	Resolution (Å)	Notes	References
	3MOM	HasA	Fe(III)-protoporphyrin-IX, imidazole	2.25	His32Ala, homo-dimer	Jepkorir et al. (2010)
	3W8M	HasA	Fe(III)-salophen	1.46		Shirataki et al. (2014)
	3W8O	HasA	Phtalocyaninato-Fe(III)	1.85		Shirataki et al. (2014)
	3WAH	HasA	Fe(III)-mesoporphyrin-IX	1.54		Shirataki et al. (2014)
	4O6Q	HasA	Fe(III)-protoporphyrin-IX, formate	0.95	Tyr75Ala	Kumar et al. (2014)
	4O6S	HasA	Fe(III)-protoporphyrin-IX, H_2O	1.32	His83Ala	Kumar et al. (2014)
	4O6T	HasA	Fe(III)-protoporphyrin-IX	1.25	His83Ala, pH 5.4	Kumar et al. (2014)
	4O6U	HasA	Fe(III)-protoporphyrin-IX	0.89	His83Ala, pH 6.5	Kumar et al. (2014)
Yersinia pestis	3MD9	HmuT	Ligand-free	1.50		Mattle, Zeltina, Woo, Goetz, and Locher (2010)
	3NU1	HmuT	Fe(III)-protoporphyrin-IX	2.50		Mattle et al. (2010)
	4G1U	HmuUV	Ligand-free	3.01	Hetero-tetramer	Woo, Zeltina, Goetz, and Locher (2012)
	4JER	HasA	Ligand-free	1.10	Tetragonal form	Kumar et al. (2013)
	4JES	HasA	Ligand-free	1.60	Hexagonal form	Kumar et al. (2013)
	4JET	HasA	Fe(III)-protoporphyrin-IX	2.20		Kumar et al. (2013)

Haemophilus influenzae	4I84	HxuA	Ligand-free	1.50	Homo-dimer	Baelen, Dewitte, Clantin, and Villeret (2013)
Porphyromonas gingivalis	3H8T	HmuY	Fe(III)-protoporphyrin-IX	1.80	Homo-tetramer	Wójtowicz et al. (2009)
Staphylococcus aureus	1XBW	IsdG	Ligand-free	1.90		Wu et al. (2005)
	2E7D	IsdH-N3	Ligand-free	2.20	Homo-dodecamer	Watanabe et al. (2008)
	2H3K	IsdH	Ligand-free	–[b]		Pilpa et al. (2006)
	2ITE	IsdA	Ligand-free	1.60	Se-Met	Grigg, Vermeiren, Heinrichs, and Murphy (2007a)
	2ITF	IsdA	Fe(III)-protoporphyrin-IX	1.90	Se-Met	Grigg et al. (2007a)
	2LHR	IsdH	Ligand-free	–[b]	Linker domain	Spirig et al. (2013)
	2K73	IsdC	Zn(II)-protoporphyrin-IX	–[b]		Villareal, Pilpa, Robson, Fadeev, and Clubb (2008)
	2MOQ	IsdB	Ligand-free	–[b]		Fonner et al. (2014)
	2O1A	IsdA	Ligand-free	1.60	Ala47-Lys188, Se-Met	Chang et al. (to be published)
	2O6P	IsdC	Fe(III)-protoporphyrin-IX	1.50		Sharp, Schneider, Cockayne, and Paoli (2007)
	2Q8P	IsdE	Fe(III)-protoporphyrin-IX	1.95	Se-Met	Grigg, Vermeiren, Heinrichs, and Murphy (2007b)

Continued

Table 1 Three-Dimensional Structures of Bacterial Haemophores—cont'd

Organism	PDB ID	Protein	Ligand	Resolution (Å)	Notes	References
	2Q8Q	IsdE	Fe(III)-protoporphyrin-IX	2.15		Grigg et al. (2007b)
	2Z6F	IsdH-N3	Fe(III)-protoporphyrin-IX	1.90	Homo-dimer	Watanabe et al. (2008)
	2ZDO	IsdG-N7A	Fe(III)-protoporphyrin-IX	1.80	Homo-dimer	Lee, Reniere, Skaar, and Murphy (2008)
	2ZDP	IsdI	Co(III)-protoporphyrin-IX, chloride	1.50	Homo-dimer	Lee et al. (2008)
	3LGM	IsdI	Fe(II)-protoporphyrin-IX	1.88	Homo-dimer	Reniere et al. (2010)
	3LGN	IsdI	Fe(II)-protoporphyrin-IX, dioxygen	1.50	Homo-dimer	Reniere et al. (2010)
	3QGP	IsdI	Fe(III)-protoporphyrin-IX, cyanide	1.80	Homo-dimer	Takayama et al. (to be published)
	3QUG	IsdH-N3	Ga(III)-protoporphyrin-IX	1.70		Moriwaki et al. (2011)
	3QUH	IsdH-N3	Mn(III)-protoporphyrin-IX	2.70		Moriwaki et al. (2011)
	3QZL	IsdA	Fe(III)-protoporphyrin-IX	1.30	Lys75Ala	Grigg, Mao, and Murphy (2011)
	3QZM	IsdA	Fe(III)-protoporphyrin-IX	1.25	His83Ala	Grigg et al. (2011)
	3QZN	IsdA	Fe(III)-protoporphyrin-IX	2.00	Tyr166Ala	Grigg et al. (2011)
	3QZO	IsdA	Fe(II)-protoporphyrin-IX	1.95		Grigg et al. (2011)
	3QZP	IsdA	Co(II)-protoporphyrin-IX	1.90		Grigg et al. (2011)

	3RTL	IsdB-N2	Fe(III)-protoporphyrin-IX	1.45		Gaudin, Grigg, Arrieta, and Murphy (2011)
	3RUR	IsdB-N2	Fe(III)-protoporphyrin-IX	1.70	Se-Met	Gaudin et al. (2011)
	3SZK	IsdH-αβHb	Fe(III)-protoporphyrin-IX	3.01	Hetero-trimer	Krishna Kumar et al. (2011)
	3VTM	IsdH-N3	In(III)-protoporphyrin-IX	2.80		Vu et al. (2013)
	3VUA	IsdH-N3	Ligand-free	1.85		Vu et al. (2013)
	4FNH	IsdI	Fe(III)-protoporphyrin-IX	1.90	Trp66Tyr, homo-dimer	Ukpabi, Takayama, Mauk, and Murphy (2012)
	4FNI	IsdI	Fe(III)-protoporphyrin-IX, cyanide	1.80	Trp66Tyr, homo-dimer	Ukpabi et al. (2012)
Bacillus anthracis	3SIK	IsdX1	Fe(III)-protoporphyrin-IX	2.15		Ekworomadu et al. (2012)
	3SZ6	IsdX1	Ligand-free	1.80		Ekworomadu et al. (2012)
	4H8P	IsdX2-N5	Fe(III)-protoporphyrin-IX	2.05		Honsa, Owens, Goulding, and Maresso (2013)
	4H8Q	IsdX2-N5	Fe(III)-protoporphyrin-IX	1.70	Gln29Thr	Honsa et al. (2013)
Listeria monocytogenes	4MYP	Hbp2-N2	Fe(III)-protoporphyrin-IX	1.80	Homo-dimer	Malmirchegini et al. (2014)
	4NLA	Hbp2-N2	Ligand-free	2.70	Homo-dimer	Malmirchegini et al. (2014)
Mycobacterium tuberculosis	3MAY	Rv0203	Ligand-free	2.50	Homo-tetramer	Tullius et al. (2011)

[a]Where not indicated, proteins are wild-type monomers.
[b]Solution NMR structure.

The affinity of haem for *S. marcescens* HasA is very high ($K_d \sim 10^{-12}$ M) increasing when Hb is in the ferric form and dissociates into αβ-dimers (Izadi et al., 1997). Kinetics of haem binding to *P. aeruginosa* HasA is biphasic. The first step corresponds to the penta-coordination of the haem–Fe atom by Tyr75, whereas the second step involves the closure of the loop carrying His32 and the hexa-coordination of the haem–Fe atom. On the other hand, kinetics of haem binding to *Y. pestis* HasA is very fast and monophasic reflecting the minimal structural reorganization occurring upon haem binding (Figs. 2 and 3) (Kumar et al., 2013; Yukl et al., 2010).

HasA-type haemophores of *S. marcescens* and *P. aeruginosa* have been reported to recover haem from Hb, Hpx, legHb, and Mb (Contreras et al., 2014; Wandersman & Delepelaire, 2012). Of note, the rate of haem transfer from Hb to HasA is limited by haem dissociation from Hb (Yukl et al., 2010). Since the HasA–Hb complex has never been observed, it has been hypothesized that HasA only binds free haem, behaving as a siderophore *sensu strictu* (Fig. 1) (Létoffé, Nato, Goldberg, & Wandersman, 1999; Létoffé et al., 2001; Wandersman & Delepelaire, 2012).

Interestingly, the homo-dimeric form of *S. marcescens* HasA, which is secreted under haem-limiting conditions, shows a swapped structural organization (Czjzek et al., 2007). The overall structure of each subunit of homo-dimeric *S. marcescens* HasA is very similar to that of monomeric HasA, but it is formed by parts belonging to the two subunits. Consequently, homo-dimeric HasA binds two haem molecules by residues His32 and Tyr75 belonging to different subunits (Fig. 2). Homo-dimeric HasA, displaying a haem affinity lower than that of the monomeric form, does not deliver the haem to the HasR. The functional meaning of HasA dimerization is intriguing, since homo-dimeric HasA (i) could act as an extracellular haem reservoir for bacteria, preventing the surplus of haem incorporation; (ii) could subtract haem from the environment; and (iii) could provide haem to the bacterium upon monomerization or by transferring the haem to the monomeric form (Czjzek et al., 2007).

2.3 The HasA–HasR Complex

HasR, belonging to the family of TonB-dependent receptors (Faraldo-Gómez & Sansom, 2003; Ferguson & Deisenhofer, 2002), actively internalizes haem through the outer membrane of Gram-negative bacteria, whereas apo-HasA persists outside the bacterium (Barjon, Wecker, Izadi-Pruneyre, & Delepelaire, 2007; Ghigo, Létoffé, & Wandersman, 1997).

HasR, displays the classical plug-and-barrel architecture of outer-membrane receptors implicated in iron uptake, consisting of a transmembrane C-terminal β-barrel (composed of 22 antiparallel β-strands) and of a N-terminal plug domain closing the β-barrel structure (Fig. 2 and Table 1) (Krieg et al., 2009).

The HasA–HasR complex formation is unaffected by the apo- or holo-form of HasA. The haem-free and haem-bound derivatives of HasA join to HasR with high affinity ($K_d \sim 10^{-9}$ M) (Létoffé et al., 2001). The HasA–HasR contact area is made up by two regions, named 1 and 2. Region 1 is characterized by loops L6, L8, and L9, and region 2 is composed by loops L2, L3, L10, and L11. The HasA–HasR recognition process has been postulated to initially occur at the contact region 1 and later at the contact region 2. It cannot be excluded that HasA binding to the contact region 1 facilitates protein–protein recognition at the contact region 2 (Barjon et al., 2007; Krieg et al., 2009; Létoffé, Debarbieux, Izadi, Delepelaire, & Wandersman, 2003).

In the HasA–HasR–haem complex, the haem is bound to HasR by axial coordination to His189 (positioned on the extracellular surface of the N-terminal plug) and His603 (sited on an extracellular loop of the β-barrel); no stabilization by hydrophobic stacking appears to occur (Fig. 2). In HasR, a large cavity under the haem may represent a path for haem translocation (Krieg et al., 2009).

When holo-HasA binds to apo-HasR, the haem is quickly transferred from HasA to HasR although the haem affinity for HasR is lower than that for HasA by about seven orders of magnitude (Benesch & Kwong, 1990; Izadi et al., 1997; Izadi-Pruneyre et al., 2006). Of note, haem transfer from holo-HasA to apo-HasR does not appear to be an energy-dependent process, but it seems to be allosterically modulated. The inspection of the three-dimensional structure of the complex built up by holo-HasA and the HasR-Ile671Gly mutant, which binds holo-HasA but does not undergo haem internalization, suggests that early steps of haem transfer are represented by the cleavage of the haem-Fe-His32 coordination bond of holo-HasA and the alteration of the geometry of the HasA loop carrying the His32 residue. Moreover, the Ile671 residue, located in loop 8 of the HasR, may come into steric clash with the haem bound to holo-HasA, thus facilitating the metal-porphyrin transfer from holo-HasA to apo-HasR (Krieg et al., 2009).

After haem transfer from holo-HasA to apo-HasR, apo-HasA remains bound to HasR. *In vivo*, apo-HasA release from HasR is an energy-driven process requiring TonB (Létoffé, Delepelaire, & Wandersman, 2004) or

HasB, the TonB orthologue specific to HasR (Paquelin, Ghigo, Bertin, & Wandersman, 2001). Energy-driven conformational changes in the HasR plug domain might lead to the rearrangement of the HasR external loops allowing apo-HasA release, which appears to occur only in the presence of haem in the medium. Moreover, haem transport through HasR might play a role in apo-HasA release (Contreras et al., 2014; Krieg et al., 2009; Wandersman & Delepelaire, 2012).

The highly pathogenic species *Y. pestis* can grow in iron-deficient media supplemented with either free haem or haem-proteins including Hb, Hb–Hp, haem–SA, haem–Hpx, and Mb (Sikkema & Brubaker, 1989). *Y. pestis* possesses the haem-protein acquisition systems Has, sharing homology with that of *S. marcescens* (Fig. 1), and Hmu. The Hmu system is composed by the HmuR outer-membrane receptor, the periplasmic binding protein HmuT, and the ABC-t HmuTUV. Both HmuR and HmuTUV are required for haem uptake and extraction from haem–SA, and Mb, while only HmuR is needed to extract haem from Hb and haem–Hpx. Lastly, the haem degradation has been hypothesized to be catalysed by HmuS (Forman, Paulley, Fetherston, Cheng, & Perry, 2010; Rossi et al., 2001).

HmuT is a bi-lobated protein binding two staked haems in the central groove. The two haem molecules are axially coordinated by different residues, i.e., Tyr70 and His167 (Fig. 3). HmuT binds the two haem molecules in discrete, sequential steps with high affinity ($K_d \sim 3 \times 10^{-8}$ M). Based on structural and functional considerations, it has been hypothesized that the cognate ABC-t HmuUV may simultaneously transfer two haem molecules per translocation cycle (Mattle et al., 2010). The homo-dimeric HmuU transporter features 20 transmembrane α-helices and the homo-dimeric nucleotide-binding HmuV is a α/β protein. The HmuU subunits bind to HmuV chains by non-covalent contacts mediated by the coupling α-helices. At the interface of the HmuU subunits, a cavity mainly formed by transmembrane α-helix 5 and the α-helix 5a is present, however it is only accessible from the periplasm being closed to the cytoplasm. A haem molecule has been modelled into the cavity without producing steric clashes. The ATP hydrolysis rate of HmuUV is stimulated by HmuT binding, reflecting a tight HmuT–HmuUV association. It has been calculated that the apparent stoichiometry of ATP consumption per transported haem is ~300:1 (Woo et al., 2012). Also in this case, transport of haem across the outer membrane appears to require energy coupling by the TonB system (Perry, Shah, Bearden, Thompson, & Fetherston, 2003).

While the haem route represents the main way for *Y. pestis* dissemination in infected animals, the importance of haem uptake in its pathogenicity appears low since mutations in both *has* and *hmu* do not affect virulence in mouse models of bubonic and pneumonic plague (Forman et al., 2010; Rossi et al., 2001).

3. THE HxuA–HxuB–HxuC HAEMOPHORE SYSTEM OF *Haemophilus influenzae*

The HxuA–HxuB–HxuC haemophore system has been described in *H. influenzae*, a bacterial species renowned for its obligate requirement of exogenous haem (the X factor) for aerobic growth in laboratory media. The corresponding gene cluster, comprising the genes *hxuA*, *hxuB*, and *hxuC*, allows the utilization of the haem–Hpx as the metal-porphyrin source (Fig. 4) (Cope et al., 1994).

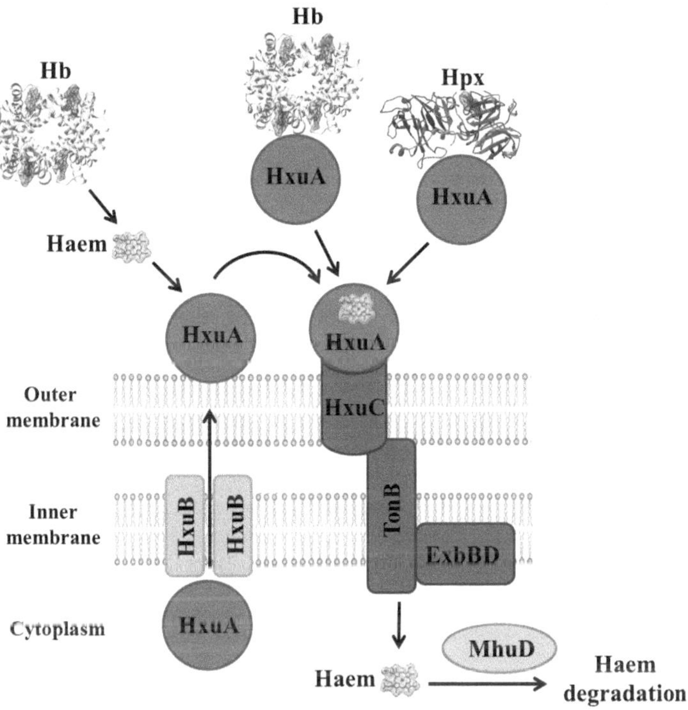

Figure 4 Haem acquisition in *H. influenzae*. HxuA is secreted by a signal peptide-dependent pathway that requires the specific pore-forming transporter HxuB. HxuA, which is either anchored at the cell surface or found at least partially in the medium supernatant, depending on the strain, binds both haem-free and haem-Hpx with high-affinity. HxuA facilitates haem uptake from the haem–Hpx complex and binds the specific outer-membrane receptor HxuC, which is also involved in haem and Hb uptake. Active haem transport across the outer membrane from the haem–Hpx complex depends on the inner membrane energy-transducing TonB–ExbB–ExbD complex. Once into the cytoplasm, haem is degraded by the haem-degrading enzyme MhuD. For details, see the text.

H. influenzae HxuA, a 100-kDa protein only present in the *Pasteurellaceae* genus, is secreted by a signal peptide-dependent pathway that requires the specific pore-forming β-barrel transporter HxuB. Of note, the HxuA–HxuB machine belongs to the so-called two-partner secretion systems (TPS). Hallmarks of TPS are the presence of (i) an N-proximal module where specific secretion signals in the substrate protein are found and (ii) a β-barrel channel homologue (Jacob-Dubuisson et al., 2000). HxuA displays the conserved N-terminal region of TPS proteins, which is pivotal for secretion. The secretion domain of HxuA folds into a β-helix domain with two extra-α-helical motifs, a four-stranded β-sheet, and a N-terminal cap (Fig. 5 and Table 1). Homo-dimeric HxuA can either be anchored at the cell surface or be found in the extracellular medium depending on the strain. HxuA binds both haem-free and the haem–Hpx complex with $K_d \sim 10^{-6}$ and $\sim 10^{-9}$ M, respectively. Also, Hb has been reported to bind to HxuA. HxuA binds the specific outer-membrane receptor HxuC, then the active haem transport across the outer membrane depends on the inner membrane energy-transducing TonB–ExbB–ExbD complex (Figs. 4 and 5) (Baelen et al., 2013; Cope, Thomas, Hrkal, & Hansen, 1998; Cope, Yogev,

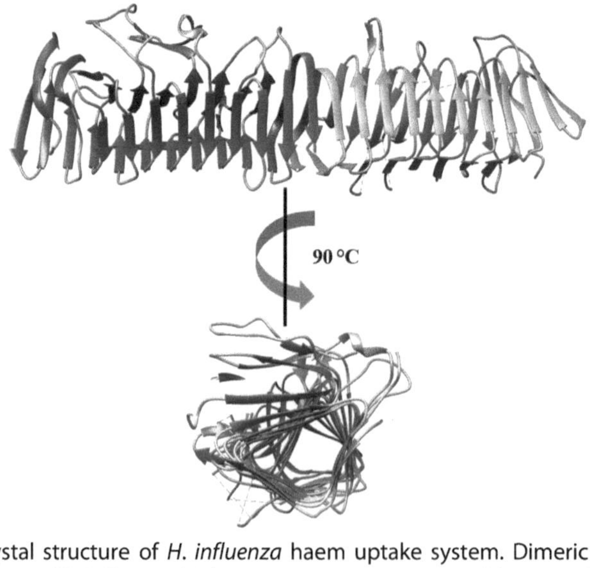

Figure 5 Crystal structure of *H. influenza* haem uptake system. Dimeric *H. influenzae* ligand-free HxuA (PDB ID: 4I84). The two pictures are rotated by 90°. For clarity, the two β-helix subunits are in forest green and chartreuse (coils in grey). *Pictures drawn by UCSF-Chimera (Pettersen et al., 2004).* (See the colour plate.)

Muller-Eberhard, & Hansen, 1995; Cope et al., 2001, 1994; Fournier, Smith, & Delepelaire, 2011; Morton et al., 2007; Wandersman & Delepelaire, 2012; Whitby, Sim, Morton, Patel, & Stull, 1997; Wong et al., 1995).

4. THE HmuY–HmuR HAEMOPHORE SYSTEM OF *Porphyromonas gingivalis*

Many Gram-negative bacteria express bipartite Hb receptors, which are pivotal to recover the haem from host haem-proteins. They consist of an outer-membrane lipoprotein and of a typical outer-membrane receptor with the plug inside the β-barrel fold (Contreras et al., 2014; Olczak, Sroka, Potempa, & Olczak, 2008; Rohde & Dyer, 2004; Wandersman & Delepelaire, 2012; Wójtowicz et al., 2009).

P. gingivalis uptakes haem through a variety of potential gateways, among which HmuY/HmuR of the *hmu* locus are the best described. Bacterial haemophores are expressed in the course of infection by *P. gingivalis*, the haem being required for microbial growth during the establishment of infection and for biofilm formation. Most importantly, *P. gingivalis* lipoprotein HmuY induces interleukin-1b, -6, -8, and -10, Bcl-2, and Fas eukaryotic proteins, as well as immunoglobulin G, suggesting that it may be an immunogenic protein, associated with host–pathogen interactions. Moreover, the *hmuY* gene sequence or its fragments may be used as one of the molecular markers of *P. gingivalis* (Carvalho-Filho et al., 2013; Contreras et al., 2014; Gao, Nguyen, & Hunter, 2010; Gmiterek et al., 2013; Olczak, Wójtowicz, Ciuraszkiewicz, & Olczak, 2010; Trindade et al., 2012, 2013; Wandersman & Delepelaire, 2012; Wójtowicz et al., 2009).

The monomeric haemophore HmuY of *P. gingivalis* is synthesized and exported to the outer membrane where it is anchored by an acylated Cys residue located at its N-terminus. The release of HmuY in the medium is induced by the cleavage of the acylated N-terminus by gingipain-K; then, HmuY undergoes tetramerization and haem uptake. Since the affinity of the haem for HmuY ($K_d \sim 3 \times 10^{-6}$ M) is lower than that of host haem-proteins ($K_d \leq 10^{-8}$ M), the haem is actively transferred to HmuY; the transfer only occurs after the hydrolysis of host haem-proteins by *P. gingivalis* proteinases, including lysine- and arginine-specific enzymes (gingipain-K (Kgp) and -R (Rgp), respectively). Then, holo-HmuY delivers the haem to its cognate outer-membrane transporter, the TonB-dependent receptor HmuR (Fig. 6). Interestingly, haem acquisition by *P. gingivalis* HmuY is facilitated

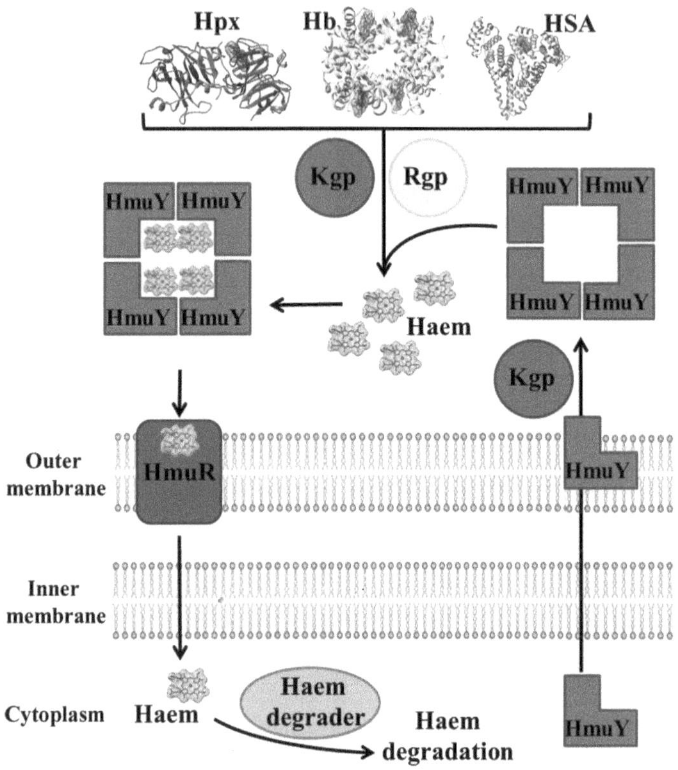

Figure 6 Haem acquisition in *P. gingivalis*. The monomeric haemophore HmuY is synthesized and exported to the outer membrane, where it is anchored by an acylated Cys residue located at its N-terminus. The release of HmuY in the medium is induced by the cleavage of the acylated N-terminus by Kgp. HmuY undergoes tetramerization and haem uptake after the hydrolysis of host haem proteins by Kgp and Rgp. The four haem groups are buried into the tetramer to be protected from competing host haem-binding proteins like SA and Hpx. The tetrameric arrangement allows haem delivery to HmuR, the TonB-dependent receptor acting as an outer-membrane transporter. Here, haem is degraded enzymatically. For details, see the text.

by the cysteine proteinase interpain A of *Prevotella intermedia*, representing a typical case of mutualism between two periodontal pathogens. In contrast, HusA, expressed on the outer membrane and released from the microorganism, rapidly binds the haem with high affinity ($K_d = 7 \times 10^{-10}$ M). Lastly, haem is enzymatically degraded (Byrne, Potempa, Olczak, & Smalley, 2013; Liu, Sroka, Potempa, & Genco, 2004; Potempa, Sroka, Imamura, & Travis, 2003; Smalley et al., 2011; Wójtowicz et al., 2009).

P. gingivalis HmuY has been reported to display a cross-like tetrameric structure made up by interactions between thumb and finger tips of monomers including β-hairpins β7β8, β9β10, β11β12, and β14β15. The four haem groups are in direct contact with each other in the centre of the tetramer. As a result, the haem groups are buried to be protected from competing host haem–Fe(III)-binding proteins (e.g. SA and Hpx). Moreover, the tetrameric arrangement allows haem storage and prevents the proteolytic cleavage of HmuY (Wójtowicz et al., 2009).

The monomer of *P. gingivalis* HmuY resembles a right hand, consisting of a roughly globular portion reminiscent of a palm, out of which emerge segments mimicking thumb and fingers. HmuY is an all β-protein constituted by 15 β-strands, the haem being sited in a hydrophobic cleft made up by strands β7, β8, β11, and β12. The haem propionates point inside the cavity; the carboxylate O atoms of the propionate substituent of the pyrrol ring *a* form a salt-bridge with the Arg79 Nη2 atom and two hydrogen bonds with Tyr80 Oη and Tyr173 Oη atoms, and the propionate of the pyrrol ring *d* is hydrogen bonded to the Thr124 Oγ1 atom. The haem–Fe atom is hexa-coordinated by axial residues His134 and His166, from strands β8 and β11, respectively (Fig. 7 and Table 1) (Wójtowicz et al., 2009).

Although the three-dimensional structure of HmuR has not been solved, the role of residues His95 and His434, as well as of the Asn-Pro-Asp-Leu and Tyr-Arg-Ala-Pro motifs in haem-binding and host haem protein utilization has been hypothesized (Liu, Olczak, Guo, Dixon, & Genco, 2006).

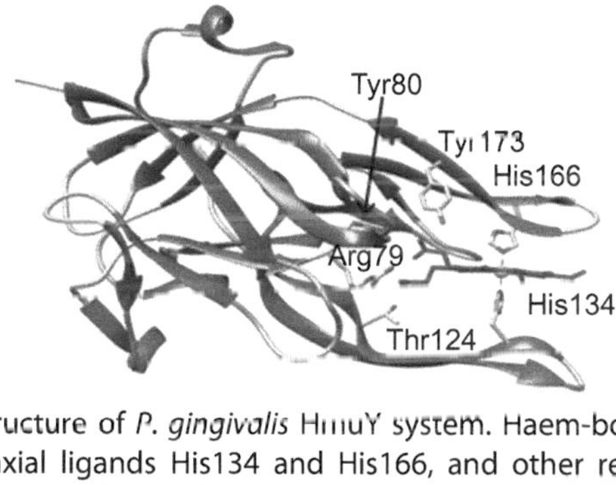

Figure 7 Crystal structure of *P. gingivalis* HmuY system. Haem-bound HmuY (PDB ID: 3H8T). The haem axial ligands His134 and His166, and other relevant residues, are shown in stick representation. Helices are in purple, strands in forest green, coils in grey, haem in red, and the haem–Fe atom in gold. *Picture drawn by UCSF-Chimera (Pettersen et al., 2004).* (See the colour plate.)

5. THE ISD HAEMOPHORE SYSTEM OF *Staphylococcus aureus*

S. aureus is a Gram-positive commensal organism which can cause severe disease when overcoming the skin barrier and invading underlying tissues. This species secretes numerous exotoxins, including several of cytolytic polypeptides capable of damaging the plasma membrane of erythrocytes, facilitating the access to host Hb (Malachowa & DeLeo, 2011; Vandenesch, Lina, & Henry, 2012). To overcome iron restriction, *S. aureus* produces two carboxylate-type siderophores, staphyloferrin A and B that can withdraw iron from transferrin and are transported into the cytoplasm as ferric complexes via ABC permeases (Beasley, Marolda, Cheung, Buac, & Heinrichs, 2011). *S. aureus* can also acquire xenosiderophores (i.e. siderophores produced by other microbial species occupying the same niche), which have been suggested to contribute to its ability to utilize exogenous chelators employed in iron overload therapy (Brock & Ng, 1983; Sebulsky, Hohnstein, Hunter, & Heinrichs, 2000). Moreover, *S. aureus* has evolved sophisticated mechanisms to obtain haem from the host and transport it across the cytoplasmic membrane, serving as paradigm for haem transport in Gram-positive species (Sheldon & Heinrichs, 2015).

The so-called iron surface determinants (Isd) are implicated in *S. aureus* haem transport, and are encoded by genes/operons that are negatively regulated by the Fur repressor protein (Hammer & Skaar, 2011). IsdA, IsdB, IsdC, and IsdH (or HarA) are the haem-binding proteins exposed on the bacterial surface. Since Gram-positive cells lack an outer membrane, these haem-binding proteins need to be covalently linked to the peptidoglycan layer by StrA and StrB sortases (Mazmanian et al., 2002). Sortases cleave specific motifs (Leu-Pro-Xxx-Thr-Gly and Asn-Pro-Thr-Gln-Asn for StrA and StrB, respectively) present in Isd haem-binding proteins and catalyse a transpeptidation reaction with a glycine of the nascent peptidoglycan peptide, allowing the proteins to be retained at the cell surface with their haem-binding domain exposed to extracellular milieu (Andrade, Ciccarelli, Perez-Iratxeta, & Bork, 2002; Contreras et al., 2014; Mazmanian et al., 2003, 2002; Wandersman & Delepelaire, 2012). Isd proteins are characterized by structurally different haem-binding clefts and functions. In fact, *S. aureus* IsdA, IsdB, IsdC, and IsdH are predominantly β-proteins acting as haem transporters, the cargo lipoprotein IsdE is mainly a α-protein, and IsdG and IsdI are α/β-proteins displaying the haem-degrading activity. In particular,

IsdA, IsdB, IsdC, and IsdH contain near-iron-transporter (NEAT) haem domains, IsdE binds haem within the N- and C-terminal globular regions, and IsdG and IsdI bind the haem in a ruled fashion facilitating the oxidation of the metal–porphyrin complex (Contreras et al., 2014; Gaudin et al., 2011; Grigg, Ukpabi, Gaudin, & Murphy, 2010; Grigg et al., 2007a, 2007b, 2011; Krishna Kumar et al., 2011; Lee et al., 2008; Moriwaki et al., 2011; Pilpa et al., 2006; Reniere et al., 2010; Sharp et al., 2007; Spirig et al., 2013; Ukpabi et al., 2012; Villareal et al., 2008; Vu et al., 2013; Wandersman & Delepelaire, 2004, 2012; Watanabe et al., 2008; Wu et al., 2005). Of note, the NEAT domains, that have been evolved by Gram-positive bacteria to extract haem from haem-protein donors, were firstly identified *in silico* by analysing the genomic neighbourhood of genes encoding siderophore transporters (Andrade et al., 2002).

It is worth emphasizing that *S. aureus* shows a repertoire of proteins involved in haem and Hb binding, transport, and degradation (Table 1) (Contreras et al., 2014; Grigg et al., 2010; Wandersman & Delepelaire, 2012), reflecting the strong microbial adaptation to haem- and Hb-rich biotopes and particularly to the human host (Pishchany et al., 2010; Skaar, Humayun, Bae, DeBord, & Schneewind, 2004).

In addition to the IsdA, IsdB, IsdC, and IsdH proteins, the *S. aureus* Isd system comprises: (i) the IsdE and IsdDF proteins, which are two components of a haem permease and (ii) the IsdI and IsdG proteins, which are cytoplasmic haem degrading enzymes (Fig. 8) (Contreras et al., 2014; Fonner et al., 2014; Gaudin et al., 2011; Grigg et al., 2007a, 2010, 2011; Krishna Kumar et al., 2011; Lee et al., 2008; Mazmanian et al., 2002; Moriwaki et al., 2011; Pilpa et al., 2006; Reniere et al., 2010; Sharp et al., 2007; Spirig et al., 2013; Tullius et al., 2011; Ukpabi et al., 2012; Villareal et al., 2008; Vu et al., 2013; Wandersman & Delepelaire, 2012; Watanabe et al., 2008; Wu et al., 2005).

IsdA and IsdC contain one NEAT domain, IsdB comprises two NEAT domains, and IsdH encompasses three NEAT domains. In contrast, IsdE, IsdG, and IsdI do not contain NEAT domains; in fact, IsdE is a canonical bilobed-binding protein belonging to family III (Grigg et al., 2007b), and IsdG and IsdI folds belong to the PG130-like protein family in which two ferredoxin-like-sheets from each monomer come together to form a β-barrel (Lee et al., 2008; Reniere et al., 2010; Ukpabi et al., 2012).

IsdA binds the haem, IsdB recognizes the haem and Hb, and IsdH binds the haem, Hb, and the Hb–Hp complex, suggesting that these NEAT domains have been evolved to recognize different targets (Dryla et al.,

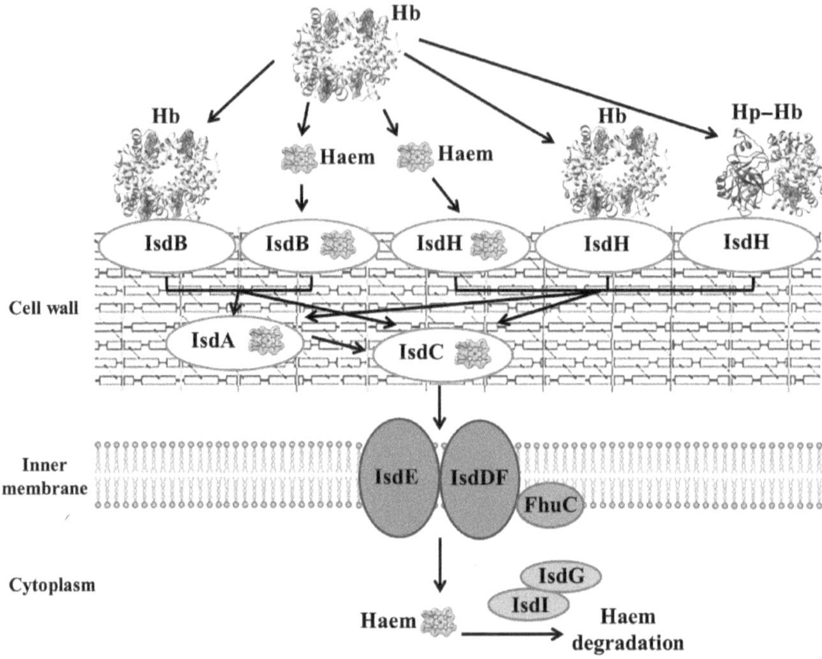

Figure 8 Haem acquisition in *S. aureus*. The *S. aureus* Isd system comprises: (i) the IsdA, IsdB, IsdC, and IsdH (or HarA) proteins, which are secreted by the secretory pathway and are anchored to the cell wall; (ii) the IsdE and IsdDF proteins, which are two components of a haem permease, and (iii) the IsdI and IsdG proteins, which are cytoplasmic haem-degrading enzymes. IsdA binds haem, IsdB recognizes haem and Hb, and IsdH binds haem, Hb, and the Hb–Hp complex. After haem transport across the cell wall, haem is transported across the inner membrane by an haem transporter composed of the haem-binding lipoprotein IsdE, and the haem permease IsdDF. FhuC has been postulated to serve as the ATPase for the Isd system. Once in the cytoplasm, haem is oxidized by the haem-degrading enzymes IsdG and IsdI. For details, see the text.

2007; Grigg et al., 2007a; Torres, Pishchany, Humayun, Schneewind, & Skaar, 2006). In particular, IsdH-NEAT domain 1 (IsdH-N1) binds Hb and the Hb–Hp complex with high affinity ($K_d \sim 2 \times 10^{-8}$ and 5×10^{-9} M, respectively) (Dryla et al., 2007; Pilpa et al., 2009), whereas IsdH-NEAT domain 3 (IsdH-N3) binds haem–Fe(III) with low affinity ($K_d = 2.8 \times 10^{-6}$ M) (Pilpa et al., 2009). The fast rate of haem transfer from the host haem-protein to IsdH-N3 has been postulated to reflect the close proximity of the haem-protein binding domains IsdH-N1 and IsdH-NEAT domain 2 (IsdH-N2) with the haem-binding region IsdH-N3 (Pilpa et al., 2009). Although IsdC is not exposed at the cell surface, it binds the haem efficiently (Marraffini & Schneewind, 2005; Mazmanian et al., 2003).

The haem-free NEAT domains IsdA, IsdC, IsdH-N1, IsdH-N2, and IsdH-N3 display a similar three-dimensional structural organization despite the low-sequence identity, the main differences being in the loops connecting the structural secondary elements. These NEAT domains display an immunoglobulin-like fold built up by eight β-strands forming the two sheets of a β-sandwich (Fig. 9 and Table 1) (Grigg et al., 2007a; Sharp et al., 2007; Villareal et al., 2008; Watanabe et al., 2008). The haem-bound NEAT domains IsdA, IsdB-N2, IsdC, and IsdH-N3 also display a similar haem-binding mode. In particular, the haem is accommodated in a hydrophobic cleft, is positioned on the β8 strand, and is covered by the N-terminal 3_{10}-helix (Grigg et al., 2007a; Pilpa et al., 2006; Sharp et al., 2007; Villareal et al., 2008; Watanabe et al., 2008). The haem–Fe atom of IsdA, IsdC, IsdH-N3, and IsdI is penta-coordinated, the proximal axial ligand of the haem–Fe atom being Tyr166, Tyr132, Tyr642, and His76, respectively. In holo-IsdB-N2, the haem–Fe atom exists in a mixture of hexa- and penta-coordinated states, residues Met362, provided by the 3_{10}-helix, and Tyr440 coordinating axially the haem-metal centre. Moreover, the haem–Fe atom of IsdE is hexa-coordinated, Met78 and His229 representing the axial ligands (Gaudin et al., 2011; Grigg et al., 2007a; Sharp et al., 2007; Watanabe et al., 2008). A conserved nearby Tyr residue (at position 170 in IsdA, 444 in IsdB-N2, 136 in IsdC, and 646 in IsdH-N3) is hydrogen bonded to the Tyr residue coordinating the haem–Fe(III) atom, likely enhancing the phenolate character of the Tyr side chain coordinating the haem–Fe atom and hence its affinity for the metal centre (Fig. 9 and Table 1) (Gaudin et al., 2011; Grigg et al., 2007a; Sharp et al., 2007; Watanabe et al., 2008). The haem propionate(s) is hydrogen bonded to Lys75, Ser82, and His83 in IsdA; Ser361 and Met362 in IsdB-N2; Ser47 in IsdC; and Ser563 and Tyr646 in IsdH-N3 (Gaudin et al., 2011; Grigg et al., 2007a; Sharp et al., 2007; Watanabe et al., 2008).

As shown in Fig. 9, the haem is bound to IsdA and IsdB-N2 at equal occupancy in two orientations when rotated by 180° along the Cα–Cγ meso axis (Gaudin et al., 2011; Grigg et al., 2007a).

IsdH and IsdB capture Hb by binding selectively to α-chains ($K_d = 1 \times 10^{-7}$ M). A pivotal role in protein–protein recognition is represented by Lys11 of Hb α-chains, which protrudes into the electronegative pocket of IsdH-N1 lined with hydroxyl and carbonyl groups from Ser130, Tyr126, Asn151, and Thr152 residues (Fig. 9). The occurrence of the Thr12 residue, at the equivalent position of Lys11, in the Hb β-chains impairs the IsdH-N1-β-chain recognition (Krishna Kumar et al., 2011).

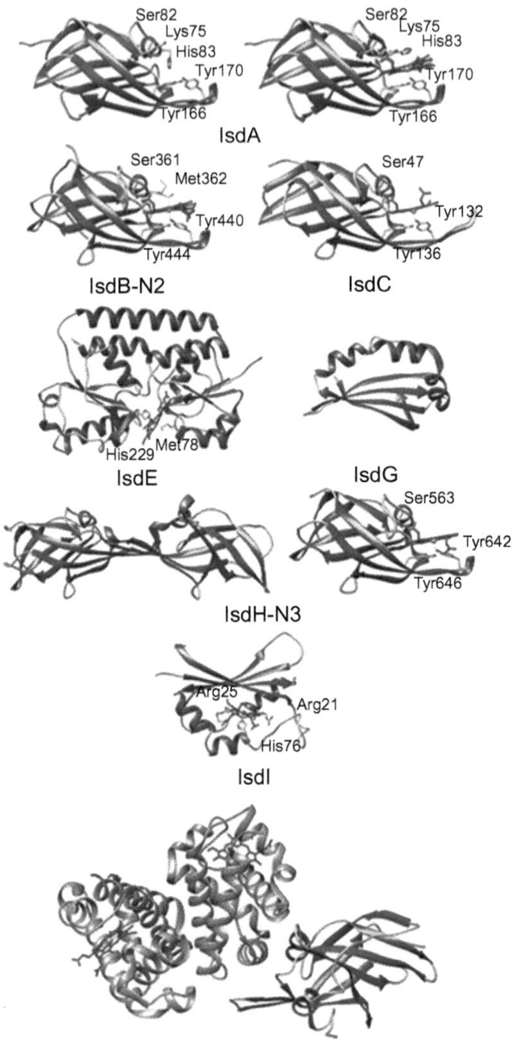

Figure 9 Crystal structures of *S. aureus* Isd system. Ligand-free (left) and haem-bound (right) IsdA (PDB ID: 2ITE and 2ITF, respectively), haem-bound IsdB-N2 (PDB ID: 3RTL), haem-bound IsdC (PDB ID: 2O6P), haem-bound IsdE (PDB ID: 2Q8Q), ligand-free IsdG (PDB ID: 1XBW), ligand-free (left) and haem-bound (right) IsdH-N3 (PDB ID: 2E7D and 2Z6F, respectively), and haem-bound IsdI (PDB ID: 3QGP). IsdH-N2 is complexed with αβ-Hb (PDB ID: 4FC3). Relevant residues are shown in stick representation. Helices are in purple, strands in forest green, coils in grey, haem in red, the haem–Fe atom in gold, and Hb in old rose. *Pictures drawn by UCSF-Chimera (Pettersen et al., 2004).* (See the colour plate.)

Isolated IsdH-N1 does not induce haem release from ferric Hb, haem dissociation occurring only in the presence of the whole IsdH or IsdB proteins. Haem transfer from Hb to IsdH-N3 is limited from the rate of haem dissociation; the rate of haem transfer increases by twofold in the presence of both IsdH-N1 and IsdH-N3 (Pilpa et al., 2009; Zhu et al., 2008).

The N- and C-terminal NEAT domains of IsdH (i.e. IsdH-N2 and IsdH-N3, respectively) display very different functions in haem extraction from Hb. In fact, IsdH-N2 binds to the Hb α–β dimer at a site distant from the haem pockets of Hb chains (i.e. between the A and E α helices of α- or β-chains). Then, the linker region positions IsdH-N3 over a haem pocket of the Hb α–β dimer to remove the haem, which would need to move only ~12 Å to be transferred from Hb chains to IsdH-N3 (Fig. 9). Of note, multiple NEAT domains cooperate to extract efficiently the haem from Hb complexed in the hetero-octameric IsdH–Hb complex (Dickson et al., 2014).

In vitro, haem is transferred unidirectionally from IsdB to IsdA and IsdC, from IsdH to IsdA and IsdC, and from IsdA to IsdC. Then, IsdC translocates the haem to the IsdE–IsdDF transmembrane complex, which in turn relocates it to the cytoplasm where the haem is degraded by IsdG and IsdI (Fig. 8) (Contreras et al., 2014; Grigg et al., 2011; Liu et al., 2008; Muryoi et al., 2008; Wandersman & Delepelaire, 2012; Zhu et al., 2008).

Haem transfer from holo-IsdA-N1 to apo-IsdC-N1 (Fig. 8) appears to reflect the occurrence of the IsdA-N1–IsdC N1 complex enhancing by four to five orders of magnitude the rate constant for the dissociation of the IsdA-N1–haem complex. Moreover, the haem affinity for IsdC-N1 is higher than that for IsdA-N1 (Grigg et al., 2011).

In the holo-IsdA–apo-IsdC-N1 handclasp-type complex, the haem-binding site of IsdC-N1 is positioned symmetrically to the haem-binding cleft of IsdA-N1. The β7–β8 loop (L8) of IsdC-N1 contacts the 3_{10}-helix of IsdA-N1, and the loop 1 of IsdC-N1 contacts the β7–β8 loop of IsdA-N1. Moreover, Pro162 to Ile164 residues present in the turn connecting strands β7 to β8 in IsdA-N1 contact residues IsdC-N1 immediately following the β1 strand. Of note, the β1 strand of IsdA-N1 contains Tyr166 representing the haem-Fe atom axial ligand. Furthermore, residues Val125 to Phe130 located in the turn connecting strands β7 and β8 of IsdC-N1 contact residues Tyr87 to Lys93 following the 3_{10}-helix of IsdA. These interactions may facilitate haem release from holo-IsdA-N1 inducing the cleavage of the Tyr166-haem-Fe axial bond by displacing the β-sheet and altering the positioning of the 3_{10}-helix which stabilizes the haem non-covalently.

Remarkably, the haem would only need to slide ~18 Å from the IsdA-N1 pocket to the IsdC-N1 cleft (Villareal et al., 2011).

Loop 8 of IsdC-N1 is pivotal for haem transfer from IsdC-N1 to IsdE (Fig. 8), a member of the α-helical backbone metal receptor superfamily, which displays a bilobed topology formed by the N- and C-terminal domains bridged by a single α-helix. In fact, loop 8 of IsdC-N1 could alter the geometry of the Thr59, Ser60, and Tyr61 residues of IsdE, which are on the edge of the haem-binding site and participate to metal-porphyrin stabilization. The value of the first-order rate constant for haem transfer is $4 \times 10^{-2}\ s^{-1}$. In holo-IsdE, the haem–Fe atom is axially bound to Met78 and His229 residues, contributed by the N- and C-terminal domains, respectively (Fig. 9). Moreover, both haem propionates are oriented approximately parallel to the metal-porphyrin-binding cleft, are essentially buried within the protein matrix, and form extensive interactions with IsdE residues. In fact, the Lys62 Nζ atom forms a hydrogen-bond to a water molecule that, in turn, is hydrogen bonded with both haem propionates. One of the haem propionates forms additional direct hydrogen bonds with the main chain amides of Val41 and Ala42. An additional water molecule-mediated interaction is formed with the OH groups of Thr40 and Thr271. The other propionate is hydrogen bonded to the Ser60 Oγ atom and to the main chain amide of Tyr61 (Abe, Caaveiro, Kozuka-Hata, Oyama, & Tsumoto, 2012; Grigg et al., 2007b).

Haem transfer from IsdE to the cytoplasm is mediated by IsdF (Fig. 8), which is suggested to represent the permease, a homo-dimeric integral membrane protein (Mazmanian et al., 2003). However, although the final component needed to drive haem uptake is an ATP hydrolase, the *S. aureus* Isd system lacks an obvious ATPase encoded by the *isd* locus (Grigg et al., 2010). As observed in several ABC-ts devoted to Fe-compound conveyance from this bacterium, FhuC has been postulated to function as the ATPase for the Isd system (Beasley et al., 2009; Speziali, Dale, Henderson, Vines, & Heinrichs, 2006). The inspection of the IsdDF-FhuC-IsdE model suggests that two Arg-rich patches present on the extracellular side of IsdF could form electrostatic interactions with Glu83 and Glu214 of IsdE, which, in turn, likely induce haem transport (Grigg et al., 2010; Hvorup et al., 2007).

Upon transfer to the cytoplasm, haem is oxidized by the haem-degrading enzymes IsdG and IsdI (Fig. 8), belonging to a haem-degradation family distinct from canonical eukaryotic and prokaryotic haem oxygenases. The metallo-porphyrins are axially penta-coordinated by His77 and His76 in IsdG and IsdI, respectively, and are buried into similar deep hydrophobic clefts. In IsdG, haem propionates are salt bridged to the guanidine groups of Arg22 and Arg26, and contact the side chain of His27 and the main chain amides of His27 and Ile29. In IsdI, propionic groups form salt bridges with the guanidine moieties of Arg21 and Arg25 and contact the side chain of Gln27 and the main chain nitrogen atoms of Gln27 and Ile28 (Fig. 9). Extensive steric hindrance induces a high distortion (i.e. ruffling) of the porphyrin rings in the IsdG-N7A–haem–Fe(III) and IsdG-N7A–haem–Co(III) complexes and places the β- and δ-*meso* carbons proximal to the haem-distal oxygen-binding site. The reaction mechanism of haem oxidation by IsdG and IsdI has been postulated to account for a reactive peroxide intermediate that proceeds with the nucleophilic oxidation of the haem β- or δ-*meso* carbon atom (Lee et al., 2008; Reniere et al., 2010; Ukpabi et al., 2012).

The contribution of haem-acquisition proteins to *S. aureus* pathogenicity has been proven by testing *isdA* and *isdB* mutants in different animal models. The most prominent role seems to be played by IsdB, since *isdB* mutants display significantly lower ability to spread to different organs of either wild type or transgenic mice expressing human Hb (Pishchany, Dickey, & Skaar, 2009; Pishchany et al., 2010). Interestingly, non-iron metalloporphyrins (e.g. Ga-protoporphyrin-IX) have also been used as Trojan horses to suppress staphylococcal growth via the inhibition of bacterial haem oxygenases (Stojiljkovic, Kumar, & Srinivasan, 1999), paving the way to the development of new haem-based iron mimetic drugs targeting bacterial iron metabolism (Minandri, Bonchi, Frangipani, Imperi, & Visca, 2014).

6. THE ISD, BslK, AND HAL HAEMOPHORE SYSTEMS OF *Bacillus anthracis*

Haem acquisition in *B. anthracis*, the causative agent of anthrax, is mediated by three different haemophore systems: Isd, *B. anthracis* S-layer homology protein K (BslK), and haem-acquisition leucine-rich repeat protein (Hal), reflecting either the very different environmental conditions occurring during the microbial life cycle or the capability to increase the overall rate of haem acquisition (Balderas, Nobles, Honsa, Alicki, & Maresso, 2012).

The IsdX1, IsdX2, and IsdC are part of the *isd* locus, an array of eight genes proposed to encode a system that acquires haem from Hb and delivers captured haem to the cell surface and across the cell membrane into the bacterial cytosol. IsdX1 and IsdX2 are secreted proteins that extract haem from Hb and deliver the haem to cell wall-bound IsdC (Balderas et al., 2012; Ekworomadu et al., 2012; Honsa et al., 2013; Tarlovsky et al., 2010).

IsdX1 and IsdX2 contain one and five NEAT domain(s), respectively. IsdX2-N1, IsdX2-N3, IsdX2-N4, and IsdX2-N5 domains bind haem–Fe(III), IsdX2-N5 displaying the highest affinity for the metal-porphyrin. All IsdX2 NEAT domains recognize Hb, but only IsdX2-N1 and IsdX2-N5 remove haem from Hb. Moreover, IsdX2-N1, IsdX2-N3, and IsdX2-N4 relocate haem to IsdC, but not IsdX2-N5 (Fig. 10) (Ekworomadu et al., 2012; Honsa, Fabian, Cardenas, Olson, & Maresso, 2011; Honsa et al., 2013).

The overall structure of IsdX1 and IsdX2-N5 (Ekworomadu et al., 2012; Honsa et al., 2013) is similar to those of *S. aureus* NEAT domains (Gaudin et al., 2011; Grigg et al., 2007a, 2007b, 2011; Krishna Kumar et al., 2011;

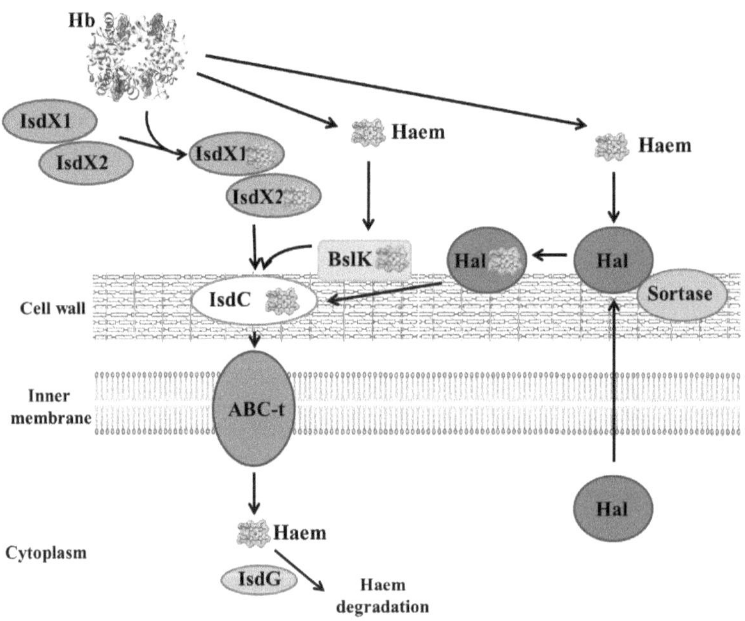

Figure 10 Haem acquisition in *B. anthracis*. The secreted haemophores IsdX1 and IsdX2 extract the haem from Hb. The haem-bound haemophores transfer haem to IsdC. Alternatively, *B. anthracis* can bind haem by the surface protein BslK to cell wall-localized IsdC. IsdC transfers the haem to ABC-t localized in the inner membrane. Also, the surface protein Hal is devoted to haem acquisition. Hal is secreted and tethered to the surface peptidoglycan by covalent attachment through a sortase-transpeptidase. After having extract the haem from Hb, Hal eventually transfers the haem to a downstream cell-wall transporter, e.g., IsdC. Finally, the haem is transported by the ABC-t into the cytoplasm where it is degraded by IsdG. For details, see the text.

Moriwaki et al., 2011; Pilpa et al., 2006; Sharp et al., 2007; Spirig et al., 2013; Villareal et al., 2008; Vu et al., 2013; Watanabe et al., 2008), consisting of an immunoglobulin-like fold built up by eight β-strands arranged in two antiparallel β-sheets of a β-sandwich. The hydrophobic environment of the haem-binding cleft of IsdX1 and IsdX2-N5 is formed by a 3_{10}-helix on one side and a β-hairpin on the other (Figs. 9 and 11) (Ekworomadu et al., 2012; Honsa et al., 2013).

In IsdX1, the axial ligand of the penta-coordinated haem–Fe atom is Tyr136; the Tyr136–haem–Fe atom coordination bond is stabilized by Tyr140, which forms a hydrogen bond with the phenolate oxygen of Tyr136. The propionate-D of haem–Fe is solvent exposed, whereas propionate-A is hydrogen bonded to the OH group of Ser53 and the backbone N atom of Arg54. The haem molecule is also stabilized by Tyr58, that weakly π-stacks with the buried haem–Fe pyrrole ring (Fig. 11 and Table 1) (Ekworomadu et al., 2012).

In IsdX2-N5, the haem–Fe atom is penta-coordinated, the axial ligand being Tyr108; the hydrogen bond between residues Tyr108 and Tyr112 renders the hydroxyl group of Tyr108 more electronegative and more prone

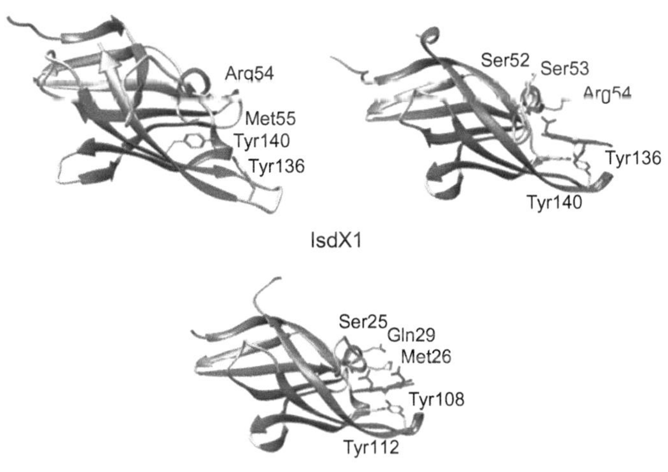

Figure 11 Crystal structures of *B. anthracis* Isd system. Ligand-free (left) and haem-bound (right) IsdX1 (PDB ID: 3SZ6 and 3SIK, respectively) and haem-bound IsdX2-N5 (PDB ID: 4H8P). The haem axial ligands of IsdX1 (Tyr136) and IsdX2-N5 (Tyr108) and other relevant residues are shown in stick representation. Helices are in purple, strands in forest green, coils in grey, haem in red, and the haem–Fe atom in gold. *Pictures drawn by UCSF-Chimera (Pettersen et al., 2004).* (See the colour plate.)

to bind the haem–Fe atom. Further stabilization of the haem is achieved through hydrogen bonding between Ser25 and haem propionate-A, together with π-stacking with Tyr30 and Tyr112 residues. The side chain of Met26 provides non-polar and weak electrostatic interactions with the haem group (Fig. 11 and Table 1) (Honsa et al., 2013).

The amino acid sequence alignment of IsdX2-N1, IsdX2-N2, IsdX2-N3, IsdX2-N4, and IsdX2-N5 indicates that Gln in the fifth position of the 3_{10}-helix is present only in IsdX2-N1 and IsdX2-N5, being Thr in IsdX2-N2, IsdX2-N3, and IsdX2-N4. In IsdX2-N5, Gln29 is exposed to the solvent, is not in contact with haem, and is positioned to recognize Hb. Haem extraction from ferric Hb is a fast process, whereas haem uptake from ferrous oxygenated Hb is very slow. Of note, the Gln29Thr mutation impairs haem uptake from ferric Hb without affecting haem binding. The Gln29 residue of IsdX2-N1 and IsdX2-N5 has been postulated to interact with the HisE7 residue of Hb. Upon destabilization of the ferric Hb–haem interface, Tyr108 could coordinate the haem–Fe atom, resulting in the completion of haem extraction (Honsa et al., 2013).

The *B. anthracis* Hal is a surface protein devoted to haem acquisition from Hb. Hal is secreted through the general secretory pathway and tethered to the surface peptidoglycan, most likely by covalent attachment by a sortase-transpeptidase (Fig. 10). The NEAT domain of Hal contains two regions common to all NEAT domains, the N-terminal 3_{10}-helix and the C-terminal β-hairpin. In addition to the NEAT domain, Hal contains 16 Leu-Leu-Arg repeats possibly involved in protein–protein recognition. Interestingly, Hal lacks the Tyr residue (i.e. mutated to Phe116) that stabilizes the coordinating Tyr112 residue axially bound to the haem–Fe atom. Moreover, Tyr56, located on the opposite side of the haem-binding pocket in the 3_{10}-helix may represent the sixth coordination ligand of haem–Fe atom. The bis-Tyr coordination of the haem–Fe atom is a peculiar feature for a bacterial protein containing a NEAT domain. The Hal–haem adduct appears to be stabilized by hydrogen bonding of haem propionate-A with Ser55. After having extract haem from Hb, Hal eventually transfers the bound haem to a downstream membrane-localized transporter, e.g., IsdC (Fig. 10) (Balderas et al., 2012).

The BslK, containing a NEAT domain, localizes non-covalently to the microbial surface, most likely by the three S-layer homology regions, and transfers the haem to the Isd system via the cell wall protein IsdC (Fig. 10). BslK binds haem–Fe with high affinity ($K_d = 1 \times 10^{-7}$ M) and transfers it rapidly to IsdC ($k \sim 6$ s^{-1}) (Tarlovsky et al., 2010).

Lastly, IsdC may next pass haem to a ABC-t, which transfers the haem to the cytoplasm where it is oxidized by haem-oxygenase IsdG (Fig. 10) (Balderas et al., 2012).

Haem and Hb uptake by NEAT domains of IsdX1, IsdX2, Hal, and BslK is pivotal for cell growth. Therefore, they could have implications for the development of therapeutics and vaccines against anthrax infection (Balderas et al., 2012; Ekworomadu et al., 2012; Fabian, Solomaha, Olson, & Maresso, 2009; Honsa et al., 2013; Tarlovsky et al., 2010).

7. THE HBP HAEMOPHORE SYSTEM OF *Listeria monocytogenes*

L. monocytogenes is a facultative intracellular pathogen endowed with the unusual feature of lacking endogenous siderophore-mediated iron transport systems, whereas being capable of acquiring xenosiderophores via ABC-ts (Jin et al., 2006), and iron supplied by host's catecholamines via reductive process (Coulanges, Andre, Ziegler, Buchheit, & Vidon, 1997). More important, *L. monocytogenes* relies on haem transport to meet with its nutritional iron requirement, both in nature and in the infected host, and has evolved multiple systems for haem binding either in the extracellular milieu or on the cell surface (Klebba, Charbit, Xiao, Jiang, & Newton, 2012).

In *L. monocytogenes*, two NEAT domains containing open reading frames encoding the haem/Hb-binding proteins 1 and 2 (Hbp1 and Hbp2, respectively) are present in a Fur regulated operon, together with genes encoding a sortase and an ABC-t. However, mutations in this locus do not appear to grossly alter the ability of *L. monocytogenes* to use haem iron and do not affect the lethality in mice (Xiao et al., 2011). Hbp1 and Hbp2 are orthologous of the staphylococcal Isd proteins, and biochemical evidence indicates that they can scavenge haem and Hb from the environment. Hbp1 is linked to the cell wall, whereas Hbp2 is either surface-displayed or secreted in the environment. Hbp1 contains one NEAT domain (Hbp1, Asp31-Ala149), whereas Hbp2 displays three NEAT domains (Hbp2-N1, Ser31-Phe170; Hbp2-N2, Thr183-Lys303; and Hbp2-N3, Leu360-Ala479). *In vitro*, isolated NEAT domains bind the haem with high affinity, K_d values ranging between 2.5×10^{-8} and 8.1×10^{-8} M. In Hbp1, the single NEAT domain mediates Hb binding, whereas in Hbp2, the first and third NEAT domains (i.e. Hbp2-N1 and Hbp2-N3, respectively) are responsible of Hb recognition; Hbp1 and Hbp2 bind Hb with $K_d \sim 10^{-9}$ M. After scavenging, the haem is transferred unidirectionally from holo-Hbp2 to apo-Hbp1. Both, holo-Hbp1 and holo-Hbp2 may transfer the haem across the cell wall to the plasma membrane. Then, the haem has been hypothesized to be transferred to ABC-t (HupDGC or Lmo2182-4) to be used directly by the bacterium or degraded by haem monooxygenase to release free iron (Fig. 12) (Klebba et al., 2012; Malmirchegini et al., 2014).

L. monocytogenes Hbp2-N2 consists of eight β-strands that form a β-sandwich. One face of the structure contains the antiparallel strands β1a-β2-β3-β6-β5b, whereas the other face includes strands β1b-β8-β7-β4-β5a. The haem is sited in the non-polar cleft located between the short 3_{10}-helix (called lip) and the β-hairpin formed by strands β7 and β8, the haem–Fe atom axial ligand being Tyr280. Surprisingly, Hbp2-N2 binds the haem even though its primary structure lacks the highly conserved fifth

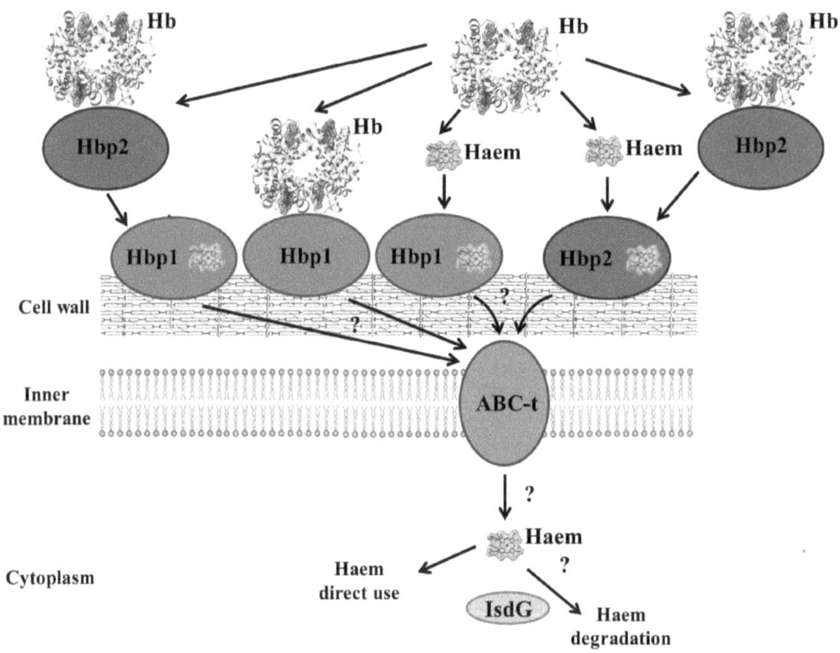

Figure 12 Haem acquisition in *L. monocytogenes*. The bacterial pathogen *L. monocytogenes* produces Hbp1 and Hbp2, which scavenge haem and Hb from the environment. Hbp1 is linked to the cell wall, and it can bind Hb and haem. Hbp2 is either surface-displayed or secreted in the environment. The secreted Hbp2 directly binds haem and Hb, and haem can be then transferred to cell wall-associated Hbp1 and Hbp2. Both, holo-Hbp1 and holo-Hbp2 may transfer haem across the cell wall to the plasma membrane by the ABC-t HupDGC or Lmo2182-4. Then, haem may be either cleaved by the haem monooxygenase IsdG or directly used. For details, see the text.

Tyr residue of the Tyr-Xxx-Xxx-Xxx-Tyr motif, thus lacking the stabilization of the tyrosinate state of the first Tyr side chain (i.e. Tyr280) coordinating firmly the haem atom, by hydrogen bonding. Although the majority of Hbp2-N2–haem interactions are non-polar (e.g. Val278, Ala282, Phe284, and Tyr289), the side chains of Ser204 and Ser205, and the backbone amide of Ser205 donate hydrogen bonds to the propionate group-A of haem–Fe(III) (Fig. 13 and Table 1) (Klebba et al., 2012; Malmirchegini et al., 2014).

Haem binding causes large conformational changes in Hbp2-N2; the major changes occur in the loop that connects strands β7 and β8, which is shortened from 10 to 4 residues and undergoes a disordered-to-ordered transition. Ordering of the β-hairpin has been postulated to be driven by haem contacts to Tyr289 and Phe284, which are packing against the metal-porphyrin in the Hbp2-N2–haem complex (Fig. 13) (Klebba et al., 2012; Malmirchegini et al., 2014).

The asymmetric unit of the Hbp2-N2–haem complex contains two copies of the adduct in which haem pockets approach one another in such a way

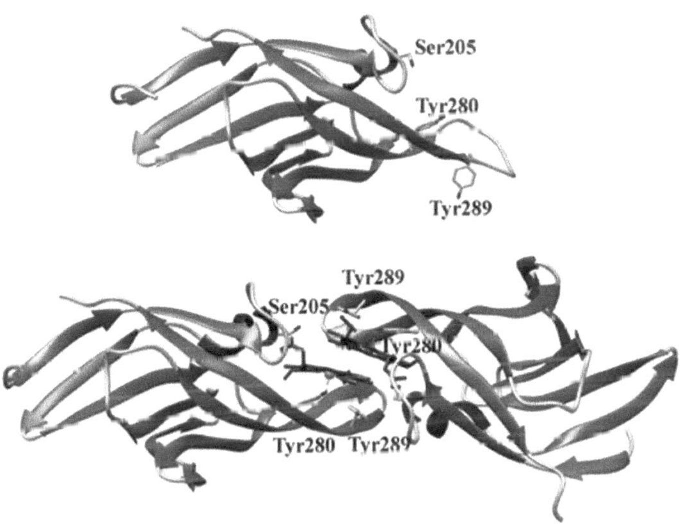

Figure 13 Crystal structures of *L. monocytogenes* Hbp2 system. Monomeric ligand-free (top) and homo dimeric haem-bound (bottom) Hbp2 (PDB ID: 4NLA and 4MYP, respectively). The haem axial ligand Tyr280 and the Tyr289 residue stabilizing the tyrosinate form of Tyr280 are shown in stick representation. Helices are in purple, strands in forest green, coils in grey, haem in red, and the haem–Fe atom in gold. *Pictures drawn by UCSF-Chimera (Pettersen et al., 2004).* (See the colour plate.)

that the β7–β8 hairpin of one protein is adjacent to the 3_{10}-helix on the opposing protein. This orientation may resemble that of the activated protein–protein complex through which Hbp2-N2 may transfer hemin to Hbp1, Hbp2-N1, and Hbp2-N3. In the transfer complex, the haem molecule bound to Hbp2-N2 is transferred only when the β7–β8 hairpin of the accepting NEAT domain simultaneously contacts also the porphyrin ring (Fig. 13 and Table 1) (Klebba et al., 2012; Malmirchegini et al., 2014). Hbp2 is implicated in haem–Fe(III) acquisition as *hbp2*-strains are impaired in their ability to import the haem and use it as a source of iron (Malmirchegini et al., 2014).

Lastly, a prominent role in haem uptake by *L. monocytogenes* is also played by the HupDGC ABC-type permease, a sortase-independent system, which translocates haem in the cytoplasm and acts as a virulence determinant of this species (Fig. 12). Haem binds to the HupDGC permease with relatively lower affinity than Hbp ($K_d \sim 10^{-8}$ M) and is transported with $K_M = 1.5 \times 10^{-9}$ M and $V_{max} = 23$ pmol/min/10^9 cells, suggesting that this system functions at high haem concentration higher than 50 nM (Klebba et al., 2012; Xiao et al., 2011).

8. THE RV0203–MmpL11–MmpL3 HAEMOPHORE SYSTEM OF *Mycobacterium tuberculosis*

M. tuberculosis is a facultative intracellular bacterium that cannot survive without iron, and has evolved multiple strategies to gain access to iron sources both intracellularly and extracellularly, including mycobactin and carboxymycobactin siderophores and one haemophore system (Fang, Sampson, Warren, Gey van Pittius, & Newton-Foot, 2015). The identification of *M. tuberculosis* genomic region responsible for haem uptake revealed the occurrence of several mycobacterial-specific proteins including the secreted haem- and Hb-binding protein Rv0203 and the transmembrane haem-transporting proteins Rv0202 and Rv0206 (also named MmpL11 and MmpL3, respectively) (Fig. 14) (Wandersman & Delepelaire, 2012).

The Rv0203 haemophore is a mycobacteria-specific protein showing a unique tetrameric all α-helical structure, consisting of a dimer of dimers, in which each monomer consists of five α-helices (Fig. 15 and Table 1). Within each monomer, α-helices 1 and 4 are linked by the Cys40-Cys114 disulphide bridge (between the N-terminus of α-helix 1 and the C-terminus of α-helix 4) and capped by α-helices 2 and 3; α-helix 5 lies perpendicularly across α-helices 1 and 4. The dimer is formed by antiparallel interactions

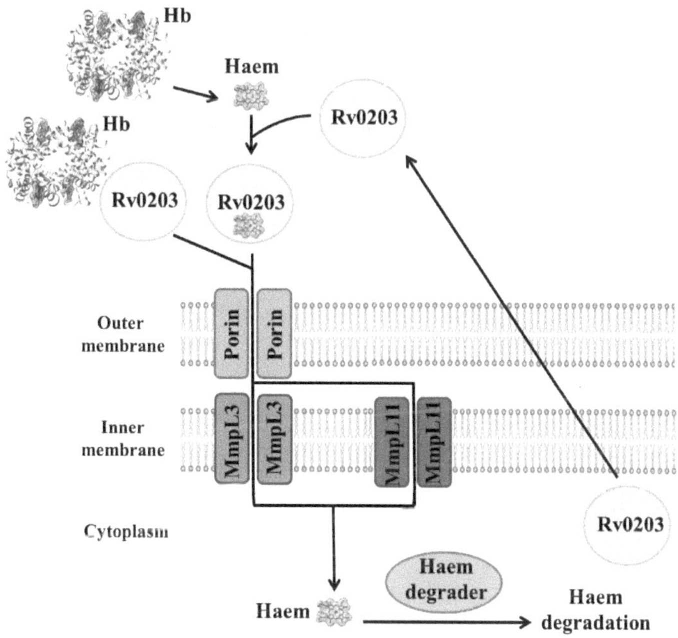

Figure 14 Haem acquisition in *M. tuberculosis*. Haem uptake is mediated by the secreted haem- and Hb-binding protein Rv0203 and the transmembrane haem-transporting porins MmpL11 and MmpL3. Once into the cytoplasm, the haem is degraded enzymatically. For details, see the text.

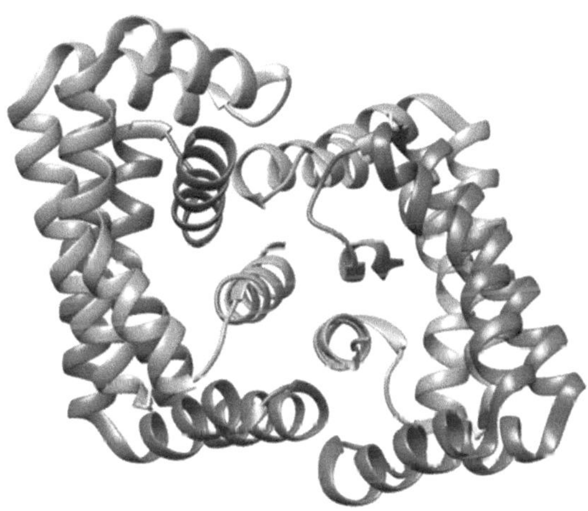

Figure 15 Crystal structure of *M. tuberculosis* haem carrier protein Rv0203. Heterotetrameric ligand-free Rv0203 (PDB ID: 3MAY). For clarity, two subunits are in purple and two subunits in hot pink. Disulphide bridges are in yellow. *Pictures drawn by UCSF-Chimera (Pettersen et al., 2004).* (See the colour plate.)

between the α-helices 1; the four α-helices of the dimer form an antiparallel α-helical sheet, with the two shorter α-helices capping each end of the dimer. Two dimers associate to form an off-tilt cage-like structure, the α-helices building up a weak hydrophobic core in the centre of the cage (Fig. 15). The largest differences between subunits occur at the loops connecting α-helices 2 and 3 and α-helices 4 and 5, causing the non-superimposition of α-helices 5. The Tyr59 and His89 residues have been postulated to coordinate the haem–Fe atom. As reported in other haemophores such as HasA (see Figs. 2 and 3, and Table 1), Tyr59 is hydrogen bonded to His63, stabilizing the highly reactive form of the tyrosinate state of Tyr59. However, some rearrangement(s) could occur to permit the formation of the Rv0203–haem complex (Tullius et al., 2011).

Haem binds to Rv0203 with high affinity, the K_d value ($=k_{off(Fe(III))}/k_{on(Fe(II)-CO)}$) being 5.0×10^{-10} M. Values of $k_{on(Fe(II)-CO)}$ and $k_{off(Fe(III))}$ are 1.3×10^{7} M^{-1} s^{-1} and $\sim 8 \times 10^{-2}$ s^{-1}, respectively. Kinetics of haem–Fe(II)–CO binding to Rv0203 and of dissociation of the Rv0203–haem–Fe(III) complex are limited by CO dissociation from haem–Fe(II)–CO and by the cleavage of the axial haem–Fe(III) bond(s), respectively (Owens, Du, Dawson, & Goulding, 2012).

MmpL11 and MmpL3 have been reported to contain 11 transmembrane α-helices, the extracellular domains E1 and E2, and the cytoplasmic domain C1. The extracellular domains E1 and E2 are located between the transmembrane α-helices 1 and 2, and at the C-terminus, respectively. The cytoplasmic domain C1 is positioned between the transmembrane α-helices 6 and 7 (Tullius et al., 2011; Owens et al., 2013).

The extracellular domains E1 and E2 have been reported to bind haem, whereas the cytoplasmic domain C1 does not recognize the metal-porphyrin (Tullius et al., 2011). In particular, the E1 domain of MmpL11 and MmpL3 binds haem with the 1:1 stoichiometry and high affinity, apparent values of K_d ($=k_{off(Fe(III))}/k_{on(Fe(II)-CO)}$) being 6.3×10^{-9} and 1.2×10^{-10} M for haem–Fe binding to MmpL11-E1 and MmpL3-E1, respectively. Values of k_{on} CO binding to MmpL11-E1 and MmpL3-E1 are similar ($k_{on} = 5.3 \times 10^{7}$ and 2.9×10^{7} M^{-1} s^{-1}, respectively). At high haem–Fe(II)–CO concentration, kinetics of haem binding to MmpL11-E1 and MmpL3-E1 are limited by the intrinsic rate of haem–Fe(II) atom coordination. Values of k_{off} for the dissociation of the MmpL11-E1–haem–Fe(III) and MmpL3-E1–haem–Fe(III) complexes are very different. Kinetics of MmpL11-E1–haem–Fe(III) dissociation are monophasic with $k_{off} = 3.4 \times 10^{-1}$ s^{-1}, whereas kinetics of MmpL3-E1–haem–Fe(III) dissociation are biphasic with values of $k_{off}=3.6\times10^{-3}$ and $4.0\times10^{-4}\ s^{-1}$. Haem binding to MmpL11-E1 and MmpL3-E1 reflects haem-dependent changes in the oligomerization state of E1 domains, which is likely a multistep process (Owens et al., 2013).

Kinetics of haem–Fe(III) transfer from holo-Rv0203 to apo-MmpL11-E1 and apo-MmpL3-E1 are biphasic, the rate of the fast step ranges between 3.3 and 5.4 s^{-1}, whereas the rate of the slow step ranges between 3.0×10^{-2} and $1.3\times10^{-1}\ s^{-1}$. Of note, the rate of the fast step is higher than the rate of passive haem–Fe(III) release from the Rv0203–haem–Fe(III) complex ($k_{off}\sim8\times10^{-2}\ s^{-1}$), which in turn corresponds to the rate of the slow step. Notably, the rates of haem–Fe(III) transfer from holo-MmpL11-E1 and holo-MmpL3-E1 domains to apo-Rv0203 are slower than those from holo-Rv0203 to apo-MmpL11-E1 and apo-MmpL3-E1 domains, suggesting that holo-Rv0203 binds preferentially apo-MmpL11-E1 and apo-MmpL3-E1 domains (Owens et al., 2013, 2012).

These observations highlight haem uptake by Rv0203 and the extracellular to intracellular haem transport by MmpL11 and MmpL3 (Fig. 14) (Owens et al., 2013; Tullius et al., 2011), outlining the role of MmpL proteins in mycobacterial metabolite import (Domenech, Reed, & Barry, 2005). Cytoplasmic haem can be degraded by the mycobacterial haem-degrading protein MhuD to provide iron and mycobilin (Owens et al., 2013; Tullius et al., 2011).

It can be speculated that *M. tuberculosis* may gain access to haem and haem-containing proteins during both intracellular and extracellular life. For instance, macrophages can degrade erythrocytes and release haem from the phagosome making it available to intracellular mycobacteria (Soe-Lin et al., 2009), whereas the contact-dependent mycobacterial haemolysin could promote Hb leakage from circulating erythrocytes (Deshpande, Khan, Bhat, & Navalkar, 1997). Accordingly, it has been demonstrated that MmpL haemophores contributes to the survival and pathogenicity of *M. tuberculosis* in a murine model of tuberculosis (Domenech et al., 2005). Therefore, paths of iron uptake may be targets of anti-mycobacterial therapeutic strategies (Tullius et al., 2011; Owens et al., 2013, 2012).

9. CONCLUSION AND PERSPECTIVES

The main role of haemophores is to provide iron to microbes. They bind free haem or extract haem from host haem-proteins in the external medium and present it to specific outer-membrane receptors. Haemophores are either tethered to the cell surface or released into the extracellular medium. However, in a few cases, they are found at both localizations.

Haemophores achieve their role in very different fashions, the Has and Isd systems representing prototypical haem uptake pathways. In the Has system, the haem acceptor protein HasA acts as both haem scavenger and transporter, the haem internalization requiring the formation of the HasA–HasR complex. Members of the Isd system bind free haem and recognize selectively Hb α-chains, then the haem is extracted from host Hb inducing the alteration of the host haem-protein conformation. The haem is transferred to other Isd haem acceptor proteins (i.e. IsdA and IsdC) and finally reaches the most deeply buried protein (i.e. IsdC) prior to be transferred to the cytoplasm. In most bacteria, it has been proposed that the degradation of Hb is a prerequisite of haem uptake and the haem enzymatic degradation is pivotal for iron release.

Although haemophores may play multiple roles in pathogenesis, the inactivation of Has systems has not been associated with the loss of virulence of *P. aeruginosa* and *Y. pestis*. However, the inactivation of the Isd, gingipain, and Hxu systems leads to a decrease, albeit modest, in the virulence of *S. aureus*, *P. gingivalis*, and *H. influenzae*, respectively.

The important role of iron and haem acquisition during infection of mammalian hosts supports the view that inhibition of such systems represents a promising strategy for new drug discovery. An understanding of the molecular mechanism by which haem-acquisition systems extract, bind, and transfer haem into bacterial cells is mandatory to accomplish this goal.

Lastly, it may be speculated that haem-bound haemophores display haem-mediated reactivity properties *in vivo*, by scavenging reactive nitrogen and oxygen species and thus protecting bacteria from oxidative and nitrosative damage. Time- and ligand-dependent properties of haemophores may represent a case for "chronosteric effects" (Ascenzi & Fasano, 2007; Ascenzi & Gianni, 2013).

ACKNOWLEDGEMENTS

This work was partially supported by the grant CAL 2015 of Roma Tre University to P.A. and by grants from the Italian Cystic Fibrosis Research Foundation (grant FFC 14/2010) and the Italian Ministry of University and Research-PRIN 2012 (project 2012WJSX8K) to P.V.

REFERENCES

Abe, R., Caaveiro, J. M., Kozuka-Hata, H., Oyama, M., & Tsumoto, K. (2012). Mapping the ultra-weak protein-protein interactions between heme transporters of *Staphylococcus aureus*. *Journal of Biological Chemistry*, *287*, 16477–16487.

Aisen, P., Enns, C., & Wessling-Resnick, M. (2001). Chemistry and biology of eukaryotic iron metabolism. *International Journal of Biochemistry and Cell Biology*, *33*, 940–959.

Alontaga, A., Rodriguez, J., Schönbrunn, E., Becker, A., Funke, T., Yukl, E. T., et al. (2009). Structural characterization of the hemophore HasAp from *Pseudomonas aeruginosa*: NMR spectroscopy reveals protein-protein interactions between Holo-HasAp and hemoglobin. *Biochemistry, 48*, 96–109.

Andrade, M., Ciccarelli, F., Perez-Iratxeta, C., & Bork, P. (2002). NEAT: A domain duplicated in genes near the components of a putative Fe^{3+} siderophore transporter from Gram-positive pathogenic bacteria. *Genome Biology, 3*. RESEARCH0047.1–RESEARCH0047.5.

Arnoux, P., Haser, R., Izadi, N., Lecroisey, A., Delepierre, M., Wandersman, C., et al. (1999). The crystal structure of HasA, a hemophore secreted by *Serratia marcescens*. *Nature Structure Biology, 6*, 516–520.

Arnoux, P., Haser, R., Izadi-Pruneyre, N., Lecroisey, A., & Czjzek, M. (2000). Functional aspects of the heme bound hemophore HasA by structural analysis of various crystal forms. *Proteins, 41*, 202–210.

Ascenzi, P., Bocedi, A., Visca, P., Altruda, F., Tolosano, E., Beringhelli, T., et al. (2005). Hemoglobin and heme scavenging. *IUBMB Life, 57*, 749–759.

Ascenzi, P., & Fasano, M. (2007). Heme-hemopexin: A 'chronosteric' heme-protein. *IUBMB Life, 59*, 700–708.

Ascenzi, P., & Gianni, S. (2013). Functional role of transient conformations: Rediscovering "chronosteric effects" thirty years later. *IUBMB Life, 65*, 836–844.

Baelen, S., Dewitte, F., Clantin, B., & Villeret, V. (2013). Structure of the secretion domain of HxuA from *Haemophilus influenzae*. *Acta Crystallographica Section F Structural Biology Communications, 69*, 1322–1327.

Balderas, M. A., Nobles, C. L., Honsa, E. S., Alicki, E. R., & Maresso, A. W. (2012). Hal is a *Bacillus anthracis* heme acquisition protein. *Journal of Bacteriology, 194*, 5513–5521.

Barjon, C., Wecker, K., Izadi-Pruneyre, N., & Delepelaire, P. (2007). Mutagenesis and molecular modeling reveal three key extracellular loops of the membrane receptor HasR that are involved in hemophore HasA binding. *Journal of Bacteriology, 189*, 5379–5382.

Beasley, F., Marolda, C., Cheung, J., Buac, S., & Heinrichs, D. (2011). *Staphylococcus aureus* transporters Hts, Sir, and Sst capture iron liberated from human transferrin by Staphyloferrin A, Staphyloferrin B, and catecholamine stress hormones, respectively, and contribute to virulence. *Infection and Immunity, 79*, 2345–2355.

Beasley, F. C., Vines, E. D., Grigg, J. C., Zheng, Q., Liu, S., Lajoie, G. A., et al. (2009). Characterization of staphyloferrin A biosynthetic and transport mutants in *Staphylococcus aureus*. *Molecular Microbiology, 72*, 947–963.

Becker, K. W., & Skaar, E. P. (2014). Metal limitation and toxicity at the interface between host and pathogen. *FEMS Microbiology Reviews, 38*, 1235–1249.

Benesch, R., & Kwong, S. (1990). The stability of the hemoglobin linkage in some normal, mutant, and chemically modified hemoglobins. *Journal of Biological Chemistry, 265*, 14881–14885.

Braun, V. (2001). Iron uptake mechanisms and their regulation in pathogenic bacteria. *International Journal of Medical Microbiology, 291*, 67–79.

Braun, V., & Hantke, K. (2011). Recent insights into iron import by bacteria. *Current Opinion in Chemical Biology, 2*, 328–334.

Brickman, T. J., Vanderpool, C. K., & Armstrong, S. K. (2006). Heme transport contributes to in vivo fitness of *Bordetella pertussis* during primary infection in mice. *Infection and Immunity, 74*, 1741–1744.

Brock, J. H., & Ng, J. (1983). The effect of desferrioxamine on the growth of *Staphylococcus aureus, Yersinia enterocolitica*, and *Streptococcus faecalis* in human serum: Uptake of desferrioxamine-bound iron. *FEMS Microbiology Letters, 20*, 439–442.

Brunori, M. (2014). Variations on the theme: Allosteric control in hemoglobin. *Federation of European Biochemical Societies Journal, 281*, 633–643.

Butler, A. (2005). Marine siderophores and microbial iron mobilization. *Biometals, 18*, 369–374.

Byrne, D. P., Potempa, J., Olczak, T., & Smalley, J. W. (2013). Evidence of mutualism between two periodontal pathogens: Co-operative haem acquisition by the HmuY haemophore of *Porphyromonas gingivalis* and the cysteine protease interpain A (InpA) of *Prevotella intermedia*. *Molecular Oral Microbiology, 28*, 219–229.

Caillet-Saguy, C., Turano, P., Piccioli, M., Lukat-Rodgers, G., Czjzek, M., Guigliarell, I. B., et al. (2008). Deciphering the structural role of histidine 83 for heme binding in hemophore HasA. *Journal of Biological Chemistry, 283*, 5960–5970.

Carvalho-Filho, P. C., Trindade, S. C., Olczak, T., Sampaio, G. P., Oliveira-Neto, M. G., Santos, H. A., et al. (2013). *Porphyromonas gingivalis* HmuY stimulates expression of Bcl-2 and Fas by human $CD3^+$ T cells. *BioMed Central Microbiology, 13*, 206.

Cescau, S., Cwerman, H., Létoffé, S., Delepelaire, P., Wandersman, C., & Biville, F. (2007). Heme acquisition by hemophores. *Biometals, 20*, 603–613.

Chiabrando, D., Mercurio, S., & Tolosano, E. (2014). Heme and erythropoiesis: More than a structural role. *Haematologica, 99*, 973–983.

Contreras, H., Chim, N., Credali, A., & Goulding, C. W. (2014). Heme uptake in bacterial pathogens. *Current Opinion in Chemical Biology, 19*, 34–41.

Cope, L., Love, R., Guinn, S., Gilep, A., Usanov, S., Estabrook, R. W., et al. (2001). Involvement of HxuC outer membrane protein in utilization of hemoglobin by *Haemophilus influenzae*. *Infection and Immunity, 69*, 2353–2363.

Cope, L., Thomas, S., Hrkal, Z., & Hansen, E. (1998). Binding of heme-hemopexin complexes by soluble HxuA protein allows utilization of this complexed heme by *Haemophilus influenzae*. *Infection and Immunity, 66*, 4511–4516.

Cope, D., Thomas, S., Latimer, J., Slaughter, C., Muller-Eberhard, U., & Hansen, E. (1994). The 100 kDa haem: Haemopexin-binding protein of *Haemophilus influenzae*: Structure and localization. *Molecular Microbiology, 13*, 863–873.

Cope, L. D., Yogev, R., Muller-Eberhard, U., & Hansen, E. J. (1995). A gene cluster involved in the utilisation of both free heme and heme:hemopexin by *Haemophilus influenzae* type B. *Journal of Bacteriology, 177*, 2644–2653.

Cornelis, P. (2010). Iron uptake and metabolism in pseudomonads. *Applied Microbiology and Biotechnology, 86*, 1637–1645.

Coulanges, V., Andre, P., Ziegler, O., Buchheit, L., & Vidon, D. J. (1997). Utilization of iron-catecholamine complexes involving ferric reductase activity in *Listeria monocytogenes*. *Infection and Immunity, 65*, 2778–2785.

Crosa, J. H. (1997). Signal transduction and transcriptional and posttranscriptional control of iron-regulated genes in bacteria. *Microbiology and Molecular Biology Reviews, 61*, 319–336.

Czjzek, M., Létoffé, S., Wandersman, C., Delepierre, M., Lecroisey, A., & Izadi-Pruneyre, N. (2007). The crystal structure of the secreted dimeric form of the hemophore HasA reveals a domain swapping with an exchanged heme ligand. *Journal of Molecular Biology, 365*, 1176–1186.

de Amorim, G. C., Prochnicka-Chalufour, A., Delepelaire, P., Lefèvre, J., Simenel, C., Wandersman, C., et al. (2013). The structure of HasB reveals a new class of TonB protein fold. *PLoS One, 8*, e58964.

Deshpande, R. G., Khan, M. B., Bhat, D. A., & Navalkar, R. G. (1997). Isolation of a contact-dependent haemolysin from *Mycobacterium tuberculosis*. *Journal of Medical Microbiology, 46*, 233–238.

Dickson, C. F., Kumar, K. K., Jacques, D. A., Malmirchegini, G. R., Spirig, T., Mackay, J. P., et al. (2014). Structure of the hemoglobin-IsdH complex reveals the molecular basis of iron capture by *Staphylococcus aureus*. *Journal of Biological Chemistry, 289*, 6728–6738.

Domenech, P., Reed, M. B., & Barry, C. E., 3rd. (2005). Contribution of the *Mycobacterium tuberculosis* MmpL protein family to virulence and drug resistance. *Infection and Immunity, 73*, 3492–3501.

Dryla, A., Hoffmann, B., Gelbmann, D., Giefing, C., Hanner, M., Meinke, A., et al. (2007). High-affinity binding of the staphylococcal HarA protein to haptoglobin and hemoglobin involves a domain with an antiparallel eight-stranded β-barrel fold. *Journal of Bacteriology, 189*, 254–264.

Ekworomadu, M. T., Poor, C. B., Owens, C. P., Balderas, M. A., Fabian, M., Olson, J. S., et al. (2012). Differential function of lip residues in the mechanism and biology of an anthrax hemophore. *Public Library of Science Pathogens, 8*, e1002559.

Fabian, M., Solomaha, E., Olson, J. S., & Maresso, A. W. (2009). Heme transfer to the bacterial cell envelope occurs via a secreted hemophore in the Gram-positive pathogen *Bacillus anthracis*. *Journal of Biological Chemistry, 284*, 32138–32146.

Fang, Z., Sampson, S. L., Warren, R. M., Gey van Pittius, N. C., & Newton-Foot, M. (2015). Iron acquisition strategies in mycobacteria. *Tuberculosis (Edinburg), 95*, 123–130.

Faraldo-Gómez, J. D., & Sansom, M. S. (2003). Acquisition of siderophores in Gram-negative bacteria. *Nature Reviews Molecular Cell Biology, 4*, 105–116.

Farhana, A., Saini, V., Kumar, A., Lancaster, J. R., Jr., & Steyn, A. J. (2012). Environmental heme-based sensor proteins: Implications for understanding bacterial pathogenesis. *Antioxidants & Redox Signaling, 17*, 1232–1245.

Ferguson, A. D., & Deisenhofer, J. (2002). TonB-dependent receptors-structural perspectives. *Biochimica et Biophysica Acta, 1565*, 318–332.

Foley, T. L., & Simeonov, A. (2012). Targeting iron assimilation to develop new antibacterials. *Expert Opinion on Drug Discovery, 7*, 831–847.

Fonner, B. A., Tripet, B. P., Eilers, B. J., Stanisich, J., Sullivan-Springhetti, R. K., Moore, R., et al. (2014). Solution structure and molecular determinants of hemoglobin binding of the first NEAT domain of IsdB in *Staphylococcus aureus*. *Biochemistry, 53*, 3922–3933.

Forman, S., Paulley, J. T., Fetherston, J. D., Cheng, Y. Q., & Perry, R. D. (2010). *Yersinia* ironomics: Comparison of iron transporters among *Yersinia pestis* biotypes and its nearest neighbor, *Yersinia pseudotuberculosis*. *Biometals, 23*, 275–294.

Fournier, C., Smith, A., & Delepelaire, P. (2011). Haem release from haemopexin by HxuA allows *Haemophilus influenzae* to escape host nutritional immunity. *Molecular Microbiology, 80*, 133–148.

Gao, J. L., Nguyen, K. A., & Hunter, N. (2010). Characterization of a hemophore-like protein from *Porphyromonas gingivalis*. *Journal of Biological Chemistry, 285*, 40028–40038.

Gaudin, C., Grigg, J., Arrieta, A., & Murphy, M. (2011). Unique heme-iron coordination by the hemoglobin receptor IsdB of *Staphylococcus aureus*. *Biochemistry, 50*, 5443–5452.

Ge, R., & Sun, X. (2014). Iron acquisition and regulation systems in *Streptococcus species*. *Metallomics, 6*, 996–1003.

Ghigo, J. M., Létoffé, S., & Wandersman, C. (1997). A new type of hemophore-dependent heme acquisition system of *Serratia marcescens* reconstituted in *Escherichia coli*. *Journal of Bacteriology, 179*, 3572–3579.

Gmiterek, A., Wójtowicz, H., Mackiewicz, P., Radwan-Oczko, M., Kantorowicz, M., Chomyszyn-Gajewska, M., et al. (2013). The unique hmuY gene sequence as a specific marker of *Porphyromonas gingivalis*. *Public Library of Science One, 8*, e67719.

Granger, J., & Price, N. M. (1999). The importance of siderophores in iron nutrition of heterotrophic marine bacteria. *Limnology and Oceanography, 44*, 541–555.

Grigg, J., Mao, C., & Murphy, M. (2011). Iron-coordinating tyrosine is a key determinant of NEAT domain heme transfer. *Journal of Molecular Biology, 413*, 684–698.

Grigg, J. C., Ukpabi, G., Gaudin, C. F., & Murphy, M. E. (2010). Structural biology of heme binding in the *Staphylococcus aureus* Isd system. *Journal of Inorganic Biochemistry, 104*, 341–348.

Grigg, J., Vermeiren, C., Heinrichs, D., & Murphy, M. (2007a). Haem recognition by a *Staphylococcus aureus* NEAT domain. *Molecular Microbiology*, *63*, 139–149.

Grigg, J., Vermeiren, C., Heinrichs, D., & Murphy, M. (2007b). Heme coordination by *Staphylococcus aureus* IsdE. *Journal of Biological Chemistry*, *282*, 28815–28822.

Hammer, N. D., & Skaar, E. P. (2011). Molecular mechanisms of *Staphylococcus aureus* iron acquisition. *Annual Review of Microbiology*, *65*, 129–147.

Henderson, D. P., & Paine, S. M. (1994). *Vibrio cholerae* iron transport systems: Roles of heme and siderophore iron transport in virulence and identification of a gene associated with multiple iron transport systems. *Infection and Immunity*, *62*, 5120–5125.

Honsa, E., Fabian, M., Cardenas, A., Olson, J., & Maresso, A. (2011). The five near-iron transporter (NEAT) domain anthrax hemophore, IsdX2, scavenges heme from hemoglobin and transfers heme to the surface protein IsdC. *Journal of Biological Chemistry*, *286*, 33652–33660.

Honsa, E. S., Owens, C. P., Goulding, C. W., & Maresso, A. W. (2013). The near-iron transporter (NEAT) domains of the anthrax hemophore IsdX2 require a critical glutamine to extract heme from methemoglobin. *Journal of Biological Chemistry*, *288*, 8479–8490.

Hulkoti, N. I., & Taranath, T. C. (2014). Biosynthesis of nanoparticles using microbes—A review. *Colloids and Surfaces B Biointerfaces*, *121*, 474–483.

Hvorup, R. N., Goetz, B. A., Niederer, M., Hollenstein, K., Perozo, E., & Locher, K. P. (2007). Asymmetry in the structure of the ABC transporter-binding protein complex BtuCD-BtuF. *Science*, *317*, 1387–1390.

Izadi, N., Henri, Y., Haladjan, J., Goldberg, M. E., Wandersman, C., Delepierre, M., et al. (1997). Purification and characterization of an extra cellular heme binding protein, HasA, involved in heme iron acquisition. *Biochemistry*, *36*, 7050–7057.

Izadi-Pruneyre, N., Huche, F., Lukat-Rodgers, G., Lecroisey, A., Gilli, R., Rodgers, K., et al. (2006). The heme transfer from the soluble HasA hemophore to its membrane bound receptor HasR is driven by protein-protein interaction from a high to a lower affinity binding site. *Journal of Biological Chemistry*, *281*, 25541–25550.

Jacob-Dubuisson, F., Kehoe, B., Willery, E., Reveneau, N., Locht, C., & Relman, D. A. (2000). Molecular characterization of *Bordetella bronchiseptica* filamentous haemagglutinin and its secretion machinery. *Microbiology*, *146*, 1211–1221.

Jepkorir, G., Rodríguez, J. C., Rui, H., Im, W., Lovell, S., Battaile, K. P., et al. (2010). Structural, NMR spectroscopic, and computational investigation of hemin loading in the hemophore HasAp from *Pseudomonas aeruginosa*. *Journal of the American Chemical Society*, *132*, 9857–9872.

Jin, B., Newton, S. M., Shao, Y., Jiang, X., Charbit, A., & Klebba, P. E. (2006). Iron acquisition systems for ferric hydroxamates, haemin and haemoglobin in *Listeria monocytogenes*. *Molecular Microbiology*, *59*, 1185–1198.

Klebba, P. E., Charbit, A., Xiao, Q., Jiang, X., & Newton, S. M. (2012). Mechanisms of iron and haem transport by *Listeria monocytogenes*. *Molecular Membrane Biology*, *29*, 69–86.

Kontoghiorghes, G. J., Kolnagou, A., Skiada, A., & Petrikkos, G. (2010). The role of iron and chelators on infections in iron overload and non-iron loaded conditions: Prospects for the design of new antimicrobial therapies. *Hemoglobin*, *34*, 227–239.

Krieg, S., Huché, F., Diederichs, K., Izadi-Pruneyre, N., Lecroisey, A., Wandersman, C., et al. (2009). Heme uptake across the outer membrane as revealed by crystal structures of the receptor-hemophore complex. *Proceedings of the National Academy of Sciences of the United States of America*, *106*, 1045–1050.

Krishna Kumar, K., Jacques, D., Pishchany, G., Caradoc-Davies, T., Spirig, T., Malmirchegini, G., et al. (2011). Structural basis for hemoglobin capture by *Staphylococcus aureus* cell-surface protein, IsdH. *Journal of Biological Chemistry*, *286*, 38439–38447.

Kumar, R., Lovell, S., Matsumura, H., Battaile, K. P., Moënne-Loccoz, P., & Rivera, M. (2013). The hemophore HasA from *Yersinia pestis* (HasAyp) coordinates hemin with a single residue, Tyr75, and with minimal conformational change. *Biochemistry*, *52*, 2705–2707.

Kumar, R., Matsumura, H., Lovell, S., Yao, H., Rodríguez, J. C., Battaile, K. P., et al. (2014). Replacing the axial ligand tyrosine 75 or its hydrogen bond partner histidine 83 minimally affects hemin acquisition by the hemophore HasAp from *Pseudomonas aeruginosa*. *Biochemistry*, *53*, 2112–2125.

Lee, W. C., Reniere, M. L., Skaar, E. P., & Murphy, M. E. (2008). Ruffling of metalloporphyrins bound to IsdG and IsdI, two heme-degrading enzymes in *Staphylococcus aureus*. *Journal of Biological Chemistry*, *283*, 30957–30963.

Létoffé, S., Debarbieux, L., Izadi, N., Delepelaire, N., & Wandersman, C. (2003). Ligand delivery by heme carrier proteins: The binding of *Serratia marcescens* hemophore to its outer membrane receptor is mediated by two distinct peptide regions. *Molecular Microbiology*, *50*, 77–88.

Létoffé, S., Delepelaire, P., & Wandersman, C. (2004). Free and hemophore-bound heme acquisitions through the outer membrane receptor HasR have different requirements for the TonB-ExbB-ExbD complex. *Journal of Bacteriology*, *186*, 4067–4074.

Létoffé, S., Deniau, C., Wolff, N., Dassa, E., Delepelaire, P., Lecroisey, A., et al. (2001). Haemophore-mediated bacterial haem transport: Evidence for a common or overlapping site for haem-free and haem-loaded haemophore on its specific outer membrane receptor. *Molecular Microbiology*, *41*, 439–450.

Létoffé, S., Ghigo, J. M., & Wandersman, C. (1994a). Iron acquisition from heme and hemoglobin by *Serratia marcescens* extracellular protein. *Proceedings of the National Academy of Sciences of the United States of America*, *91*, 9876–9880.

Létoffé, S., Ghigo, J. M., & Wandersman, C. (1994b). Secretion of the *Serratia marcescens* HasA protein by an ABC transporter. *Journal of Bacteriology*, *176*, 5372–5377.

Létoffé, S., Nato, F., Goldberg, M. E., & Wandersman, C. (1999). Interactions of HasA, a bacterial haemophore, with haemoglobin and with its outer membrane receptor HasR. *Molecular Microbiology*, *33*, 546–555.

Létoffé, S., Redeker, V., & Wandersman, C. (1998). Isolation and characterisation of an extracellular heme binding protein from *Pseudomonas aeruginosa* that shares function and sequence similarities with the *Serratia marcescens* HasA hemophore. *Molecular Microbiology*, *28*, 1223–1234.

Liu, X., Olczak, T., Guo, H. C., Dixon, D. W., & Genco, C. A. (2006). Identification of amino acid residues involved in heme binding and hemoprotein utilization in the *Porphyromonas gingivalis* heme receptor HmuR. *Infection and Immunity*, *74*, 1222–1232.

Liu, X., Sroka, A., Potempa, J., & Genco, C. A. (2004). Coordinate expression of the *Porphyromonas gingivalis* lysine-specific gingipain proteinase, Kgp, arginine-specific gingipain proteinase, RgpA, and the heme/hemoglobin receptor, HmuR. *Biological Chemistry*, *385*, 1049–1057.

Liu, M., Tanaka, W., Zhu, H., Xie, G., Dooley, D., & Lei, B. (2008). Direct hemin transfer from IsdA to IsdC in the ironregulated surface determinant (Isd) heme acquisition system of *Staphylococcus aureus*. *Journal of Biological Chemistry*, *283*, 6668–6676.

Llamas, M. A., Imperi, F., Visca, P., & Lamont, I. L. (2014). Cell-surface signaling in *Pseudomonas*: Stress responses, iron transport, and pathogenicity. *FEMS Microbiology Reviews*, *38*, 569–597.

Malachowa, N., & DeLeo, F. R. (2011). *Staphylococcus aureus* survival in human blood. *Virulence*, *2*, 567–569.

Malmirchegini, G. R., Sjodt, M., Shnitkind, S., Sawaya, M. R., Rosinski, J., Newton, S. M., et al. (2014). Novel mechanism of hemin capture by Hbp2, the hemoglobin-binding hemophore from *Listeria monocytogenes*. *Journal of Biological Chemistry*, *289*, 34886–34899.

Maresso, A., Garufi, G., & Schneewind, O. (2008). *Bacillus anthracis* secretes proteins that mediate heme acquisition from hemoglobin. *PLoS Pathogens, 4*, e1000132.

Marraffini, L., & Schneewind, O. (2005). Anchor structure of staphylococcal surface proteins. V. Anchor structure of the sortase B substrate IsdC. *Journal of Biological Chemistry, 280*, 16263–16271.

Mattle, D., Zeltina, A., Woo, J. S., Goetz, B. A., & Locher, K. P. (2010). Two stacked heme molecules in the binding pocket of the periplasmic heme-binding protein HmuT from *Yersinia pestis*. *Journal of Molecular Biology, 404*, 220–231.

Mazmanian, S. K., Skaar, E. P., Gaspar, A. H., Humayun, M., Gornicki, P., Jelenska, J., et al. (2003). Passage of heme-iron across the envelope of *Staphylococcus aureus*. *Science, 299*, 906–909.

Mazmanian, S. K., Ton-That, H., Su, K., & Schneewind, O. (2002). An iron-regulated sortase anchors a class of surface protein during *Staphylococcus aureus* pathogenesis. *Proceedings of the National Academy of Sciences of the United States of America, 99*, 2293–2298.

Minandri, F., Bonchi, C., Frangipani, E., Imperi, F., & Visca, P. (2014). Promises and failures of gallium as an antibacterial agent. *Future Microbiology, 9*, 379–397.

Moriwaki, Y., Caaveiro, J. M., Tanaka, Y., Tsutsumi, H., Hamachi, I., & Tsumoto, K. (2011). Molecular basis of recognition of antibacterial porphyrins by heme-transporter IsdH-NEAT3 of *Staphylococcus aureus*. *Biochemistry, 50*, 7311–7320.

Morton, D., Seale, T., Madore, L., VanWagoner, T., Whitby, P., & Stull, T. (2007). The haem-haemopexin utilization gene cluster (hxuCBA) as a virulence factor of *Haemophilus influenzae*. *Microbiology, 153*, 215–224.

Morton, D. J., Smith, A., Ren, Z., Madore, L. L., VanWagoner, T. M., Seale, T. W., et al. (2004). Identification of a haem-utilization protein (Hup) in *Haemophilus influenzae*. *Microbiology, 150*, 3923–3933.

Muryoi, N., Tiedemann, M., Pluym, M., Cheung, J., Heinrichs, D., & Stillman, M. (2008). Demonstration of the ironregulated surface determinant (Isd) heme transfer pathway in *Staphylococcus aureus*. *Journal of Biological Chemistry, 283*, 28125–28136.

Nairz, M., Schroll, A., Sonnweber, T., & Weiss, G. (2010). The struggle for iron—A metal at the host-pathogen interface. *Cellular Microbiology, 12*, 1691–1702.

Neilands, J. B. (1995). Siderophores: Structure and function of microbial iron transport compounds. *Journal of Biological Chemistry, 270*, 26723–26726.

Olczak, T., Sroka, A., Potempa, J., & Olczak, M. (2008). *Porphyromonas gingivalis* HmuY and HmuR: Further characterization of a novel mechanism of heme utilization. *Archives of Microbiology, 189*, 197–210.

Olczak, T., Wójtowicz, H., Ciuraszkiewicz, J., & Olczak, M. (2010). Species specificity, surface exposure, protein expression, immunogenicity, and participation in biofilm formation of *Porphyromonas gingivalis* HmuY. *BMC Microbiology, 10*, 134.

Owens, C. P., Chim, N., Graves, A. B., Harmston, C. A., Iniguez, A., Contreras, H., et al. (2013). The *Mycobacterium tuberculosis* secreted protein Rv0203 transfers heme to membrane proteins MmpL3 and MmpL11. *Journal of Biological Chemistry, 288*, 21714–21728.

Owens, C., Du, J., Dawson, J., & Goulding, C. (2012). Characterization of heme ligation properties of Rv0203, a secreted heme-binding protein involved in *Mycobacterium tuberculosis* heme uptake. *Biochemistry, 51*, 1518–1531.

Paquelin, A., Ghigo, J. M., Bertin, S., & Wandersman, C. (2001). Characterization of HasB, a *Serratia marcescens* TonB-like protein specifically involved in the haemophore dependent haem acquisition system. *Molecular Microbiology, 42*, 995–1005.

Paulley, J. T., Anderson, E. S., & Roop, R. M., 2nd. (2007). *Brucella abortus* requires the heme transporter BhuA for maintenance of chronic infection in BALB/c mice. *Infection and Immunity, 75*, 5248–5254.

Perry, R. D., Shah, J., Bearden, S. W., Thompson, J. M., & Fetherston, J. D. (2003). *Yersinia pestis* TonB: Role in iron, heme, and hemoprotein utilization. *Infection and Immunity, 71*, 4159–4162.

Pettersen, E. F., Goddard, T. D., Huang, C. C., Couch, G. S., Greenblatt, D. M., Meng, E. C., et al. (2004). UCSF Chimera—A visualization system for exploratory research and analysis. *Journal of Computational Chemistry, 25*, 1605–1612.

Pilpa, R., Fadeev, E., Villareal, V., Wong, M., Phillips, M., & Clubb, R. (2006). Solution structure of the NEAT (NEAr Transporter) domain from IsdH/HarA: The human hemoglobin receptor in *Staphylococcus aureus*. *Journal of Molecular Biology, 360*, 435–447.

Pilpa, R., Robson, S., Villareal, V., Wong, M., Phillips, M., & Clubb, R. (2009). Functionally distinct NEAT (NEAr Transporter) domains within the *Staphylococcus aureus* IsdH/HarA protein extract heme from methemoglobin. *Journal of Biological Chemistry, 284*, 1166–1176.

Pishchany, G., Dickey, S. E., & Skaar, E. P. (2009). Subcellular localization of the *Staphylococcus aureus* heme iron transport components IsdA and IsdB. *Infection and Immunity, 77*, 2624–2634.

Pishchany, G., McCoy, A., Torres, V., Krause, J., Crowe, J. J., Fabry, M., et al. (2010). Specificity for human hemoglobin enhances *Staphylococcus aureus* infection. *Cell Host and Microbe, 8*, 544–550.

Potempa, J., Sroka, A., Imamura, T., & Travis, J. (2003). Gingipains, the major cysteine proteinases and virulence factors of *Porphyromonas gingivalis*: Structure, function and assembly of multidomain protein complexes. *Current Protein and Peptide Science, 4*, 397–407.

Reniere, M. L., Ukpabi, G. N., Harry, S. R., Stec, D. F., Krull, R., Wright, D. W., et al. (2010). The IsdG-family of haem oxygenases degrades haem to a novel chromophore. *Molecular Microbiology, 75*, 1529–1538.

Rohde, K., & Dyer, D. (2004). Analysis of haptoglobin and hemoglobin-haptoglobin interactions with the *Neisseria meningitidis* TonB-dependent receptor HpuAB by flow cytometry. *Infection and Immunity, 72*, 2494–2506.

Rossi, M. S., Fetherston, J. D., Létoffé, S., Carniel, E., Perry, R. D., & Ghigo, J. M. (2001). Identification and characterization of the hemophore-dependent heme acquisition system of *Yersinia pestis*. *Infection and Immunity, 69*, 6707–6717.

Rossi, M. S., Paquelin, A., Ghigo, J. M., & Wandersman, C. (2003). Haemophore-mediated signal transduction across the bacterial cell envelope in *Serratia marcescens*: The inducer and the transported substrate are different molecule. *Molecular Microbiology, 48*, 1467–1480.

Runyen-Janecky, L. J. (2013). Role and regulation of heme iron acquisition in gram-negative pathogens. *Frontiers in Cellular and Infection Microbiology, 3*, 55.

Sawatzki, G., Hoffmann, F. A., & Kubanek, B. (1983). Acute iron overload in mice: Pathogenesis of *Salmonella typhimurium* infection. *Infection and Immunity, 39*, 659–665.

Schröder, I., Johnson, E., & de Vries, S. (2003). Microbial ferric iron reductases. *FEMS Microbiology Reviews, 27*, 427–447.

Sebulsky, M. T., Hohnstein, D., Hunter, M. D., & Heinrichs, D. E. (2000). Identification and characterization of a membrane permease involved in iron-hydroxamate transport in *Staphylococcus aureus*. *Journal of Bacteriology, 182*, 4394–4400.

Sharp, K., Schneider, S., Cockayne, A., & Paoli, M. (2007). Crystal structure of the heme-IsdC complex, the central conduit of the Isd iron/heme uptake system in *Staphylococcus aureus*. *Journal of Biological Chemistry, 28*, 10625–10631.

Sheldon, J. R., & Heinrichs, D. E. (2015). Recent developments in understanding the iron acquisition strategies of gram positive pathogens. *FEMS Microbiology Reviews, 39*, 592–630. http://dx.doi.org/10.1093/femsrev/fuv009.

Shirataki, C., Shoji, O., Terada, M., Ozaki, S., Sugimoto, H., Shiro, Y., et al. (2014). Inhibition of heme uptake in *Pseudomonas aeruginosa* by its hemophore ($HasA_p$) bound to synthetic metal complexes. *Angewandte Chemie International Edition in English, 53,* 2862–2866.

Sikkema, D. J., & Brubaker, R. R. (1989). Outer membrane peptides of *Yersinia pestis* mediating siderophore-independent assimilation of iron. *Biology of Metals, 2,* 174–184.

Skaar, E., Humayun, M., Bae, T., DeBord, K., & Schneewind, O. (2004). Iron-source preference of *Staphylococcus aureus* infections. *Science, 305,* 1626–1628.

Smalley, J., Byrne, D., Birss, A., Wojtowicz, H., Sroka, A., Potempa, J., et al. (2011). HmuY haemophore and gingipain proteases constitute a unique syntrophic system of haem acquisition by *Porphyromonas gingivalis. PLoS One, 6,* e17182.

Soe-Lin, S., Apte, S. S., Andriopoulos, B. Jr, Andrews, M. C., Schranzhofer, M., Kahawita, T., et al. (2009). Nramp1 promotes efficient macrophage recycling of iron following erythrophagocytosis *in vivo. Proceedings of the National Academy of Sciences of the United States of America, 106,* 5960–5965.

Speziali, C. D., Dale, S. E., Henderson, J. A., Vines, E. D., & Heinrichs, D. E. (2006). Requirement of *Staphylococcus aureus* ATP-binding cassette-ATPase FhuC for iron-restricted growth and evidence that it functions with more than one iron transporter. *Journal of Bacteriology, 188,* 2048–2055.

Spirig, T., Malmirchegini, G. R., Zhang, J., Robson, S. A., Sjodt, M., Liu, M., et al. (2013). *Staphylococcus aureus* uses a novel multidomain receptor to break apart human hemoglobin and steal its heme. *Journal of Biological Chemistry, 288,* 1065–1078.

Stojiljkovic, I., Kumar, V., & Srinivasan, N. (1999). Non-iron metalloporphyrins: Potent antibacterial compounds that exploit haem/Hb uptake systems of pathogenic bacteria. *Molecular Microbiology, 31,* 429–442.

Taabazuing, C. Y., Hangasky, J. A., & Knapp, M. J. (2014). Oxygen sensing strategies in mammals and bacteria. *Journal of Inorganic Biochemistry, 133,* 63–72.

Tai, S. S., Lee, C. J., & Winter, R. E. (1993). Hemin utilization is related to virulence of *Streptococcus pneumoniae. Infection and Immunology, 61,* 5401–5405.

Tarlovsky, Y., Fabian, M., Solomaha, E., Honsa, E., Olson, J. S., & Maresso, A. W. (2010). A *Bacillus anthracis* S-layer homology protein that binds heme and mediates heme delivery to IsdC. *Journal of Bacteriology, 192,* 3503–3511.

Torres, V., Pishchany, G., Humayun, M., Schneewind, O., & Skaar, E. (2006). *Staphylococcus aureus* IsdB is a hemoglobin receptor required for heme iron utilization. *Journal of Bacteriology, 188,* 8421–8429.

Trindade, S. C., Olczak, T., Gomes-Filho, I. S., de Moura-Costa, L. F., Vale, V. C., Galdino-Neto, M., et al. (2013). *Porphyromonas gingivalis* HmuY-induced production of interleukin-6 and IL-6 polymorphism in chronic periodontitis. *Journal of Periodontology, 84,* 650–655.

Trindade, S. C., Olczak, T., Gomes-Filho, I. S., Moura-Costa, L. F., Cerqueira, E. M., Galdino-Neto, M., et al. (2012). Induction of interleukin (IL)-1β, IL-10, IL-8 and immunoglobulin G by *Porphyromonas gingivalis* HmuY in humans. *Journal of Periodontal Research, 47,* 27–32.

Tullius, M. V., Harmston, C. A., Owens, C. P., Chim, N., Morse, R. P., McMath, L. M., et al. (2011). Discovery and characterization of a unique mycobacterial heme acquisition system. *Proceedings of the National Academy of Sciences of the United States of America, 108,* 5051–5056.

Ukpabi, G., Takayama, S. J., Mauk, A. G., & Murphy, M. E. (2012). Inactivation of the heme degrading enzyme IsdI by an active site substitution that diminishes heme ruffling. *Journal of Biological Chemistry, 287,* 34179–34188.

Vandenesch, F., Lina, G., & Henry, T. (2012). *Staphylococcus aureus* hemolysins, bi-component leukocidins, and cytolytic peptides: A redundant arsenal of membrane-damaging virulence factors? *Frontiers in Cellular and Infection Microbiology*, *2*, 12.

Vanderpool, C. K., & Armstrong, S. K. (2003). Heme-responsive transcriptional activation of *Bordetella bhu* genes. *Journal of Bacteriology*, *185*, 909–917.

Villareal, V., Pilpa, R., Robson, S., Fadeev, E., & Clubb, R. (2008). The IsdC protein from *Staphylococcus aureus* uses a flexible binding pocket to capture heme. *Journal of Biological Chemistry*, *283*, 31591–31600.

Villareal, V., Spirig, T., Robson, S., Liu, M., Lei, B., & Clubb, R. (2011). Transient weak protein-protein complexes transfer heme across the cell wall of *Staphylococcus aureus*. *Journal of the American Chemical Society*, *133*, 14176–14179.

Visca, P., Imperi, F., & Lamont, I. L. (2007). Pyoverdine siderophores: From biogenesis to biosignificance. *Trends in Microbiology*, *15*, 22–30.

Vraspir, J. M., & Butler, A. (2009). Chemistry of marine ligands and siderophores. *Annual Review of Marine Science*, *1*, 43–63.

Vu, N. T., Moriwaki, Y., Caaveiro, J. M., Terada, T., Tsutsumi, H., Hamachi, I., et al. (2013). Selective binding of antimicrobial porphyrins to the heme-receptor IsdH-NEAT3 of *Staphylococcus aureus*. *Protein Science*, *22*, 942–953.

Wandersman, C., & Delepelaire, P. (2004). Bacterial iron sources: From siderophores to hemophores. *Annual Review of Microbiology*, *58*, 611–647.

Wandersman, C., & Delepelaire, P. (2012). Haemophore functions revisited. *Molecular Microbiology*, *85*, 618–631.

Watanabe, M., Tanaka, Y., Suenaga, A., Kuroda, M., Yao, M., Watanabe, N., et al. (2008). Structural basis for multimeric heme complexation through a specific protein-heme interaction: The case of the third neat domain of IsdH from *Staphylococcus aureus*. *Journal of Biological Chemistry*, *283*, 28649–28659.

Whitby, P. W., Sim, K. E., Morton, D. J., Patel, J. A., & Stull, T. L. (1997). Transcription of genes encoding iron and heme acquisition proteins of *Haemophilus influenzae* during acute otitis media. *Infection and Immunity*, *65*, 4696–4700.

Wilks, A., & Burkhard, K. A. (2007). Heme and virulence: How bacterial pathogens regulate, transport and utilize heme. *Natural Product Reports*, *24*, 511–522.

Wittenberg, J. B., & Wittenberg, B. A. (1990). Mechanisms of cytoplasmic hemoglobin and myoglobin function. *Annual Review of Biophysics and Biophysical Chemistry*, *19*, 217–241.

Wójtowicz, H., Guevara, T., Tallant, C., Olczak, M., Sroka, A., Potempa, J., et al. (2009). Unique structure and stability of HmuY, a novel heme-binding protein of *Porphyromonas gingivalis*. *PLoS Pathogens*, *5*, e1000419.

Wolff, N., Izadi-Pruneyre, N., Couprie, J., Habeck, M., Linge, J., Rieping, W., et al. (2008). Comparative analysis of structural and dynamic properties of the loaded and unloaded hemophore HasA: Functional implications. *Journal of Molecular Biology*, *376*, 517–525.

Wong, J., Patel, R., Kendall, D., Whitby, P., Smith, A., Holland, J., et al. (1995). Affinity, conservation, and surface exposure of hemopexin binding proteins in *Haemophilus influenzae*. *Infection and Immunity*, *63*, 2327–2333.

Woo, J. S., Zeltina, A., Goetz, B. A., & Locher, K. P. (2012). X-ray structure of the *Yersinia pestis* heme transporter HmuUV. *Nature Structural and Molecular Biology*, *19*, 1310–1315.

Wu, R., Skaar, E. P., Zhang, R., Joachimiak, G., Gornicki, P., Schneewind, O., et al. (2005). *Staphylococcus aureus* IsdG and IsdI, heme-degrading enzymes with structural similarity to monooxygenases. *Journal of Biological Chemistry*, *280*, 2840–2846.

Xiao, Q., Jiang, X., Moore, K., Shao, Y., Pi, H., Dubail, I., et al. (2011). Sortase independent and dependent systems for acquisition of haem and haemoglobin in *Listeria monocytogenes*. *Molecular Microbiology*, *80*, 1581–1597.

Yukl, E., Jepkorir, G., Alontaga, A., Pautsch, L., Rodriguez, J., Rivera, M., et al. (2010). Kinetic and spectroscopic studies of hemin acquisition in the hemophore HasAp from *Pseudomonas aeruginosa*. *Biochemistry*, *49*, 6646–6654.

Zhu, H., Xie, G., Liu, M., Olson, J., Fabian, M., Dooley, D., et al. (2008). Pathway for heme uptake from human methemoglobin by the iron-regulated surface determinants system of *Staphylococcus aureus*. *Journal of Biological Chemistry*, *283*, 18450–18460.

CHAPTER FOUR

The Haemoglobins of Algae

Eric A. Johnson, Juliette T.J. Lecomte[1]
T.C. Jenkins Department of Biophysics, Johns Hopkins University, Baltimore, Maryland, USA
[1]Corresponding author: e-mail address: lecomte_jtj@jhu.edu

Contents

Abstract

In the last few years, advances in algal research have identified the participation of haemoglobins in nitrogen metabolism and the management of reactive nitrogen and oxygen species. This chapter summarises the state of knowledge concerning algal haemoglobins with a focus on the most widely used model system, namely, *Chlamydomonas reinhardtii*. Genetic, physiologic, structural, and chemical information is compiled to provide a framework for further studies.

ABBREVIATIONS

CtrHb haeme domain of *Chlamydomonas eugametos* LI637 haemoglobin
EPR electron paramagnetic resonance
FHb flavohaemoglobin
GCS globin-coupled sensor

Advances in Microbial Physiology, Volume 67
ISSN 0065-2911
http://dx.doi.org/10.1016/bs.ampbs.2015.08.003

Hb haemoglobin
IDP intrinsically disordered protein
LI light induced
M myoglobin like
Mb myoglobin
NMR nuclear magnetic resonance
NOD nitric oxide dioxygenase
NOS nitric oxide synthase
NR nitrate reductase
nsHb nonsymbiotic haemoglobin
ORF open reading frame
PDB protein data bank
Pgb protoglobin
PTM post-translational modification
RNS reactive nitrogen species
ROS reactive oxygen species
S sensor
SDgb single-domain globin
sHb symbiotic haemoglobin
SSDgb sensor single-domain globin related to the GCS globin domain
T truncated
TAP Tris–Acetate–Phosphate

1. INTRODUCTION

In 1805 British, French, and Spanish fleets met at the Battle of Trafalgar. In this battle, the mighty ships did not trade shots at a distance, but engaged at close quarters; men boarded enemy vessels with heavy sail and rigging tangled between ships. The crews had to protect themselves and their ships not only from the enemy, but also from the tonnes of sailing equipment that rained from above. The Battle of Trafalgar has become a definition for the term melee.

Consider then the daily environment for the unicellular microalgae. The individual cell exists within an aqueous environment cohabitated with a wide assortment of other species, some benign, some not. The cell must defend itself against a host of organisms while protecting itself from the dangerous cellular machinery that drives its metabolism and the energy unleashed from capture of sunlight. The daily life of an algal cell is a type of melee, and algae possess weapons for this battle, perhaps not essential for basic metabolism, but important for survival.

Algae are, by definition, photosynthetic. They are capable of obligate or facultative autotrophic growth. The interior of a photosynthetic cell can be a punishing environment. Water splitting (as an example) generates diatomic oxygen while sending electrons through complexes of manganese, iron, and other metals (Yruela, 2013). These complexes can interact with dioxygen, giving rise to reactive species that are destructive to biomolecules (Kornas, Kuźniak, Ślesak, & Miszalski, 2010; Ludwig & Bryant, 2012). And the frequency of such reactions is at the whim of the solar radiation at any particular instant, being both continually variable and generally unpredictable. These cells are also in constant competition with other cells in the surrounding environment for access to this solar radiation. Added to this is the fact that algae generally comprise the bottommost layer of a food chain, making all other heterotrophic species their predators.

This chapter attempts to bring some perspective to the relatively new field of algal haemoglobin research. The update builds on several recent reviews, in particular, those of Vinogradov, Bailly, et al. (2013) and our own (Johnson & Lecomte, 2013), and put forth the hypothesis that haemoglobins play key roles in cellular protection—helping the algal cell participate in a daily melee for its own survival.

1.1 Haemoglobin Superfamily and Subfamilies

The word haemoglobin invariably conjures up the pigment of red blood cells responsible for diatomic oxygen transport in the circulatory system of animals. At the molecular level, haemoglobin is traditionally envisioned as a tetramer of α helical proteins, each wrapped around a *b* haeme molecule. The haeme cofactor, an iron–porphyrin derivative synthesised in most organisms by complex and ancient pathways (Fujita, Tsujimoto, & Aoki, 2015; Layer, Reichelt, Jahn, & Heinz, 2010; Obornik & Green, 2005), is used for multiple purposes other than reversible dioxygen binding, and many proteins contain domains rich in α helices that fold to form a hydrophobic pocket capable of incorporating this cofactor (Landfried, Vuletich, Pond, & Lecomte, 2007). The superfamily of haemoglobins comprises a subset of these haemeproteins, identified as fulfilling several criteria: a haemoglobin is generally longer than 100 residues, has pairwise alignment with a known haemoglobin yielding an Expect value *E* larger than 0.005, has a histidine at topological position F8 to ligate the haeme iron (proximal histidine), and aligns well over helices B through G (Vinogradov, Tinajero-Trejo, Poole, & Hoogewijs, 2013). By these criteria, the haemoglobin superfamily now counts over 32,000 members and shows remarkable diversity of primary, supersecondary, and quaternary structures.

As the haemoglobin superfamily grows with the frequent inclusion of new sequences, periodic bioinformatic analyses refine and complicate its history. We now know that the proteins can be separated into three families, designated M for myoglobin-like, S for sensor, and T for truncated (Vinogradov, Tinajero-Trejo, et al., 2013) (Table 1). M proteins have a fold referred as 3-on-3 because of the topology of their helical sandwich (Fig. 1A). They occur as isolated domains (SDgbs, for example, myoglobin) or as two-domain flavohaemoglobins (FHbs). S proteins also adopt the 3-on-3 fold and comprise the globin-coupled sensors (GCS), protoglobins (Pgb), and sensor single-domain globins (SSDgb). T proteins, in contrast, have a smaller, truncated, domain folded in a 2-on-2 topology (Fig. 1B). Within the T family (collectively TrHbs), there are three groups or subfamilies called TrHb1, TrHb2, and TrHb3 (the 1, 2, 3 numbers have recently replaced the N, O, P nomenclature) (Vuletich & Lecomte, 2006; Wittenberg, Bolognesi, Wittenberg, & Guertin, 2002). M, S, and T proteins are distributed in bacteria, whereas to date there is no example of eukaryotic S protein and no example of archaeal M protein (Vinogradov, Tinajero-Trejo, et al., 2013).

Table 1 Haemoglobin Nomenclature Advocated by Vinogradov and Coworkers (Vinogradov, Tinajero-Trejo, et al., 2013)

Family Name	M (Myoglobin-Like Globins)	S (Sensor Globins)	T (Truncated Globins)
Fold	3-on-3	3-on-3	2-on-2
Bacteria	FHb (flavohaemoglobins) SDgb (single-domain globins related to FHbs)	GCS (globin-coupled sensors) Pgb (protoglobins) SSDgb (sensor single-domain globins)	TrHb1 (group I or N) TrHb2 (group II or O) TrHb3 (group III or P)
Archaea		HemAT Pgb SSDgb	TrHb1
Eukaryotes	FHb SDgb		TrHb1 TrHb2

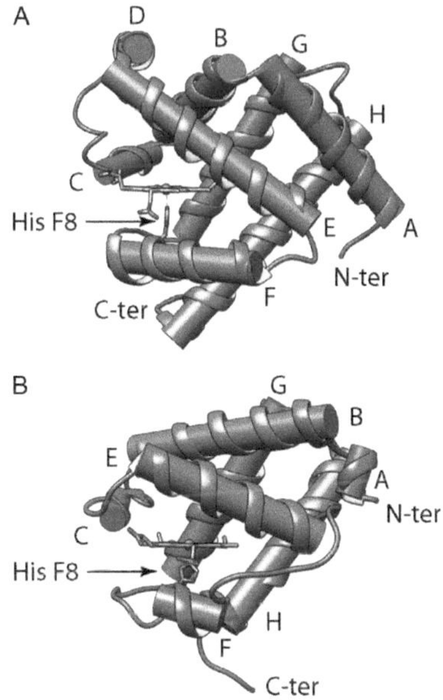

Figure 1 The two main topologies of the haemoglobin superfamily. (A) The 3-on-3 fold illustrated by sperm whale myoglobin (1JP6) (Urayama, Phillips, & Gruner, 2002). The α-helical sandwich is formed by the A, B, E, F, G, and H helices depicted in blue. (B) The 2-on-2 fold illustrated by the haeme domain of *Chlamydomonas eugametos* LI637 (1DLY) (Pesce et al., 2000). The α-helical sandwich is formed by the B, E, G, and H helices depicted in blue. Haemes and proximal histidines are represented with sticks. The proximal histidine is denoted as F8 (eighth position in the F helix) by alignment with the canonical 3-on-3 haemoglobins (Perutz, 1979). (See the colour plate.)

Phylogenetic analysis of haemoglobin apoprotein genes places their first appearance at least three billion years ago (Vinogradov et al., 2007). Algae, in contrast, began to arise during the Great Oxidation Event that occurred with the advent of photosynthesis on Earth about two billion years ago (Holland, 2006). Modern algae are also influenced by the endosymbiotic events that lead to the mitochondria and chloroplast found in eukaryotic photosynthesisers (Yoon, Hackett, Ciniglia, Pinto, & Bhattacharya, 2004). The appearance of haemoglobins would seem to predate these other events, and evolutionarily divergent groups of haemoglobins have sometimes been found within the same algae. This has led to speculation that the endosymbiotic events that formed the origins of the mitochondria and chloroplasts brought haemoglobin genes from the different original organisms together within the same eukaryotic genome (Vinogradov, Fernandez, Hoogewijs, & Arredondo-Peter, 2011). This horizontal transfer of haemoglobin genes during evolution has intertwined them and given a polyphyletic nature to many algae.

1.2 Relationship to Cyanobacteria

The term alga is applied to a vast array of organisms without a known common ancestor. Simple, photosynthetic aquatic organisms lacking stems, roots, or complex structures are generally grouped together as algae. A prokaryotic alga is often considered a misnomer, as the term algae typically refer only to eukaryotes. However, cyanobacteria are commonly known as blue-green algae and they posses chlorophyll, split water to liberate electrons for photosynthesis, and are considered to be related to ancient bacteria that gave rise to the chloroplast in eukaryotes. The prokaryotes, specifically the cyanobacteria, are at the foundation of our understanding of haemoglobins in algae.

Not all cyanobacteria contain haemoglobins. A recent survey speculates that roughly 50% of known species do not possess any haemoglobin gene (Vinogradov, Tinajero-Trejo, et al., 2013). However, among those species possessing one or more haemoglobins, there is a remarkable variety, with all three families of globins represented. Haemoglobins from the T lineage far outnumber those from M and S families, and it is interesting that among the TrHbs in cyanobacteria only the TrHb1 group is represented (Vinogradov, Tinajero-Trejo, et al., 2013). Also interesting is that TrHb2s dominate in plants even though a cyanobacterium was probably the original endosymbiont producing the chloroplast (Fig. 2). The remainder of the haemoglobins found in cyanobacteria are classified as SDgbs (from the M family) and SSDgbs (from the S family). Although plants and algae may have SDgbs, none have been found to contain SSDgb genes. This suggests that while the modern chloroplast may have derived from ancient cyanobacteria, the modern counterparts of these prokaryotes possess a different genetic composition.

Eukaryotes arose with the advent of membrane specialisation, possibly when larger cells formed invaginations to provide more surface area for the cytoplasm (Vothknecht & Westhoff, 2001). This compartmentalisation was linked to the acquisition of an aerobic proteobacterium that would eventually form the mitochondrion. The evident monophylogeny associated with the mitochondrion suggests that this endosymbiosis was unique and the resulting cell was able to survive while all its contemporaries have

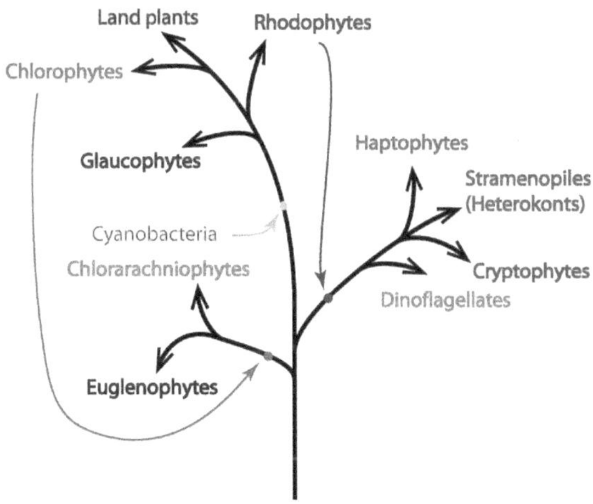

Figure 2 Schematic representation of the evolution of algae. Algae can be loosely categorised into two groups. The primary endosymbionts: glaucophytes, rhodophytes, chlorophytes, and land plants (embryophytes) arose following the acquisition of a cyanobacterium. The second group is the secondary endosymbionts; these species are predicted to have occurred following the endosymbiosis of a primary endosymbiont. Both the chlorophyte algae (green (dark grey in the print version) line) and the rhodophyte algae (red (dark grey in the print version) line) are predicted to have participated separately. Lengths of lines are not meant to be representative of evolutionary timelines.

gone extinct (Richards & van der Giezen, 2006). Photosynthetic eukaryotes diverged from this lineage when a photosynthetic cyanobacterium became incorporated into the cell, giving the organism not only the powerhouse of a mitochondrion but solar harvesting power as well. Unlike for the mitochondrion, monophylogeny associated with the advent of chloroplasts is not definitive. This has led investigators to propose that the rise of photosynthetic eukaryotes from their nonphotosynthetic ancestors (the primary endosymbiotic event) was not unique, but rather occurred several times (Vinogradov et al., 2011). Owing to this lack of monophylogeny, different photosynthetic eukaryotes contain haemoglobin genes that align in distinct groupings, such as the TrHb1 genes, which appear to form two subgroups TrHb1-1 and TrHb1-2 (Vuletich & Lecomte, 2006). Because the haemoglobin genes of algae can inform us about the evolution of these organisms, eukaryotic algae, and cyanobacteria are related through this direct evolutionary link.

The role of endosymbiosis in eukaryotic algae grows beyond the initial cyanobacterial acquisition. Figure 2 shows the branching of photosynthetic organisms that followed the primary endosymbiosis, leading to both red and green algae as well as the glaucophytes and the land plants (embryophytes). Several arrows move backwards down the tree. These represent secondary endosymbiotic events, nonphotosynthetic eukaryotic cells acquiring photosynthetic eukaryotic cells, thereby creating new species within these taxa and dividing them further into photosynthetic and nonphotosynthetic organisms. This secondary endosymbiosis has occurred with both the green and red algae (Archibald & Keeling, 2002). A hallmark of this process is the encapsulation of the photosynthetic membranes. Cyanobacteria perform photosynthesis on the plasma membrane of the cell (Bryant & Frigaard, 2006), while in primary endosymbionts the photosynthetic thylakoids are encased within two membranes, the inner and outer envelopes (Vothknecht & Westhoff, 2001). Secondary endosymbionts carry multiple membrane layers around their photosynthetic membranes (Delwiche, 1999).

In addition to encapsulation of the chloroplast, the nucleus from the acquired organism is pared down, with gene transfers leading to the eventual loss of the endosymbiont nucleus, or at least its reduction to a structure known as a nucleomorph (Moore & Archibald, 2009). As additional genetic information is acquired about the varied array of algal species displayed in Fig. 2, we will gain improved understanding of how the exchange of haemoglobin genes occurred in these organisms. Thus, while haemoglobin genes can inform about the evolution of algae, the path between cyanobacteria and algae has confounding loops and curves instead of straight lines.

1.3 Relationship to Plants

Land plants are highly multicellular organisms and differ from algae in the diversification of their tissues. As such the presence of haemoglobins in land plants must be considered in terms of protein localisation. There are three different types of haemoglobins in plants. One of these belongs to the T family, with TrHb2 as the only group so far identified (Vázquez-Limón, Hoogewijs, Vinogradov, & Arredondo-Peter, 2012). The other two types of land plant haemoglobins belong to the M family and are classified as symbiotic and nonsymbiotic (Hoy & Hargrove, 2008). Here, the symbiosis refers to the relationship some plants maintain with soil microbes, specifically leguminous plants and the root nodules formed in concert with α- and β-proteobacteria. These nodules allow the plant and the bacteria to enter into a nitrogen-fixing synergy responsible for the conversion of atmospheric dinitrogen into the ammonia essential for plant growth.

Symbiotic haemoglobins (sHbs), many of which are referred to as leghaemoglobins, are only expressed within the root nodules. These proteins are known for their ability to bind O_2, an attribute that is thought to allow root nodules to maintain the low oxygen content that is an essential requirement for the N_2-fixation process. sHbs are found only in nitrogen-fixing plants and are believed to have evolved from their companion nonsymbiotic haemoglobins (nsHbs) (Gopalasubramaniam et al., 2008). nsHbs are of two types; type 1 binds dioxygen very tightly whereas type 2 has a lower affinity, comparable to that of sHbs. Both kinds of proteins participate in nitric oxide metabolism (Dordas, 2009). nsHbs are expressed throughout the plant, and are found across the entire plant kingdom. Gene expression studies of nsHbs have demonstrated that plant expresses these genes when under stress, particularly nutrient stress, salt stress, and temperature stress (Hill, 2012). Based on these results it is thought that nsHbs have been essential for plants since their initial colonisation of the terrestrial surface, as these environmental stressors would have been common and frequent.

Both the symbiotic and nonsymbiotic haemoglobins possess the 3-on-3 fold and belong to the M family. They are also similar to the 3-on-3 proteins that are found in algae. However, although these appear to be the dominant types of haemoglobins found in plants, at present T proteins dominate over M proteins in algae. Conversely, TrHb have not been found in many plants, and those that are found belong to the TrHb2 subfamily whereas algae tend to possess TrHb1 genes. Land plants are believed to have diverged from the algae ancestors roughly 500 million years ago (Vázquez-Limón et al., 2012). During that time it appears that plants have relied on haemoglobins using a 3-on-3 fold from the M family, while algae have instead used the 2-on-2 fold of the T family more extensively.

2. DISTRIBUTION OF HAEMOGLOBINS IN ALGAE

In 1992, an experiment to study the entrainment of algae by light cycles uncovered a series of genes that are rapidly expressed or repressed upon transitioning cultures of *Chlamydomonas eugametos* from periods of growth in the dark to periods of growth in the light (Gagné & Guertin, 1992). While this work revealed a circadian oscillator in the alga, it also uncovered several genes of unknown function. Further study of one of these genes (Couture, Chamberland, St-Pierre, Lafontaine, & Guertin, 1994; Couture et al., 1999) recognised it as having strong homology to a group of haemoglobins identified in ciliates by Keilin (Keilin & Ryley, 1953). This group of proteins was sparsely represented in databases when the *Chlamydomonas* version was found. In the intervening years since that discovery, an ever-increasing number of algal haemoglobins have been identified, and algal research has shifted from extensive use of *C. eugametos* to the model organism *Chlamydomonas reinhardtii*. In this section, we review the haemoglobins of algae collectively grouped in the Chlorophyta division with a focus on the *Chlamydomonas* model.

2.1 The *Chlamydomonas* Model

The fresh water alga *C. reinhardtii* was described in 1888 and named for the Ukrainian botanist Ludwig Reinhard, who had published a description of the alga he originally classified as *Chlamydomonas pulvisculus*. Later, this organism was reclassified by P.A. Dangeard as *C. reinhardtii* (though it is still debatable if his name was spelled with a 't') (Harris, 2009). *C. reinhardtii* is a unicellular, chlorophyte flagellate physically characterised by two anterior flagella of equal lengths, a proteinaceous cell wall, and a single large chloroplast occupying the majority of the cell volume. *C. reinhardtii* belongs to the fresh water genus *Chlamydomonas*, which is found universally in nature. Isolates have been recovered from areas stretching from the equatorial to the polar regions of the planet, with samples obtained from water, soil, damp walls, mattress dust (Harris, 2009), and even from atmospheric testing at over a kilometre in altitude (Brown, Larson, & Bold, 1964).

All laboratory strains of *C. reinhardtii* are believed to have originated from a single zygote isolated from a potato field in Amherst, Massachusetts in 1945 by G.M. Smith. A key event in this early history occurred in the 1950s when the strain was sent to at least three separate laboratories where they were propagated independently for many years. One of these laboratories noted the inability of *C. reinhardtii* to grow with nitrate as a sole nitrogen source, a property resulting from two unlinked mutations to the *NIT1* and *NIT2* genes necessary for nitrate assimilation. While this set of mutations was not seen in the other samples of the alga, this particular lineage (sometimes referred to as the 'Ebersold stocks' or strain 137c) was used to generate a great many of the subsequent variant strains of *C. reinhardtii*. From its early days, the development of *C. reinhardtii* strains has been linked to nitrogen metabolism, a salient fact for haemoglobin research.

Today, *C. reinhardtii* has become a central model organism for the study of photosynthesis (Erickson, Wakao, & Niyogi, 2015), cell cycles (Cross &

Umen, 2015), cell development (Wood & Rosenbaum, 2015), the evolution of multicellularity (Hallmann, 2011), and photo/chemotaxis (Noordally & Millar, 2015). It is currently heavily studied as a source for biofuels (Scranton, Ostrand, Fields, & Mayfield, 2015) and industrial feedstocks (Scaife et al., 2015), and it has been instrumental in establishing the field optogenetics (Lórenz-Fonfría & Heberle, 2014).

The adoption of *C. reinhardtii* as a model organism for haemoglobin research was prompted by the property that, unlike *C. eugametos*, this alga is not an obligate autotroph and can survive without photosynthesis if given acetate as a carbon source (Harris, 2001). Such flexible metabolism makes *C. reinhardtii* useful in creating specific variants defective in key metabolic pathways. *C. reinhardtii* has had its nuclear, chloroplast, and mitochondrial genomes fully sequenced (Merchant et al., 2007), with its nuclear genome annotated and readily available. *C. reinhardtii* also benefits from available mutagenesis techniques that allow the genetic manipulation of all of these genomes (Bock, 2015; Jinkerson & Jonikas, 2015).

While there is increasing amounts of genetic information about algae from many taxa beyond chlorophytes, there are few species that have been so thoroughly adapted to laboratory study as *C. reinhardtii*. As a result, despite the many species containing haemoglobins, much of the biochemical research on haemoglobins in algae has been performed in *C. reinhardtii*. As a functional understanding of Hbs in algae is obtained, it is hoped that this information will be transferable to other species, similar in method to the research currently using *C. reinhardtii* as a testbed for biofuel and bioindustrial applications (Scranton et al., 2015).

The study of truncated haemoglobins in *C. reinhardtii* has begun only recently. One of the first goals has been to determine how many truncated haemoglobins are found within the alga's genome. Although there is agreement about the types and locations of these genes, some variation does exist in the current databases. The haemoglobins of *C. reinhardtii* are all from the T family and belong to the TrHb1 group; they are dispersed across four chromosomes (4, 7, 14, and 16) and vary in length from approximately 130 amino acids to over 900.

2.1.1 THB1, THB2, THB3, *and* THB4

The complete nuclear genome of *C. reinhardtii* was published in 2007 (Merchant et al., 2007), and early in the annotation it was apparent that four genes had significant similarity to known bacterial truncated haemoglobins. These genes were given the prefix *THB* and their respective products the names THB1 to THB4. Subsequently several additional genes were identified that contained domains aligning with truncated haemoglobins (Hemschemeier et al., 2013), but these genes also have significant upstream or downstream regions. We have attempted to combine the individual information from different genomic analyses into a table listing all the currently available nomenclature and database notations (Table 2). In our discussions, we have chosen to adopt the Hemschemeier designations for THB5 through THB12 (Hemschemeier et al., 2013) along with the original THB1 through THB4 names. Figure 3 contains the amino acid sequences of all the hypothetical haemoglobin domains of *C. reinhardtii.*

The *THB1* gene encodes a 136 amino acid protein with strong homology to the cyanobacterial GlbNs (Fig. 3). A key attribute of the *THB1* gene is its control via the NIT2 transcription factor (Johnson et al., 2014). The *NIT2* gene, a key regulator for the assimilation of nitrate by the cell, signals a possible inclusion of THB1 in this process. The gene product of *THB1* is also the only TrHb to be positively identified as an expressed protein within *C. reinhardtii.* In a study of the transport mechanisms of *C. reinhardtii* flagella (Lechtreck et al., 2009), a knockout mutation of the flagellar transport component known as bbs4 was found to alter the composition and quantity of the protein within the flagella. During a screen of these proteins using identification through mass spectrometry of peptide fragments isolated following gel electrophoresis, it was found that THB1 was present in the knockout *bbs4-1* strain, but completely absent from the *g1* background strain. Subsequent analysis identified a key alteration in the strains, where the background *g1* strain appeared to have acquired a spontaneous mutation within the *NIT2* gene preventing any expression of THB1 within the entire cell, not just the flagella (Johnson et al., 2014). The apparent accumulation of THB1 in the flagella of the *bbs4-1* strain was actually the failure to accumulate THB1 in the background *g1* strain. Nevertheless, the identification THB1 during these experiments was the first confirmation of the presence of the protein within *C. reinhardtii.*

THB1 has also been the first TrHb to be purified from *C. reinhardtii* (Johnson & Lecomte, 2014). The protein is present in the cytosol and could be isolated using a liquid nitrogen freeze-fracture technique followed by enrichment using anion chromatography and size exclusion chromatography. The native protein was found associated with a molecule of *b* haeme, and mass spectrometry confirmed N-acetylation of the amino-terminus. The partially purified protein also exhibited optical properties similar to

Table 2 The Haemoglobins of *C. reinhardtii*

	Phytozome Notation			UniProt	NCBI Notation		
		Length	TrHb Domain			Length	TrHb Domain
THB1	Cre14.g615400	136	10–129	A8JAR4	XP_001699005	136	10–129
THB2	Cre14.g615350	147	27–147	A8JAR3	XP_001699004	147	27–146
THB3	Cre04.g218800	297	148–263	A8ISQ0	XP_001692257	218	93–218
THB4	Cre04.g218750	210	55–174	A8ISP8	XP_001692256	188	21–149
THB5	Cre07.g351100	959	62–182, 203–326, 379–503, 571–717	A8IU82	XP_001692484	837	27–154, 180–301, 344–480
THB6	Cre16.g654250.t3.1	385	134–261	A8J9B4	XP_001698210	497	284–411
THB7	Cre16.g661000	292	54–246	A8J8P3	XP_001697939	224	53–181
THB8	Cre16.g661200	519	46–166	A8J8P7	XP_001697866	519	46–165
THB9	Cre16.g661250	712	53–176	A8J8P8[a]	XP_001697867	952	53–176[a]
THB10	Cre16.g661300	483	53–175	A8J8P8[a]	XP_001697867	952	650–757[a]
THB11	Cre16.g662750	395	51–171	A8J8S5	XP_001697881	270	85–205
THB12	Cre16.g663000	432	46–166	A8J8T3	XP_001697958	240	34–183

[a]A8J8P8 was not originally identified by Vinogradov and coworkers but contains two TrHb domains that align to the TrHb domains of THB9 and THB10 individually.
As adapted from Huwald, Schrapers, Kositzki, Haumann, and Hemschemeier (2015) and Vinogradov et al. (2011). Information obtained from the Phytozome database 10.2 and the National Center for Biotechnology.

```
                          B10                        E10
SYN6803   TLYEKLGGTTAVDLAVDKFYERVLQDDRIKHFFA-DVDMAKQRAHQKAFLTYAFGGTDKYDG-------
  LI637   SLFAKLGGREAVEAAVDKFYNKIVADPTVSTYFS-NTDMKVQRSKQFAFLAYALGGASEWKG-------
  LI410   SLFAKLGGREAVEAAVDKFYNKVVADPTVSVFFS-KTDMKVQRSKQFAFLAYALGGAAEWKG-------
            |A|  |-------B-------|    |-C-|    |-------E------|
   THB1   SLYSRMGGEAAVEKAVDVFYERIVADPQLAPFFA-NVDMKKQRRKQVAFMTYVFGGSGAYEG-------
   THB2   SIFDRLGGEAAVEKAVDVFYERIVADPQLAPFFE-GVDMRTQRRKQQAFMTYAFGGATGYTG-------
   THB3   CVYESMGGWETLSAAVESLYNRMRGDGRCASLFS-EGNEQQLKTHMLEFLTCALGGKARFAS-------
   THB4   TLHAKLGGAAAVAATVDVFYKKLMNDPDLEPFFR-GVDMVTLIAKQNRFLAYAFGATTHYHG-------
  THB5A   GIVERLGGADRVHDVIQAFYRKLFQNAMVRHYFE-NLSSDRMRSKQMKFMRYIMGGPGTYTG-------
  THB5B   PLVAKLGGPEPVQGIITAFYRRLFSSEELKYFFV-GLSTERLRTMQFRFMRYLLAGPGWYATT--------G
  THB5C   GLVRRVGGPEAAYELVRRLYSRLFADEVLKGFFA-GLSVERLKTKQLRFMRYLLGGPPAYAAA------TGG
  THB5D   SIYERLGGEEVLCDLVDMWAGKLGSDPALAHIYSGGVDMPRQRAHQLAFLRQAFGGSGGGGASRPGSA-VGA(ins1)
   THB6   HKRIKNSEEFSMKFLVDKFYGKVLKIKEIKPFFD-GVDMHKLKYHQEALMLLVFGGQELLDDELPGLS----
   THB7   TLLAKLGGNDILKQVAERWYAKLKRDERLGTFLA-DQDTIHLRARLSAYLANLFGQKHDPMAALLRKQ-GRC(ins2)
   THB8   TLLKRIGGEKGMKDVVEMFYRKIYANERLRHFLK-DKDVSLLRSKQAAFMTWLFGPSTATYT---------G
   THB9   PLLDRMGGADYVKRVAERFYHKLYADEQLAPFLG-EHDVVSLRAKQNAFLIWAFGPPNRPYP---------G
  THB10   PLLARVGGIDNVKRVVEGFYRRLYADERMDKYLS-GFDMPYLRAKQSAFMGWLFGPASVPYS---------G
  THB11   PLLQRVGGLDVVKKVVELFYRKLYADPQLIKYLH-DQDPMHLRAKQSMFVSWLFGPPNVPYT---------G
  THB12   PLLQRVGGSDVVKHVVELFYRHLYADPSLIKFLH-DQEMTHLRAKQSAFISWLFGPPNVPYT---------G

               F8
SYN6803   ---RYMREAHKELVENHGLNGEHFDAVAEDLLATLKEMGVPEDLIAE-VAAVAGAPA-----HKRDVLNQ--------
  LI637   ---KDMRTAHKDLVPH--LSDVHFQAVARHLSDTLTELGVPPEDITD-AMAVVASTR-----TEVLNMPQQ-------
  LI410   ---KDMRTAHKDLVPH--LTDVHFQAVVRHLSDTLAELGVTPGDIAD-AMAVVASTK-----TEVLNMPRQQGAESNR
              |-----F----|   |-------G-------|    |------------H------------|
   THB1   ---RDLGASHRRLIREQGMNHHHFDLVAAHLDSTLQELGVAQELKAE-AMAIVASAR-----P--LIFGT--------
   THB2   ---RDLAAAHRRLIRDKGLKEEHFDMVAGHLAGTLQLLGIGADLVDE-VIALVATTK-----P--VIFERV-------
   THB3   ---STLLTSQRDQLRQHGFGVGQFDILLQHMKAVLDEIGVQQETAAS-AIALLRCHK-----YLF-ERRQ--------
   THB4   ---KDIVMGHAHLIINRGLNLTHFDKVAGHFVDSLKEMGVGQELIDE-AAGVLIGVR-----PLFDP---E-------
  THB5A   ----NLRCIHAQMVKEEGLDKAKFQVVIGMLLDTFKELNVPQDVVSE-ILGHVESAK-----SAIFTPQP--------
  THB5B   --GGNMRCVHARMIAEGGLDLPKFQAVTGMLRAACDEHLVSPPLIRE-IMLNVKAAE-----QAIFTPGP--------
  THB5C   TGGANLRCPHLKLMVKQPAAAAAAGADD---------------DLHRS-STGGTEEGGGGASTAKAAGGGRDVLDGTK
  THB5D   VGAMRLEIAAAHLARDKALAPAQFERIADHLTACLKKEGASPELVAE-VRAMQGPGR-----CLLWPPSS--------
   THB6   ---ADLRLIHLHLLF-EGLNLSHWQIFADTFCETLDELPQIPADVKAHAKAYMRATK-----QH-FRPI---------
   THB7   GRCHHLRRVHLRLIQQMGFSPADFDAGLGYFRESLQELGAPEWLVDE-VMLKLVEVK-----DLVFDDPD--------
   THB8   --GRSVRCAHLRMIKQRGFSEDDFLLGMSCFEEAMRDYGAPQRLTNE-IMAKILPFK-----DIIFSP-
   THB9   --GRHLRTAHLRLIKQKGFTVEHFEMGLKHFEEALKEMDAPPSVVRD-ALMKLRPYK-----TVVFTPSSA-
  THB10   --GKHLRTVHLKMIKQQGFSPEDFEVGMDYFEESMKELGAPETIISE-VMSKIRPFK-----AVIFTPNA-
  THB11   --GKSVRIAHLRIIKQRGFSPEDFDLGMKYFEEAMTELGAPEVLRGE-VMRRMLPYK-----DAIFTP-
  THB12   --GKSVRIAHLRIIKQRGFSPEDFDLGMKYFAESMRELGAPEILINE-VLRRITPYK-----DAIFTP-

  THB5D ins1   A-C-------------------------------------------PALG---GTAGAAGGM
  THB7  ins2   ATSQALQHQRSSVLQHQPTSNGLALHSSQPSLVTSRAPSHAMSRLPPSYSLHHQATSSRVGA
```

Figure 3 The amino acid sequences of *C. reinhardtii* haemoglobins. For comparison, the top three sequences are for the cyanobacterium *Synechocystis* sp. PCC 6803 (UniProt identifier P73925), the globin domain of *C. eugametos* LI637 (UniProt identifier Q08753), and the globin domain of *C. eugametos* LI410 (UniProt identifier P52334). Only the globin domains are included. The secondary structure of THB1 (Rice et al., 2015) is indicated above its own sequence. THB5D and THB7 contain long insertions in the EF loop (ins1 and ins2, respectively). These inserts have been extracted from the domain sequences and are listed at the bottom of the figure. Residues B10, E10, and F8 are highlighted along with the location of the Gly-Gly motifs characteristic of TrHb1s. (See the colour plate.)

those of the recombinant version of THB1, although the yield of purified protein from the cell was too low for extensive analysis.

Currently, THB1 is the most extensively studied of these four initially identified TrHbs, but some information is known about the other three proteins as well. In the recent paper by Sanz-Luque that featured THB1, the relative gene expression of all four proteins was examined in cells grown in media containing different sources of nitrogen (Sanz-Luque et al., 2015). Cells were initially grown in medium containing ammonium as the sole nitrogen source but then washed and switched to medium containing some combination of ammonium and nitrate. Through this screen, *THB2* was seen to undergo a substantial up-regulation in expression when cells were switched to medium containing nitrate, either with or without ammonium. A similar increase was not seen in the expression levels of either *THB3* or *THB4*. Thus it appears that as with *THB1*, the *NIT2* promoter also controls the expression of *THB2*, but not of *THB3* or *THB4*. This is in conflict with the recent report of Ciaccio et al. (2015), which interprets the same study as evidence for *NIT2* controlling the expression of *THB4*.

The gene expression study of *THB1* and *THB2* shows that although both are under the control of *NIT2*, expression levels are much smaller for *THB2* than for *THB1*, specifically when cells are grown in a medium containing nitrite and an exogenous nitric oxide producer (Sanz-Luque et al., 2015). These two types of additive have different effects. The presence of nitrite is expected to increase the cytosolic nitrite levels and promote the production of nitric oxide via the endogenous nitrate reductase (NR) (Sakihama, Nakamura, & Yamasaki, 2002), whereas exogenous nitric oxide producers such as diethylammonium (Z)-1-(*N*,*N*-diethylamino)diazen-1-ium-1,2-diolate (also known as DEA-NONOate) or *S* nitrosoglutathione, decompose slowly and release nitric oxide into the medium as part of this process. The presence of nitric oxide appears to have opposite effects on *THB1* and *THB2* gene expression of Sanz-Luque et al. (2015). The expression of *THB1* increases, while the expression of *THB2* decreases, signalling that the two proteins may play different roles in the regulation of nitrogen metabolism within the cell.

Less is known about THB3 and THB4. THB3 has an unusual TrHb1 sequence. In fact, it does not qualify as a haemoglobin according to the criteria mentioned in Section 1.1 because it lacks a histidine at position F8 (Fig. 3). *THB3* is located on chromosome 4 of the *C. reinhardtii* nuclear genome and is found near the location of *THB4*, suggesting these genes may have resulted from a duplication event. The absence of proximal histidine and the possibility of gene duplication suggest that THB3 is an inactive version of THB4. However, it is intriguing that some level of detection, albeit unchanging, was seen in the same screen performed for *THB1* and *THB2* (Sanz-Luque et al., 2015). Also intriguing is the presence of a histidine at predicted position E10, which is known to serve as an axial ligand to the iron in certain cyanobacterial proteins including GlbN from *Synechocystis* sp. PCC 6803 and *Synechococcus* sp. PCC 7002 (Couture et al., 2000; Scott et al., 2002). Furthermore, although His F8 is considered a hallmark of haemoglobins, a small percentage of sequences have been identified as aligning with TrHbs, but lacking this key residue (Vinogradov, Tinajero-Trejo, et al., 2013). Thus, it is likely that THB3 is in fact a functional TrHb representing yet an additional subclass of proteins with unique structural and chemical properties.

Existing information for THB4 is comprised solely of genetic evidence. Unfortunately, there is some confusion about the exact sequence for the *THB4* gene. Information from the NCBI reference sequence (XP_001692256) states that the protein product is 188 residues in length with a TrHb domain extending from residue 37 to residue 149. This is different from the sequence given in the Phytozome database (Cre04.g218750), which presents the THB4 protein with a length of 214 residues and a TrHb domain extending from residue 59 to residue 178. As is apparent from Fig. 4A, the two sequences are identical for the majority of the THB4 product; however, they differ in their amino-terminal extensions. The NCBI

A

```
NCBI       THB4   --------------------------MQSTPQANAAPANVNGSVSSAGQSKSAATTRSTL 34
Phytozome  THB4   MAAVNNGAVPGNGANPLYGAELQGDGVKSTPQANAAPANVNGSVSSAGQSKSAATTRSTL 60

                                  B10                         E10
NCBI       THB4   HAKLGGAAAVAATVDVFYKKLMNDPDLEPFFRGVDMVTLIAKQNRFLAYAFGATTHYHGK 94
Phytozome  THB4   HAKLGGAAAVAATVDVFYKKLMNDPDLEPFFRGVDMVTLIAKQNRFLAYAFGATTHYHGK 120
                       F8
NCBI       THB4   DIVMGHAHLIINRGLNLTHFDKVAGHFVDSLKEMGVGQELIDEAAGVLIGVRPLFDPERY 154
Phytozome  THB4   DIVMGHAHLIINRGLNLTHFDKVAGHFVDSLKEMGVGQELIDEAAGVLIGVRPLFDPERY 180

NCBI       THB4   KGKVDVEKIEKAATTPQSKPAEAGQDAPACCTIM 188
Phytozome  THB4   KGKVDVEKIEKAATTPQSKPAEAGQDAPACCTIM 214
```

B

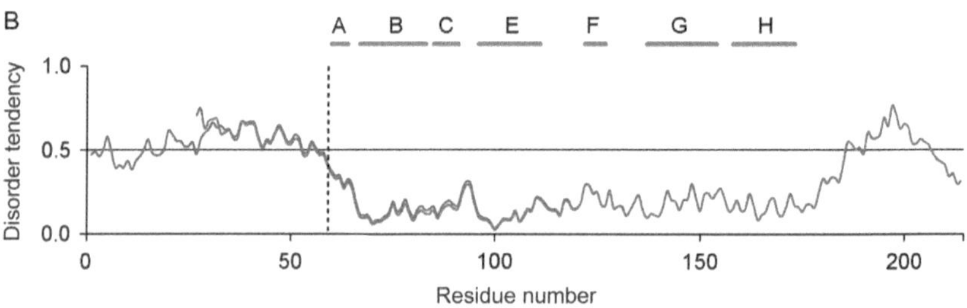

Figure 4 (A) Sequence alignment of two possible *THB4* gene products. (B) The IUPred profile (Dosztanyi, Csizmok, Tompa, & Simon, 2005) for THB4 (UniProt identifier A8ISP8) (long disorder calculation). The locations of helices in THB1 are marked. (See the colour plate.)

sequence is 25 residues shorter than the Phytozome sequence. This could be without much significance except that the predicted behaviours of the two products are different. PredAlgo is a computer programme designed to identify protein sequences with similarity to known algal organelle targeting sequences (Tardif et al., 2012). If both THB4 predicted protein sequences are analysed using this programme, the shorter NCBI sequence yields an amino-terminus with a high probability for possessing a chloroplast targeting sequence (dotted line in Fig. 4B). The Phytozome sequence, which contains the peptide, but places 25 additional amino acids in front of it, does not. It is interesting that the PredAlgo outcome is based solely upon targeting sequences and not TrHb structure, yet the signal peptide would end almost exactly at the beginning of the first turn of the A helix of THB4. If the NCBI sequence is correct, then we would expect THB4 to localise to the *C. reinhardtii* chloroplast, and given its size and sequence, THB4 would be the most likely homologue to the original LI637 isolated from *C. eugametos* (Couture et al., 1994). It would be intriguing if this possibility were verified, as it would allow the existing studies of LI637, performed decades ago, to be extended using the *C. reinhardtii* model organism. Thus far these sequences are strictly predicted, and no evidence has yet been produced to favour one interpretation over the other.

2.1.2 THB8

THB1 through *THB4* are genes that encode small, single-domain proteins very similar to the cyanobacterial GlbNs (Scott et al., 2002; Scott & Lecomte, 2000). Closer inspection of the nuclear genome of the *C. reinhardtii* has uncovered several additional, larger genes containing domains that resemble known TrHbs (Hemschemeier et al., 2013). Expression analysis of these genes from *C. reinhardtii* has found that many are translated, with detectable levels of the corresponding mRNA present in the cell. The first experimental publication investigating TrHbs in *C. reinhardtii* studied one of these TrHb-domain proteins, *THB8* (Hemschemeier et al., 2013). The expression of *THB8* was highlighted in a RNA-Seq analysis of *C. reinhardtii* following growth under anaerobic or hypoxic growth, which determined that *THB8* underwent a >1000-fold increase in expression following hypoxia. This massive up-regulation during anaerobic growth appears to be unique to *THB8* among all TrHb genes in *C. reinhardtii*, although *THB7* does show mild up-regulation under similar conditions. This investigation was coupled with the creation of a genetic knockdown strain of *C. reinhardtii*, where interfering RNA was used to reduce the abundance of THB8 protein within the cell. The reduction of THB8 within the cell during anaerobiosis caused a decrease in the expression of CYG12, a guanylyl cyclase enzyme responsible for nitric oxide sensing within the cell (Winger, Derbyshire, Lamers, Marletta, & Kuriyan, 2008). The synchronicity of the expression of these two genes suggests a role for THB8 in nitric oxide signalling pathways of *C. reinhardtii*, an inference further strengthened when the growth of these knockdown strains was studied under hypoxic conditions. Strains of *C. reinhardtii* with reduced levels of THB8 grew poorly when subjected to hypoxic environment, in contrast to the wild-type controls. While the addition of nitric oxide scavengers or exogenous nitric oxide did not recover the hypoxic growth of these strains, treatment with nitric oxide did stimulate gene expression in the wild-type cells. The role of THB8 is still unclear; however, as we will discuss in later sections the involvement in nitric oxide signalling pathway suggests that TrHbs may fill specific enzymatic roles for balancing nitric oxide chemistries in the cell.

2.1.3 THB7

As discussed in the previous section, T haemoglobins have a distinctive architecture conveyed by a set of four α-helices with specific relative orientation (Fig. 1). Within this structure are several key residues that are essential for thermodynamic stability and for high-affinity association of the haeme group with the apoprotein. Sequence alignments allow us to predict if a gene will encode a TrHb and to which subfamily (TrHb1, TrHb2, or TrHb3) this protein will belong (Vuletich & Lecomte, 2006). However, in some cases, alternative alignments appear plausible and some debate will exist about the correct assignment of a predicted domain until the protein is characterised experimentally. THB7 provides a relevant example. Domain definition for this protein has gone through several iterations. In 2011, Vinogradov et al. identified the protein as a TrHb2 (Vinogradov et al., 2011), but one that clustered with TrHb1s when placed in a phylogenetic tree. Compared to TrHb1s, TrHb2s have a tryptophan at position G8 (eighth residue in helix G). They also have a short, somewhat conserved, addition to the EF loop forming a helix (helix Φ). These features are not apparent in THB7. It is likely that THB7 is actually a TrHb1, which would explain its standing in phylogenetic analyses. In addition, the original THB7 alignment places an arginine at position F8 as noted later by Hemschemeier and coworkers (Huwald et al., 2015). Subsequent alignment by the same authors has proposed that the TrHb domain within THB7 is split into two segments with a 54 amino acid linker region separating them. This alignment introduces a long loop between the E and F helices, but does place a histidine residue at position F8.

During the expression analysis that highlighted the large increase in transcripts of *THB8*, it was noted that several other genes showed mild up-regulation, among them *THB7*. Hemschemeier et al. subsequently constructed recombinant material using cDNA created from *C. reinhardtii* RNA and their anticipated THB7 sequence (Huwald et al., 2015). The recombinant material is able to bind haeme, but its optical spectrum displays a main absorption band (Soret) unusually broad and shifted to higher energies, reminiscent of guanylate cyclase spectra. Optical spectra cannot identify the type or position (e.g. B10, E10, or F8) of iron ligands with certainty. Although it has now been established that *THB7* is expressed in the cell, without further study, the domain definition and haeme pocket properties of THB7 remain uncertain.

2.1.4 THB5

The TrHb genes in *C. reinhardtii* are clustered throughout the nuclear genome. *THB1* and *THB2* are both on chromosome 14, and *THB3* and *THB4* are found on chromosome 4. All of the other TrHbs are found on chromosome 16 except for *THB5*, which is by itself on chromosome 7. The *THB5* gene is large and is the only TrHb gene in *C. reinhardtii* to contain multiple predicted TrHb domains. In the initial description of the gene, three TrHb domains were located within the coding sequence. Revised annotations in both the NCBI reference sequence (XP_001692484.1) and the Phytozome locus (Cre07.g351100) now list four domains. The peptide sequences for the four domains from *THB5* are shown in Fig. 3. The fourth predicted TrHb domain (THB5D) does not have a proximal histidine and lacks a tyrosine at position B10, putting into doubt the nature of this domain of the protein. The expression of *THB5* did not appear to be influenced by hypoxia. THB5 has not been studied as a recombinant protein.

2.1.5 THB6, THB9, THB10, THB11, *and* THB12

The remaining five TrHb genes are all found on chromosome 16. Currently two of these genes, *THB9* and *THB10*, carry a single UniProt assignment (A8J8P8) and may represent one large gene as opposed to two smaller ones. As noted above, the RNA-seq experiment that uncovered the hypoxic expression of *THB8* also found several milder changes in expression. Among those was *THB10*, which demonstrated a 3.5-fold change relative to the 1.0-fold change seen for *THB9*, although at the zero time point both genes gave the same 6.1-fold increase. Hemschemeier and coworkers (Huwald et al., 2015) created a recombinant version of the THB10 protein using cDNA based upon RNA extracted from *C. reinhardtii*, demonstrating that the gene is actively expressed. It is still unknown if *THB9* and *THB10* represent individual genes, or subdomains of a larger gene. In the same study, a recombinant version of THB11 was also created from cDNA, implying that the *THB11* gene is actively expressed in the cell. *THB12* was seen to give a slight (threefold) increase in expression upon hypoxic growth, but to date no further investigations have been published on this gene.

The *THB6* gene expression increased 13.6-fold upon hypoxic growth induction, but returned to normal after 6 h. The predicted THB6 protein possesses extended amino and carboxyl-terminal regions. It is also unusual in that the amino region contains a RAKFDSPEAYSAYMAGLGKM sequence that is repeated six times. A search with BLAST did not reveal homology to known repeat proteins, and it is unknown what implication this appendage could have for function.

2.2 Relating the Haemoglobins of *Chlamydomonas* to Other Algal Haemoglobins

The green algae, or chlorophytes, are represented by *C. reinhardtii* as well as many other species and are found in virtually all biomes of the planet. Members of the Chlorophyte family have been studied for centuries, but sequencing of their genomes has lagged behind higher plants as well as prokaryotes such as cyanobacteria. Partial information is available for several chlorophytes, but currently there are only three completely sequenced genomes, those of *C. reinhardtii* (Merchant et al., 2007), *Volvox carteri* (Prochnik et al., 2010), and *Chlorella variabilis* (Blanc et al., 2010). The approach we take in this section will be to use our discussion of *C. reinhardtii* as a foundation, since it forms the bulk of our present knowledge of algal haemoglobins, and gradually pull back our focus to encompass a greater variety of the known algal species. One aspect that has become apparent from *C. reinhardtii* is the repeated linkage between algal haemoglobins and nitrogen metabolism.

It should be reiterated that algae have three distinct genomes. Besides the nuclear genome, each cell has genetic information stored in their mitochondria and chloroplasts. The sequencing of both the mitochondrial and chloroplastic genomes of *C. reinhardtii* has been completed. No haemoglobin genes were found in either; these genes are present only in the nuclear genome. The 12 TrHbs within the genome of *C. reinhardtii* do have similarity to genes found in other algae, and these algae are a growing area of interest in understanding the function of these proteins. Not all members of the algal community portrayed in Fig. 2 are known to possess haemoglobins, but many do. A list of organisms with known Hb genes is given in Table 3. In many cases, we know only the sequences of these haemoglobin genes, and the presence of the expressed proteins in these algae remains in question.

2.2.1 *Chlorophyta*

2.2.1.1 *Chlamydomonas eugametos*

The study of algal haemoglobins began in *C. eugametos* with investigations of light-induced (LI) genes *LI410* and *LI637*. The sequencing of the *C. eugametos* genome has never been completed, so it is not possible at present to determine how many and what type of haemoglobins are present in the full complement of its genes. However, the predicted protein sequence for these two genes align with the known TrHbs of *C. reinhardtii* (Fig. 3), though they do appear to be more similar to each other than to any other algal TrHb. They group with a small subset of algal TrHb that are single-domain proteins with minimal N- and C-terminal extensions, such as THB1, THB2, and THB4 (assuming an N-terminal targeting peptide). Although subsequent experimentation failed to find substantial expression of *LI410*, the *LI637* gene was linked to protein expression (Couture et al., 1994). The expressed protein, LI637, was localised to the chloroplast of the algae and the recombinant form of the protein was crystallised. This crystal structure represented the first structure of a truncated haemoglobin, as summarised in Section 3.2.1.

2.2.1.2 *Volvox carteri*

Chlamydomonas is considered a volvocine algae, a class of algae comprised of several species of increasingly complex multicellularity. The main representatives of this family are *C. reinhardtii*, *Gonium pectorale*, *Pandorina morum*, *Eudorina unicocca*, *Pleodorina californica*, and *V. carteri* (Hallmann, 2011). The volvocine family has been studied extensively to understand the evolution of multicellularity in organisms, and while the most complex species of this family, *V. carteri*, is closely related to *C. reinhardtii*, the two have evolved separately for approximately 200 million years (Herron, Hackett,

Table 3 List of Algae Known to Contain Haemoglobin Genes

Phylum	Genus (#Species)	References	Hb Family
Chlorophyte	*Asterochloris*	a	T and M
	Bathycoccus	a	M
	Botryococcus	a	T
	Chlamydomonas (2)	a	T
	Chlorella	a	T
	Coccomyxa	a	T
	Micromonas (2)	a	T and M
	Ostreococcus (2)	a	M
	Pyramimonas	a	T and M
	Polytomella	a	T
	Volvox	a	T
Rhodophyte	*Chondrus*	a	T
	Cyanidioschyzon	a	M
	Galdieria	a	M
	Porphyridium	a	T
Glaucophyte			
Chlorarachniophyte			
Euglenophyte			
Haptophyte	*Emiliania*	e	M
Dinoflagellate	*Symbiodinium*	d	M
Cryptophyte	*Guillardia*	c	M
Stramenopile	*Aureococcus*	a	T and M
	Chattonella	b	T
	Ectocarpus	a	M
	Heterosigma	b	T
	Nannochloropsis (2)	a	M
	Phaeodactylum	a	T and M
	Thalassiosira (2)	a	T and M

Data from a, Vinogradov et al. (2011); b, Stewart and Coyne (2011); c, Curtis et al. (2012); Smith, Vinogradov, and Hoogewij (2014); d, Rosic, Leggat, Kaniewska, Dove, and Hoegh-Guldberg (2013); and e, Read et al. (2013).

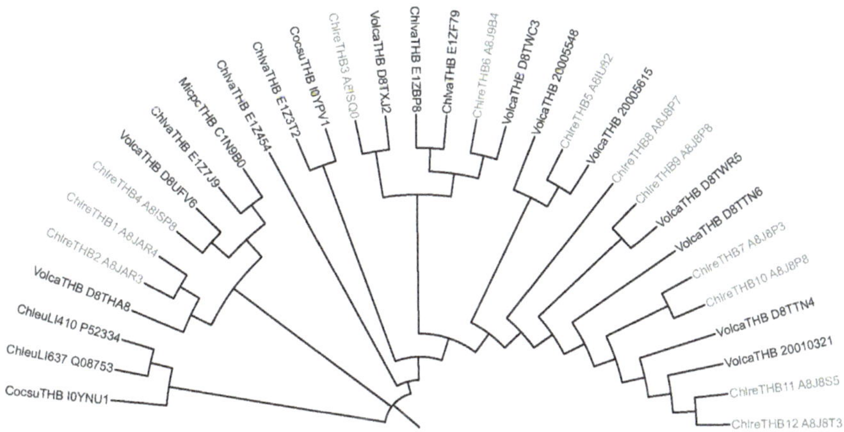

Figure 5 Phylogenetic tree for chlorophytic haemoglobins prepared with iTOL software (Letunic & Bork, 2007, 2011). Proteins are from *C. reinhardtii* (Chlre, blue font), *C. eugametos* (Chleu), *V. carteri* (Volca), *C. variabilis* (Chlva), *Micromonas pusilla* (Micpc), and *Coccomyxa subellipsoidea* (Cocsu). (See the colour plate.)

Aylward, & Michod, 2009). In 2010, the completed genome of *V. carteri* was published (Prochnik et al., 2010) and has since become an important companion to the existing *C. reinhardtii* genome for the study of green algae.

Within the *V. carteri* genome are 10 genes containing putative TrHb domains. Seven of these domains carry UniProt identifiers; the additional three, included in Fig. 5, also appear to encode TrHb domains, although two of the predicted proteins have unusual regions outside of the domain. Specifically, the gene product VolcaTHB_20005548 contains a region directly after the TrHb domain with 79 amino acids comprised only of glycines and serines, and the gene product VolcaTHB_20010321 contains a high percentage of alanine and valine following the TrHb domain.

Of particular significance is that all of the TrHb genes located in *V. carteri* are homologous to genes within *C. reinhardtii*, which suggests that the conservation of TrHbs among closely related species is high. Several interesting aspects of the alignment between these two genomes bear mentioning. Recent work has focused on understanding the relative functions for the *C. reinhardtii* TrHbs THB1 and THB2, both of which are under the control of the NIT2 transcription factor. However, in *V. carteri* there appears to be only one corresponding TrHb gene, the product of which (UniProt identifier D8THA8) exhibits 67% and 61% identity with THB1 and THB2. *C. reinhardtii* and *V. carteri* share similar nitrogen assimilation pathways (Gruber, Goetinck, Kirk, & Schmitt, 1992). This raises the intriguing possibility that the conditions necessitating two TrHbs are absent in *V. carteri*, or that the presence of the second gene in *C. reinhardtii* is redundant.

Also of interest is the similarity between *C. reinhardtii* THB4 (A8ISP8) and *V. carteri* D8UFV6. In the previous section, the THB4 predicted protein was discussed in light of a potential chloroplast targeting peptide at the beginning of its sequence. Figure 4B denotes the location of the cleavage point with a dotted line. The *V. carteri* predicted protein does not possess a targeting peptide domain, but instead appears to begin at approximately the same location as the predicted TrHb1 domain of THB4 (Fig. 6). Starting from that location, the two predicted proteins share startlingly high-sequence identity, with over 80% of the amino acids being shared between the two. This is especially surprising as the sequence identity between *C. reinhardtii* and *V. carteri* appears to be higher than the identity between either of them and the LI637 protein form *C. eugametos*, the closest known homologue of THB4. Having the identity between species of different genus higher than

```
                                                                                   B10
tr|A8ISP8|A8ISP8_CHLRE   MQSTPQANAAPANVNGSVSSAGQSKSAATTRSTLHAKLGGAAAVAATVDVFYKKLMNDPD
tr|D8UFV6|D8UFV6_VOLCA   ---------------------------MTARATLHAKLGGAAAVAATVDIFYKKLMGDPD
                                                    *:*:******************:****** ***

                                        E10                     F8
tr|A8ISP8|A8ISP8_CHLRE   LEPFFRGVDMVTLIAKQNRFLAYAFGATTHYHGKDIVMGHAHLIINRGLNLTHFDKVAGH
tr|D8UFV6|D8UFV6_VOLCA   LAPFFHGVDMVTLIAKQNRFLAYAFGATSHYHGKDIVMGHAHLIINRGLNLSHFDKVAGH
                         * ***:**********************:**********************:********

tr|A8ISP8|A8ISP8_CHLRE   FVDSLKEMGVGQELIDEAAGVLIGVRPLFDPERYKGKVDVEKIEKAATTPQSKPAEAGQD
tr|D8UFV6|D8UFV6_VOLCA   FVDSLKEMGVTQELIDDAAGVLIGVRPLFDPERYKGKIDVEKIEKAAVVPQTKDTEGVKE
                         ********** *****:*******************:*********..**:* :*. ::

tr|A8ISP8|A8ISP8_CHLRE   APACCTIM
tr|D8UFV6|D8UFV6_VOLCA   TPACCTVM
                         :*****:*
```

```
tr|A8ISQ0|A8ISQ0_CHLRE   -----------------------------MPKVDGTPSALDRTAAISAATESLIASLED
tr|D8TXJ2|D8TXJ2_VOLCA   MAQDDVNTIHQAVTDWHEKLLSLPAMHGVLLHGPRQAPPPEERGAAVALAAQHLATCLED
                                                     :     :*   :* **:: *:: * :.***

tr|A8ISQ0|A8ISQ0_CHLRE   KQLPDGLIDDILSVVDVFTTLCGLSGPSPSDSFGAGGS---------TRSGAPLNPAKLK
tr|D8TXJ2|D8TXJ2_VOLCA   KNLPDGLINDLLSVCETFTALCGVSLPNNS--LGALSSPAHGALWVGWGSNRPLKAAKVK
                         *:******:*:*** :.**:***:* *. *  :** .*           *  **: **:*

                                                                          B10
tr|A8ISQ0|A8ISQ0_CHLRE   LLKAATGLPVHTTSIGRRASASYCVYESMGGWETLSAAVESLYNRMRGDGRCASLFSEGN
tr|D8TXJ2|D8TXJ2_VOLCA   LLKAATGLPVHTTSIGRRASASYCVYESMGGWETLSAAVESLYNRMRNDQRCEVMFADAN
                         *********************************************** * **  :*::.*

                             E10                        F8
tr|A8ISQ0|A8ISQ0_CHLRE   EQQLKTHMLEFLTCALGGKARFASSTLLTSQRDQLRQHGFGVGQFDILLQHMKAVLDEIG
tr|D8TXJ2|D8TXJ2_VOLCA   EQNLKTHTLEFLTCALGGKARFASSTLLTSQRDQLRNHGFGAGQFDALLQHMKAVMDELG
                         **:**** ***************************:****.**** ********:**:*

tr|A8ISQ0|A8ISQ0_CHLRE   VQQETAASAIALLRCHK-----------------------------------
tr|D8TXJ2|D8TXJ2_VOLCA   IHPEVAASAIALLRCHKYMFEKRPSDDGEEEAEEELSPTPTSASSPTAEASG
```

Figure 6 The sequences of *V. carteri* proteins homologous to THB4 (top) and THB3 (bottom). (See the colour plate.)

between species in the same genus is a striking affirmation for the inclusion of *C. reinhardtii* into the volvocine family. If THB4 does in fact possess a targeting peptide in *C. reinhardtii* the loss of the peptide in *V. carteri* may be explained by the specialisation of the different cell types in this multicellular organism. Further work will be needed to explore if alterations to TrHbs may be part of the evolution of multicellularity.

Another interesting comparison can be made between the *C. reinhardtii* predicted protein THB3 and *V. carteri* D8TXJ2 (Fig. 6). As discussed in the previous section, the THB3 protein sequence lacks a histidine at the critical location on the F helix (F8) but has a histidine on the distal side of the haeme that may be able to serve as a ligand to the iron. While both proteins have rather dissimilar amino-terminal regions, further into the sequence (beyond position 117 in THB3, position 111 in D8TXJ2) they become remarkably similar. This similarity extends to the same replacement of histidine at position at F8 and the same placement of a histidine on the distal side (E10) of the haeme. It is therefore likely that if THB3 is a functional protein with unusual haeme binding characteristics, the function was conserved in *C. reinhardtii* and *V. carteri*.

2.2.1.3 Other Chlorophytes

As we move beyond the core group within chlorophytes, less information becomes available. The nuclear, mitochondrial, and chloroplastic genomes of *C. variabilis*, a photobiont of *Paramecium bursaria*, and a species of green algae more distantly related to *C. reinhardtii* than *V. carteri*, have been completed (Blanc et al., 2010; Orsini et al., 2015). Genetic information is also available for a few additional species. Within this remaining genetic information we do find several haemoglobin genes.

Figure 5 presents available TrHb genes in chlorophytes arrayed on a rudimentary phylogenetic tree. It is interesting that this tree has two main branch points, a lower branch that segregates the TrHb genes encoding typically smaller, single-domain TrHbs such as THB1, THB2, and THB4, and an upper branch that contains the larger genes encoding TrHb domains within them. Within the lower branch the *C. eugametos* LI genes cluster together, but in general a small single-domain TrHb1 is found in most of the algae for which significant genetic data are available. The presence of these genes throughout chlorophytes suggests a conserved function.

The majority of the putative chlorophytic haemoglobins belong to the T family, principally TrHb1s. Nevertheless, there are also some instances of genes from the M family. The corresponding hypothetical gene products are single-domain proteins with the 3-on-3 fold (SDgbs). No SDgbs have been identified in *C. reinhardtii* or *V. carteri*; however, species such as *Bathycoccus prasinos* and *Ostreococcus tauri* appear to contain a single SDgb gene. It is worth noting that, although little is known about the SDgb genes of *Ostreococcus*, two species of this alga possess a functional nitric oxide synthase (NOS) (Foresi et al., 2010). This enzyme produces nitric oxide that may be metabolised by the algal haemoglobin; however, according to a published analysis of the three available species of *Ostreococcus*, only two have haemoglobin genes, and one of the species, *Ostreococcus lucimarinus*, known to have an NOS gene does not contain a haemoglobin gene (Vinogradov, Bailly, et al., 2013).

2.2.2 Photosynthetic Eukaryotes Beyond Green Algae

Aside from the chlorophyte, only a few pieces of genetic information are available for the remaining members of the algal primary endosymbionts. There is no current evidence for a haemoglobin gene in any species of Glaucophyte. Four species of red algae, belonging to the taxon Rhodophyta, do possess haemoglobin genes. *Cyanidioschyzon merolae* (Vinogradov et al., 2006) and *Galdieria sulphuraria* (Vinogradov et al., 2011) both have a single SDgb gene, while *Porphyridium purpureum* (Bhattacharya et al., 2013) and *Chondrus crispus* (Vinogradov et al., 2011) each have multiple T family genes, all belonging to the TrHb1 group. Beyond the primary endosymbionts, information is scarcer yet.

2.2.2.1 *Symbiodinium*

Symbiotic dinoflagellates are unicellular algae that form tight mutualistic symbiosis with other marine organisms. As such they are in a position to exert a great deal of influence on the cells around them, as well as be influenced by their partners. A recent publication reports the discovery of two sets of haemoglobin genes in the dinoflagellate *Symbiodinium* sp. (type C3) (Rosic et al., 2013). This alga is an endosymbiont of coral that can secrete up to 50% of its photosynthetic output to its host, mainly as glycerol, and in return the coral provides a stable habitat for the alga (Paracer & Ahmadjian, 2000). The symbiosis is thought to enhance greatly the survival of the coral (Davy, Allemand, & Weis, 2012), and the evolution of the symbiosis is hypothesised to have lead to the explosion of coral seen during the Triassic period (Muscatine et al., 2005). The coral–dinoflagellate symbiosis is especially important to survival during periods of environmental stress such as elevated water temperature.

The two putative haemoglobins, Hb-1 and Hb-2, occur as four isoforms each. The isoforms display high-sequence identity within each set whereas Hb-1s and Hb-2s are at most 21% identical. A single open reading frame (ORF) was found within *Hb-1*, coding for a 174-amino acid product. Two ORFs were identified within Hb-2, coding for products of 247 and 198 amino acids. Interestingly, both Hb-1 and Hb-2 are 3-on-3 SDgbs. BLAST searches relate Hb-1 to globins of bacteria, brown algae, and chlorophytes, whereas Hb-2 is cytoglobin-like and clusters with SDgbs from fish and other metazoans.

Examination of gene expression found that both Hb-1 and Hb-2 appear to participate in stress responses at the RNA level, with Hb-2 showing a more pronounced effect than Hb-1. Both haemoglobins demonstrate up-regulation when the symbiotic coral–dinoflagellate pair is exposed to either nutrient or thermal stress. Because it has been shown that nitric oxide can be used in signalling within the coral–dinoflagellate pair (Safavi-Hemami, Young, Doyle, Llewellyn, & Klueter, 2010), it is possible that these responses are linked to nitric oxide signalling. Another possibility is that the haemoglobins protect the symbionts as excessive nitric oxide has been found to collapse similar cnidarian–dinoflagellate symbionts (Perez & Weis, 2006). Given the role of nitric oxide as both an endogenous signal and as a potent reactive molecule, the tight interactions necessary for these symbionts may require a combination of these functions.

2.2.2.2 *Heterosigma akashiwo*

Stramenopiles (also called the heterokonts) are a vast and complex group of organisms that consist of both photosynthetic and nonphotosynthetic cell types as well as unicellular and multicellular species (Andersen, 2004). Within this group are the raphidophytes, photosynthetic organisms known to cause toxic algal bloom or 'red tides' responsible for large-scale fish mortality (Jeong, 2011). Two members of the raphidophytes, *H. akashiwo* and *Chattonella subsalsa*, have been confirmed to carry unique versions of an algal Hb (Stewart & Coyne, 2011). The genomes of these organisms have not been sequenced fully and gene expression studies have not been carried out systematically, but there is strong evidence, at least with *H. akashiwo*, that one haemoglobin gene is expressed and active in the cell.

What makes the haemoglobins of *H. akashiwo* and *C. subsalsa* unique is their presence as a domain within a NR. Although the evidence for this arrangement in *C. subsalsa* is limited to a partial gene sequence, the information from *H. akashiwo* is complete. In *H. akashiwo* there are two genes that encode a NR, one prototypical and one containing the inserted haemoglobin domain. This insert aligns with the TrHb family, specifically the TrHb1 subfamily, and is located in one of the two known hinge regions of NR. The architecture leads to a possible functional hypothesis for these proteins. Raphidophytes are known to make large amounts of nitric oxide under normal growth conditions (Kim et al., 2008). This overproduction of nitric oxide has been linked to the organisms' lethality particularly with fish, although the method of this toxicity is unclear (Shen, Xu, Chiang, & Au, 2011). Exposure to nitric oxide was shown to cause an increase in the expression of the Hb-containing NR gene. The presence of a haemoglobin embedded within a NR could give the cell a competitive advantage for nitrogen metabolism, especially when nitrogen is a limiting nutrient (Davidson et al., 2012), as explained in Section 4.2.3.

2.2.2.3 Other Secondary Endosymbionts

One recent addition to the list of secondary endosymbionts with genetic information is the coccolithophore, *Emiliania huxleyi* (Read et al., 2013). This phytoplankton appears to have at least a single SDgb-type haemoglobin gene within its genome (XP_005794579.1). Although the alignment of the predicted protein to known globin sequences is low, analysis of the sequence by I-TASSER (Yang et al., 2015) suggests that the protein adopts the 3-on-3 fold. The majority of haemoglobins found in the partial genetic sequences of the remaining algae also appear to fit with the 3/3 fold of M family. As there is still only a relative handful of known algae genomic sequences available, out of the possibly thousands of extant species, it is possible to only draw general conclusions. Within the haemoglobins thus far there is very little consistency in the genetic make-up, including number of genes per organism or distribution of haemoglobin families (M and T) within a single organism. In general, the stramenopiles appear to be more variable in the composition of both SDgbs and TrHbs, but it is possible that is an artefact due to the larger number of sequences available for those organisms.

3. THE STRUCTURE OF ALGAL HAEMOGLOBINS

The ultimate goal of much haemoglobin research is to characterise the physical properties of the proteins and to relate these to function. This section summarises essential observations on two known algal haemoglobin structures, both of which belong to the TrHb1 subfamily of proteins. This small number of solved structures and the absence of structures for algal

TrHb2 and M family globins leads us to consider the benefits and limitations of modelling approaches.

3.1 The Folds of Algal Haemoglobins

The three-dimensional structures of globins found in unicellular algae are expected to resemble closely one of two folds: the 3-on-3 α-helix sandwich of *Physeter catodon* (sperm whale) myoglobin (Mb, M family) or the 2-on-2 α-helix sandwich of truncated haemoglobins (TrHb, T family) depicted in Fig. 1. The algal 2-on-2 proteins belong to two distinct groups, TrHb1, represented by the haeme domain of *C. eugametos* LI637 haemoglobin (1DLY) (Pesce et al., 2000), and TrHb2, represented distantly by a haemoglobin from the actinobacterium *Thermobifida fusca* (2BMM) (Bonamore et al., 2005), and a haemoglobin from *Arabidopsis thaliana* (4C0N) (Reeder & Hough, 2014). TrHb3 genes form the most recent branch of the T family and have not been found in algae. Suitable models for the 3-on-3 architecture are provided by the class 1 nonsymbiotic haemoglobin from *A. thaliana* (3ZHW (Mukhi et al., 2013)) and globin I from *Methylacidiphilum infernorum* (Hell's Gate HbI, 3S1I) (Teh et al., 2011). The 3-on-3 fold, when fused to a flavin-binding domain, is likely to mimic the FHb from *Escherichia coli* (1GVH) (Ilari, Bonamore, Farina, Johnson, & Boffi, 2002). Figure 7 contains ribbon diagrams of the models, with the 2-on-2 and 3-on-3 sets of helices emphasised. Table 4 lists the representative structures used in the text.

As illustrated by numerous and distant amino acid sequences, algal haemoglobins show high diversity in the haemoglobin domains proper. In addition, they are decorated with an extraordinary variety of N- and C-terminal extensions. In several instances, two or more haemoglobin domains are organised in succession in a single polypeptide chain and separated by stretches of variable length. Chimeric proteins such as the FHb illustrated in Fig. 7D are also expected. However, in many cases, the appended residues do not form a recognisable domain.

Analyses with programmes such as IUPred (Dosztanyi et al., 2005) and DisEMBL (Linding et al., 2003) suggest that many extensions and linkers have low propensities for adopting a folded structure. Figures 4B and 8 present the results of the analysis for *C. reinhardtii* THB4 and THB8, respectively. In Fig. 8, the N-terminal domain and haemoglobin domain stand out for their low disorder propensity, whereas the long C-terminal extension appears to favour unstructured conformations as found in intrinsically

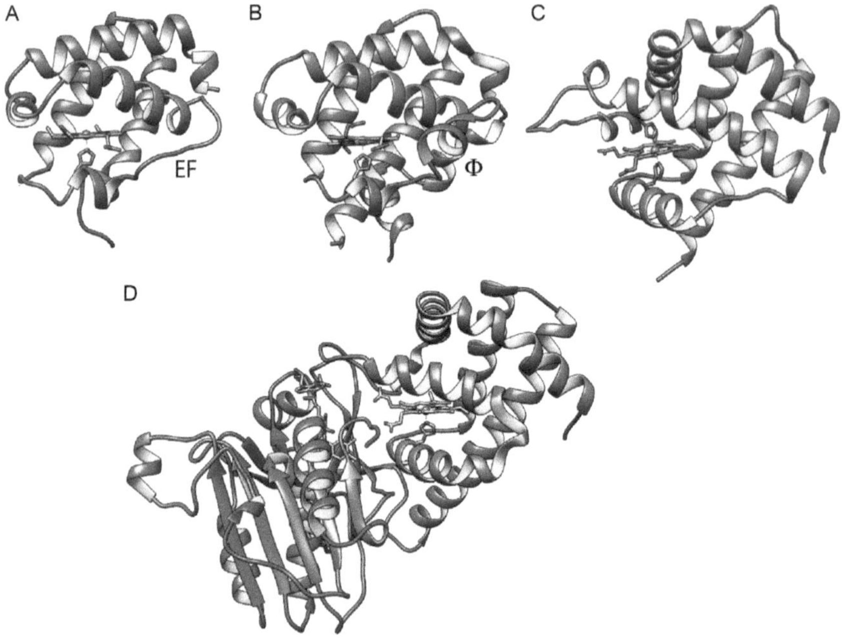

Figure 7 Ribbon structures of representative haemoglobins. (A) TrHb1: *C. eugametos* LI637 haemoglobin domain (1DLY) (Pesce et al., 2000). Distinctive features are the short A and F helices and the long EF loop. (B) TrHb2: *Arabidopsis thaliana* GLB3 (4C0N) (Reeder & Hough, 2014). The TrHb2 architecture includes an extended EF loop folded into a helix denoted by Φ. (C) SDgb: *Arabidopsis thaliana* AHb1 (3ZHW) (Mukhi et al., 2013). This protein is an example of *bis*-histidine coordination of the haeme iron. (D) FHb: *Escherichia coli* Hmp (1GVH) (Ilari et al., 2002). The long H-helix (33 residues) allows the C-terminal flavin domain to make close contact with the globin domain, which guarantees efficient electron transfer from the FAD molecule to the haeme group. The helices forming the 2-on-2 and 3-on-3 topologies are highlighted in blue. Haemes and proximal histidines are represented with sticks. (See the colour plate.)

disordered proteins (IDPs) (Oldfield & Dunker, 2014). One can speculate that if such extensions survive in the mature proteins, they serve purposes typical of IDPs, for example, the formation of complexes with functional partners. Figure 8 also contains the IUPred profile of THB1 for comparison. As often seen in haemoglobins (Landfried et al., 2007), the F helix of THB1 has a higher tendency for disorder than the rest of the domain. A disordered haeme cavity in the apoprotein state is thought to prevent local hydrophobic collapse that would retard haeme binding and the formation of a stable holoprotein.

Table 4 Representative Structures of Haemoglobin Domains

Source	Protein	PDB ID	Resolution (Å)	Oxidation State, Distal Ligand, PTM	References
Chlamydomonas eugametos	CtrHb (TrHb1)	1DLY	1.8	Ferric, cyanide	Pesce et al. (2000)
Synechocystis sp. PCC 6803	GlbN (TrHb1)	1RTX	1.8	Ferric, histidine, haeme PTM[a]	Hoy, Kundu, Trent, Ramaswamy, and Hargrove (2004)
Synechocystis sp. PCC 6803	GlbN (TrHb1)	1S69	1.68	Ferric, cyanide, haeme PTM	Trent, Kundu, Hoy, and Hargrove (2004)
Synechococcus sp. PCC 7002	GlbN (TrHb1)	4MAX	1.45	Ferric, histidine, haeme PTM	Wenke, Lecomte, Heroux, and Schlessman (2014)
Synechococcus sp. PCC 7002	GlbN (TrHb1)	4L2M	2.25	Ferric, cyanide, haeme PTM	Wenke et al. (2014)
Chlamydomonas reinhardtii	THB1 (TrHb1)	4XDI	1.89	Ferric, lysine	Rice et al. (2015)
Arabidopsis thaliana	GLB3 (TrHb2)	4C0N	1.77	Ferric, hydroxide	Reeder and Hough (2014)
Thermobifida fusca	Tf-TrHb (TrHb2)	2BMM	2.48	Ferric, acetate	Bonamore et al. (2005)
Arabidopsis thaliana	AHb1 (SDgb)	3ZHW	2.22	Ferric, histidine	Mukhi et al. (2013)
Methylacidiphilum infernorum	HGbI (Sdgb)	3S1I	1.77	Ferrous, oxy	Teh et al. (2011)
Escherichia coli	Hmp (FHb)	1GVH	2.19	Ferric	Ilari et al. (2002)

[a]PTM: post-translational modification by which the haeme is covalently attached to the protein.

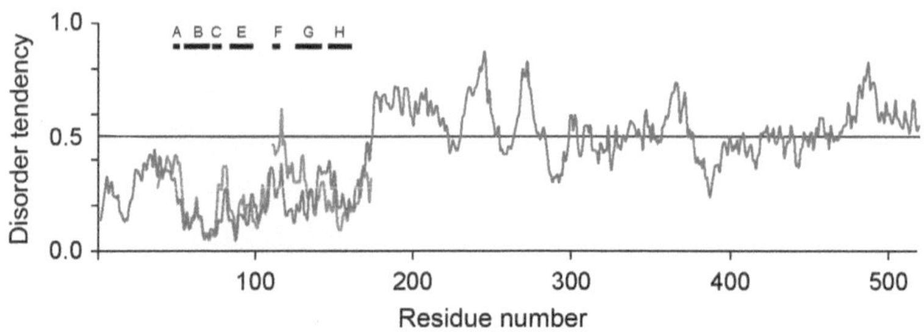

Figure 8 The IUPred profile (Dosztanyi et al., 2005) of THB1 (UniProt identifier A8JAR4, magenta) and THB8 (UniProt identifier A8J8P7, blue) (long disorder calculation). The haemoglobin domain is shaded in light blue. The folded region of THB8 is readily recognised. (See the colour plate.)

3.2 The Structure of CtrHb and THB1

3.2.1 CtrHb

Section 2.2.1 introduced *C. eugametos* LI637, a light-induced protein targeted to the chloroplast. The original spectroscopic studies of the haeme domain (CtrHb) noted an unusual feature of this protein in the absence of exogenous ligand (Couture et al., 1994, 1999; Das et al., 1999). In the ferrous and ferric states and within a pH range close to physiological, CtrHb appeared to coordinate the iron not only with the proximal histidine but also with another protein residue (Couture et al., 1999; Das et al., 1999). With a combination of mutagenesis, resonance Raman, and electron paramagnetic resonance (EPR) data, the distal ligand was identified as Tyr B10 in a configuration stabilised by Lys E10 (Das et al., 1999). This novel arrangement was a manifestation of a flexible haemoglobin fold.

Ferric CtrHb with cyanide bound to the iron (CtrHb-CN) was crystallised successfully by Bolognesi and colleagues, who proceeded to solve the structure of the protein to a resolution of 1.8 Å (Pesce et al., 2000). The landmark study presented the first view of a member of the truncated haemoglobin family. The details of the topology (Fig. 7) became apparent, but more importantly, attributes characteristic of TrHb1 in association with a strong ligand could now be visualised, and it became possible to design experiments to test specific structure–function hypotheses within the whole TrHb1 subfamily.

Figure 9 shows a portion of the haeme cavity of CtrHb-CN to emphasise a salient set of interactions. In an arrangement expected to hold for other ligands capable of hydrogen bonding, the N atom of the cyanide ligand is hydrogen bonded by the hydroxyl hydrogen of Tyr B10. Gln E7 and

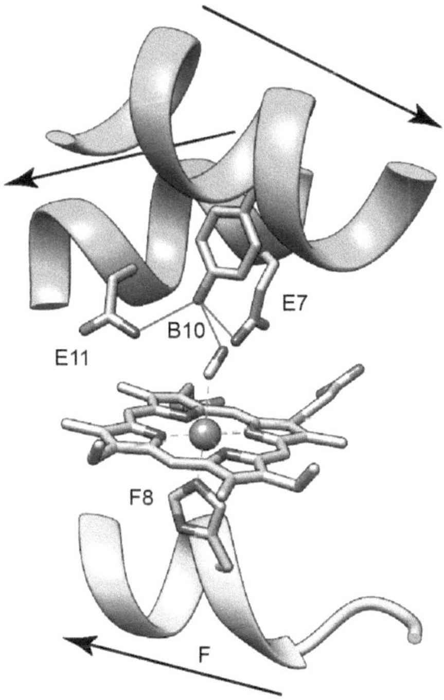

Figure 9 The haeme cavity of CtrHb (1DLY) (Pesce et al., 2000), showing portions of helices B, E, and F and the hydrogen bonds stabilising the bound cyanide. Tyr B10, Gln E7, Gln E11, and His F8 are depicted with sticks. The direction of the helices is marked with an arrow. (See the colour plate.)

Gln E11 engage in complementary interactions. The importance of Tyr B10 and the distal hydrogen-bond network in controlling the kinetics of ligand release from non-vertebrate globins has long been recognised (Bolognesi, Bordo, Rizzi, Tarricone, & Ascenzi, 1997). In TrHb1s, it has been specifically probed by amino acid replacements with the result that k_{off} for dioxygen increases and the ferrous protein undergoes ready autooxidation to the ferric state (Couture et al., 1999; Igarashi, Kobayashi, & Matsuoka, 2011; Ouellet, Milani, Couture, Bolognesi, & Guertin, 2006).

A second trait of CtrHb-CN is the presence of internal cavities. Unlike many globins and globular proteins, CtrHb-CN is not well packed. It has distinct channels, wide enough to allow passage of small diatomic ligands, connecting the haeme cavity and the bulk solvent. These passageways are present when an exogenous ligand is bound and are thought to facilitate the diffusion of additional substrates into the active site (Milani et al., 2004; Savard et al., 2011). Channels, pores, and tunnels are observed in the majority of TrHb1s with exogenous ligand bound. In contrast, TrHb2s do not display such characteristic features (Nardini, Pesce, Milani, & Bolognesi, 2007).

3.2.2 THB1

The sequence and phylogenetic analysis of THB1 places this protein in Group I of the TrHb family. As a result, the three-dimensional structure was expected to reflect a version of the 2-on-2 topology resembling closely CtrHb. In the absence of exogenous ligand, absorbance data showed THB1 to have low-spin character at neutral pH both in the ferric ($S=½$) and the ferrous ($S=0$) states (Johnson et al., 2014) in an indication that a distal residue serves as an axial ligand to the iron. By analogy with CtrHb, this residue was initially assumed to be Tyr B10 (Tyr29). However, replacement of Tyr29 with phenylalanine, a residue incapable of coordination, did not produce the expected change in ligation state or pH response, eliminating Tyr B10 as the wild-type sixth ligand to the iron.

An alternative possibility was provided by comparison to GlbN from *Synechocystis* and *Synechococcus*. Although these two proteins have structures matching that of CtrHb when in the ferric, cyanide-bound state (Trent et al., 2004; Wenke et al., 2014), they use His E10 to coordinate the iron in the absence of a strong ligand and achieve this particular hexacoordination with a conformational rearrangement. Figure 10 shows a superposition of the *Synechococcus* GlbN backbones in the *bis*-histidine and cyanomet states to illustrate the extent of the distortion. This is best appreciated by comparison to sperm whale myoglobin, for which the backbone structures in the deoxy state (1BZP) and the carbonmonoxy state (1BZR) are practically

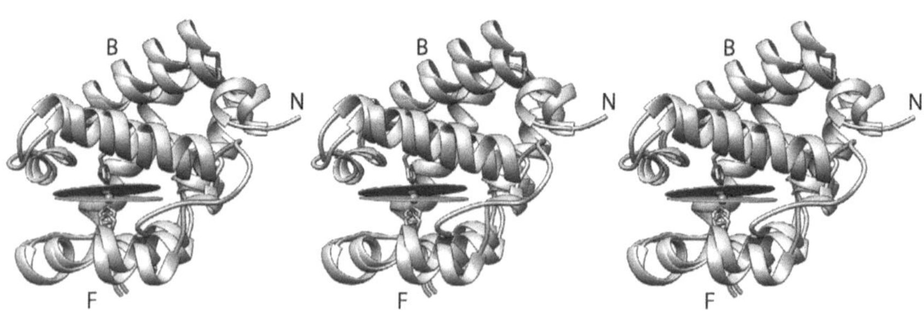

Figure 10 Superimposition of *Synechococcus* sp. PCC 7002 ferric GlbN in the *bis*-histidine state (4MAX, beige) and cyanomet state (4L2M, cyan) (Wenke et al., 2014). The left pair of the stereo triptych is for wall-eyed viewing. The haeme groups are represented by a disc with iron atom as a ball. The proximal histidines are below the haeme planes. (See the colour plate.)

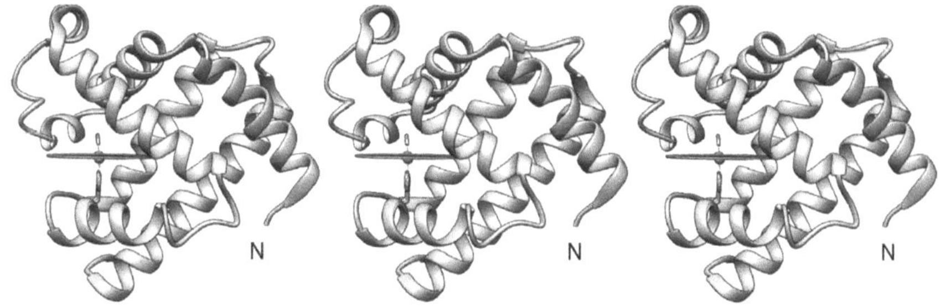

Figure 11 Superimposition of sperm whale myoglobin in the deoxy state (1BZP, beige) and the carbonmonoxy state (1BZR, cyan) (Kachalova, Popov, & Bartunik, 1999). The left pair of the stereo triptych is for wall-eyed viewing. The haeme groups are represented by a disc with iron atom as a ball. The proximal histidines are below the haeme planes. (See the colour plate.)

indistinguishable (Fig. 11). The flexibility of the fold exhibited by the cyanobacterial proteins suggested that a residue of the E helix, specifically Lys E10 in THB1, could reach the iron without causing prohibitive steric clashes.

Lysine coordination is supported by several lines of evidence. Replacement of Lys E10 (Lys53) with alanine transforms the absorbance spectrum of ferric THB1 into that of a complex possessing a water molecule (low pH) or hydroxide ion (high pH) as the distal ligand. In the proton nuclear magnetic resonance (NMR) spectrum of the ferrous state, signals can readily be attributed to a lysine in close contact with the haeme group (Johnson et al., 2014), and the chemical shifts of the terminal NH_2 (^1H and ^{15}N) are consistent with the formation of a Fe–N bond. Finally, the electron density map obtained by X-ray diffraction of the ferric protein shows a well-defined lysine ligand at position E10 (Rice et al., 2015) with geometry matching that of the ferrous state in solution. Haeme coordination in THB1 is illustrated in Fig. 12.

Because of the relatively high acidity constant of an exposed amino group (~10.5) compared to an imidazole group (~6.5), at neutral pH the histidine-lysine coordination is energetically more costly than the *bis*-histidine coordination. The specific chemical properties imparted to the protein by the histidine-lysine scheme remain to be fully determined. However, both ferric and ferrous states exhibit the same hexacoordination, which is likely to facilitate electron transfer and be a key component of an activity requiring a change in oxidation state in the absence of bound substrate or product (Johnson et al., 2014; Preimesberger, Pond, Majumdar, & Lecomte, 2012).

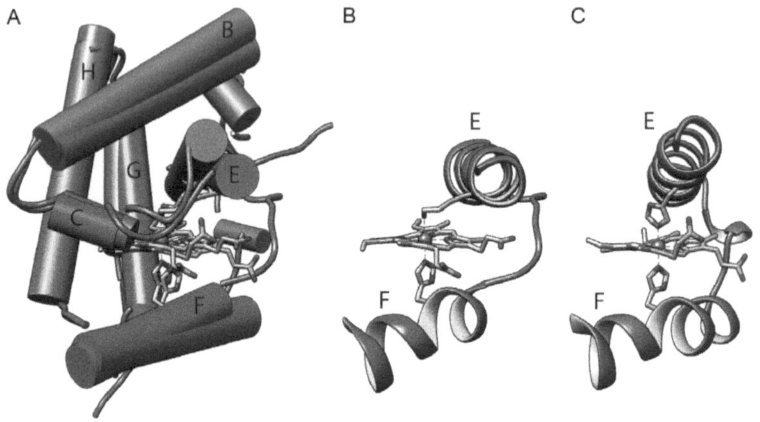

Figure 12 Superimposition of the structures of THB1 (orange) (4XDI) (Rice et al., 2015) and *Synechococcus* sp. PCC 7002 GlbN (blue, 4MAX) (Wenke et al., 2014) in the absence of exogenous ligand. (A) Cylinder diagrams emphasising the reorientation of helices. (B) The E and F helices of THB1 showing the histidine-lysine coordination. (C) The E and F helices of GlbN, showing the *bis*-histidine coordination. (B) and (C) are in the same orientation as in (A). (See the colour plate.)

The distal lysine of ferric THB1 is readily displaced by cyanide to form the cyanomet complex. Although no structure of this form is available yet, NMR data confirm that the hydrogen-bond network exhibited by CtrHb-CN and involving Tyr B10 is formed (Johnson et al., 2014). Thus, it is likely that the structures of CtrHb-CN and THB1-CN match closely and that a conformational change occurs upon binding of exogenous ligand.

A concise approach to the representation of haemoglobin topology focuses on the relative orientation of the B, E, G, and H helices and provides additional insight into packing and fold variability. The polar coordinate diagram of Fig. 13 includes data for selected TrHb1s and TrHb2s. In this analysis, an axis is calculated for crossing helices. The crossing parameters for each pair of helices are represented in two dimensions by a vector forming the crossing angle with the ordinate axis and having length equal to the distance between helical axes (Lesk & Chothia, 1980). Three clusters corresponding to the BE, BG, and GH interfaces are readily identified in the TrHb plot. The conformational change associated with the replacement of His E10 by cyanide in GlbN is represented with arrows (2–3 and 5–6). THB1 (4), with its His-Lys ligation, has a BE crossing intermediate between *bis*-histidine and cyanomet GlbNs illustrating again the flexibility of the TrHb1 fold. The same analysis carried out on a few 3-on-3 haemoglobins (Fig. 14) reveals how the BE topology varies between T and M proteins.

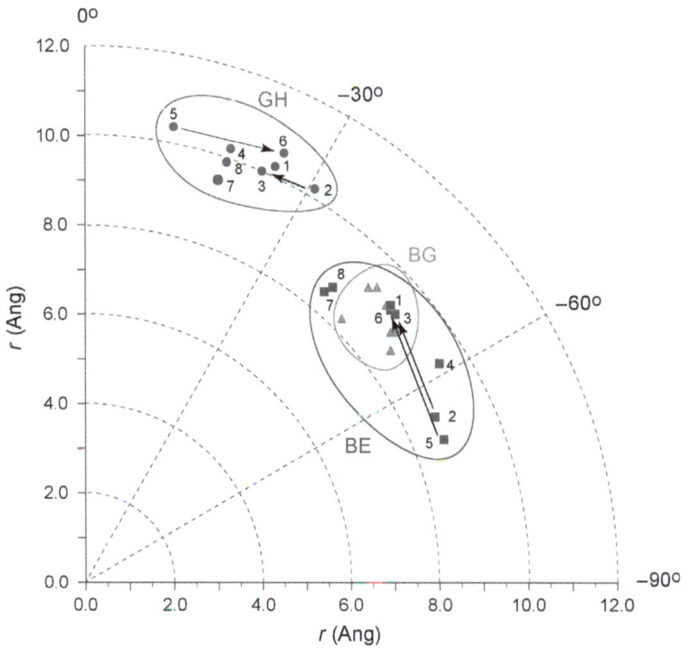

Figure 13 Helix crossing parameters for selected TrHbs. Helical axes were calculated with UCSF Chimera (Pettersen et al., 2004) for helices defined by DSSP (Kabsch & Sander, 1983). The distance *r* is obtained by finding the vector perpendicular to a pair of axes. Crossing angles were obtained by projection into a plane perpendicular to the *r* vector. A negative sign is given to pairs for which the near helix is rotated clockwise. The data points are for CtrHb-CN (1DLY, 1), *Synechocystis bis*-histidine GlbN (1RTX, 2), *Synechocystis* GlbN-CN (1S69, 3), *C. reinhardtii* histidine-lysine THB1 (4XDI chain A, 4), *Synechococcus bis*-histidine GlbN (4MAX chain C, 5), *Synechococcus* GlbN-CN (4L2M chain A, 6), *Arabidopsis thaliana* GLB3-OH (4CON, 7), and *Thermobifida fusca* TrHb-acetate (2BMM, 8). (See the colour plate.)

Sequence alignments (including those in Fig. 3) suggest that this surprising histidine-lysine coordination occurs in other proteins as well. For example, CtrHb possesses Lys E10 and NMR data collected on ferrous CtrHb reveal the presence of signals corresponding well with those of the coordinating Lys E10 in THB1 (Rice et al., 2015). Thus, it is likely that under some conditions, Lys E10 also serves as the sixth ligand of CtrHb. In contrast to the histidine lysine ligation state, the histidine-tyrosine ligation state such as proposed for CtrHb has so far eluded crystallisation. In an additional demonstration of the flexibility of the TrHb fold and versatile coordination on the distal side of the haeme, Bolognesi, Verde, and coworkers observed that *Pseudoalteromonas haloplanktis* TAC125 TrHb2 crystallises in the ferric state

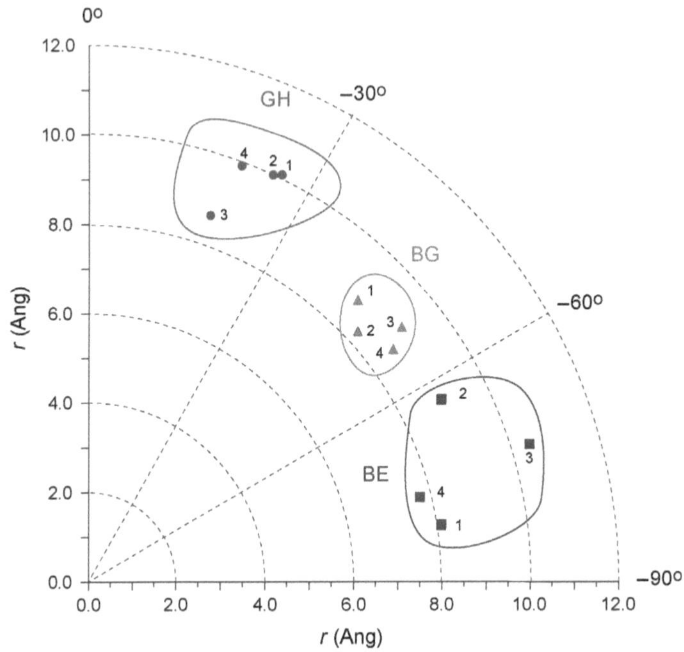

Figure 14 Helix crossing parameters for selected 3-on-3 haemoglobins (see Fig. 13 for details). The data points are for *Arabidopsis thaliana bis*-histidine AHb1 (3ZHW, 1), *Methylacidiphilum infernorum* Hb1-O_2 (3S1I, 2), *Escherichia coli* flavohaemoglobin (1GVH, 3), and *Physeter catodon* Mb-O_2 (1MBO, 4). (See the colour plate.)

with a distal water molecule embedded in a network of hydrogen bonds anchored by His CD1 (Giordano et al., 2015). However, in solution the protein shows signs of endogenous hexacoordination, with resonance Raman and EPR signatures attributed to an axial tyrosine, presumably Tyr B10. The structures currently available indicate that a substantial structural rearrangement is necessary for this tyrosine to reach the iron.

3.3 Structural Predictions

When presented with a new globin amino acid sequence, the task of modelling the structure in the absence of experimental data is straightforward if one is satisfied with a low-resolution approximation focused on the haeme domain. This is because of the large number of sequences and structures that allows one to categorise proteins within the 2-on-2 or the 3-on-3 families and the robustness of the two types of fold. The three-dimensional structure of extensions at termini or in various inter-helical regions is more difficult to predict, mostly because secondary or tertiary structure propensities do not inform on thermodynamic or kinetic stability with sufficient reliability. The effect of the cellular environment (particular conditions of pH or presence of partner proteins) adds to this uncertainty.

A distinctive aspect of haemoglobin structural modelling compared to the modelling of globular protein that do not contain a cofactor or prosthetic group is quite obviously the ability to predict accurately the details of the haeme cavity. A larger problem, however, is to ascertain the identity of the cofactor itself. The 3-on-3 globin fold is related to the α-helical fold of phycocyanin (Pastore & Lesk, 1990) and other tetrapyrrole-binding proteins. As a result, the Conserved Domain Database of the National Center for Biotechnology Information (NCBI) groups these proteins together within the 'globin-like superfamily'. This sole classification is applied by BLAST searches to some proteins distantly related to known globins. 'Globin-like' hits should therefore be inspected carefully (and preferably experimentally) before including them with true globins. The ambiguity is a special concern for the proteins of photosynthetic organisms because numerous pigments have molecular properties related to *b* haeme. Sobering instances include the binding of non-haeme tetrapyrrole derivatives by myoglobin (Boxer & Wright, 1979; Pröll, Wilhelm, Robert, & Scheer, 2006; Wright & Boxer, 1981) (which highlights the need for information on cellular localisation) and the binding of haeme by modified phycocyanins (Isogai & Ishida, 2009). Furthermore, the globin fold can be rearranged to yield a compact protein unable to bind a cofactor (Murray, Delumeau, & Lewis, 2005). Thus, in our studies of THB1, we purified the protein from the *Chlamydomonas* cell in order to confirm the presence of the haeme group (Johnson & Lecomte, 2014). Unfortunately extracting haemoglobins from algae cultures may not always be practical.

At the core of haemoglobin chemistry is the haeme group and its central iron, and not surprisingly, iron coordination is an essential determinant of reactivity (Ozaki, Roach, Matsui, & Watanabe, 2001). In typical haemoglobins, the proximal histidine is a non-labile ligand over a broad range of pH values, and the sixth coordination site of the ferrous iron is vacant, occupied loosely by a water molecule or satisfied by an oxygen atom from dioxygen. In a growing number of globins across prokaryotes and eukaryotes, a protein residue coordinates the iron on the distal side as long as no strong ligand (e.g. cyanide for the ferric state and carbon monoxide for the ferrous state) is present in the medium. These proteins, referred to as hexacoordinate haemoglobins, may behave differently from their pentacoordinate relatives. *Synechococcus* GlbN, *Synechocystis* GlbN, *A. thaliana* AHb1, and THB1 are all examples of hexacoordinate globins.

So far, all studied algal and cyanobacterial haemoglobins, whether 'hexacoordinate' or not, are capable of binding diatomic ligands. The Protein Data Bank (PDB) contains a large number of TrHb and SDgb structures with bound cyanide (ferric state), azide (ferric state), dioxygen (ferrous state), or carbon monoxide (ferrous state). These structures reveal the protein–ligand interactions that control the affinity for various ligands. Combined with the degeneracy of the globin fold (i.e. structures with low-sequence identity can adopt the same fold), it is not surprising that for a given sequence, an existing structure (template) is available that generates minor clashes resolved by side chain rotameric state adjustments. It should be stressed that details of hydrogen bonding, protonation states of cavity residues and the orientation of side chain amides, which require the inclusion of electrostatic terms or quantum mechanical calculations, are less reliable than global fold and identification of helix starts and stops.

Obtaining an acceptable model of the bound state does not guarantee that the protein is incapable of adopting an alternative structure. Helix repositioning and hinge motions at inter-helical regions driven by the ligation of a distal residue have been well documented in TrHb1s (Section 3.2.2) and is apparent in the helix crossing diagram of Fig. 13. Thus, modelling THB1 with either a TrHb1 in the *bis*-histidine state (e.g. *Synechococcus* GlbN, 4MAX) or containing an exogenous ligand (e.g. *Synechococcus* GlbN-CN, 4L2M) returns two distinct but acceptable structures. According to X-ray data, the actual structure deviates slightly from both. If backbone root-mean-squared deviation is used to assess the relationship, THB1 appears to be intermediate between these two extremes (Wenke et al., 2014). Modelling of THB2, which also has a lysine at E10 but for which there is no structure yet, yields plausible structures with the same two templates. Preliminary experiments with this protein indicate partial coordination of the lysine at neutral pH. Thus, not only is the preparation of acceptable models dependent on the availability of relevant templates in the PDB, but the relative population of these hypothetical structures cannot be predicted without high level computational work.

Finally, discrimination among pentacoordinate, water- or hydroxide bound, and endogenous hexacoordinate species poses a particular challenge for modelling exercises. Threading programmes such as I-TASSER (Yang et al., 2015) and other Web-based programmes (e.g. SWISS-MODEL (Biasini et al., 2014)) do not address this attribute well. Predicted coordination states should be regarded as questionable until confirmed with experimental data.

4. THE CHEMISTRY OF ALGAL HAEMOGLOBINS

The haeme within haemoglobins is a reactive molecule that can *a priori* convey a range of enzymatic activity to the holoprotein. Although trends may eventually emerge from combined studies of different systems, at this time little distinguishes algal haemoglobins from their relatives of non-photosynthetic organisms. They are capable of the same chemistry, specifically described in the following sections, and they appear to play roles similar to those envisioned in cyanobacteria, plants, and non-vertebrates (Vinogradov & Moens, 2008). The existence of multiple globins within the same cell illustrates that a full understanding of the complex interplay of globin functions will require detailed characterisation of the context in which these proteins operate: their localisation, regulation, expression levels under various stresses, flux of relevant metabolites, chemical properties, etc.

In hexacoordinate haemoglobins, the stability of the bond linking the iron to the distal residue (a histidine, lysine, or perhaps, tyrosine) may be such that decoordination is slow compared to the on-rate of the physiological ligand, therefore constituting the rate-determining step. In other instances, decoordination may be relatively fast and have no influence on the binding reaction. In addition to the kinetic aspect of endogenous hexacoordination, coordination of a residue on the distal side modulates the redox potential of the iron. For example, *bis*-histidine proteins tend to have a lower redox potential than their mono-histidine counterparts (Kakar, Hoffman, Storz, Fabian, & Hargrove, 2010). The rate of autooxidation (the conversion of ferrous to ferric state caused by dioxygen (Brantley, Smerdon, Wilkinson, Singleton, & Olson, 1993)) also appears to be enhanced in *bis*-histidine globins. Hexacoordination has other consequences, for example, a distal histidine may protect the protein from oxidative damage (Nothnagel et al., 2011). Lysine coordination in THB1 appears less effective than histidine for this purpose (unpublished observation). Both histidine and lysine can coordinate the ferrous and ferric iron and therefore facilitate electron transfer (Johnson et al., 2014; Preimesberger et al., 2012).

As illustrated in Fig. 10, binding of an exogenous ligand to the iron of *Synechococcus* GlbN causes a large conformational change. A similar rearrangement is observed in *Synechocystis* GlbN, but the extent and character of the structural transition are not shared by all hexacoordinate proteins (Kakar et al., 2010). It is possible that such transition modifies the affinity for a partner or relays a signal; however, no experimental data exist at this stage to support interaction of a SDgb with another protein.

Limited information is available regarding the physiological role of haemoglobins in unicellular algae. This predicament is common in non-vertebrate haemoglobin research for multiple reasons. Foremost, algal haemoglobins do not appear to be essential proteins. *In vivo* studies involve genetic manipulations that are technically complex and are further complicated by redundant pathways and the need to identify conditions under which a clear phenotype is observed. Parallel *in vitro* studies have utility, but to recapitulate cellular activity they must address the properties of the mature protein, consider relevant partners, and discriminate among adventitious and physiological reactions. As a result of concentrated efforts, the characterisation of algal haemoglobins has advanced over the last few years and has focused increasingly on the management of reactive oxygen species (ROS) and reactive nitrogen species (RNS). Management of nitric oxide, in particular, is a recurring theme. Nevertheless, the obstacles described above have prevented definitive functional assignment.

In this section, we review briefly the main reactions that contribute to the likely functions of algal haemoglobins. They are oxygen binding, nitric oxide dioxygenation, and nitrite reduction. Excellent articles are available for in-depth chemical insight (Bowman, McLean, Poole, & Fukuto, 2011; Gardner, 2012; Hughes, 2008; Marti et al., 2006) and the interested reader is referred to those.

4.1 Reversible Dioxygen Binding

Reversible dioxygen binding to the ferrous iron is a common property of haemoglobins, even if their function is unrelated to that process. For example, ferric HbI from *Lucina pectinata* is a H_2S transporter, but is capable of O_2 binding in the ferrous state (Pietri et al., 2006). GLB-26 of *Caenorhabditis elegans*, a *bis*-histidine globin, is so far the only exception (Kiger et al., 2011; Tilleman et al., 2012). GLB-26 is thought to serve as an electron transport protein, a rare, if not unique, function for a haemoglobin. Roles related to dioxygen utilisation, however, dominate the haemoglobin superfamily and are considered most ancient (Gardner, 2012). *C. elegans* GLB-26 and *L. pectinata* remind us not to ignore the possibility of other activities and to consider the ferric state as well as the ferrous state as an important player.

Dioxygen binding by haemoglobin results in a complex with ferric-superoxide character (Wittenberg, Wittenberg, Peisach, & Blumberg, 1970), as represented by

$$Fe(II) - O_2 \rightarrow Fe(III) - O_2^-$$

The propensity for transfer of an electron from the ferrous iron to the distal oxygen is responsible for the reactivity of oxy globins towards nitric oxide. Dissociation of the superoxide anion from the globin leaves the globin in the ferric state (autooxidation) and is controlled in large part by distal interactions as mentioned in Section 3.2.1.

The affinity of algal haemoglobins for dioxygen has yet to be measured systematically. *C. eugametos* CtrHb has a slow dissociation rate constant caused by the distal hydrogen-bond network shown in Fig. 9 (Couture et al., 1999). Combined with the low concentration of the protein in the cell, a scavenging role seems unlikely. Two proposals advanced by Guertin and coworkers (Couture & Guertin, 1996) are a protective role and delivery of dioxygen to a terminal oxidase involved in chlororespiration (Nixon, 2000).

4.2 Nitric Oxide Dioxygenation and Nitrite Reduction

4.2.1 Sources of Nitric Oxide

A recurring theme in the proposals for haemoglobin function in algae (and now in all other organisms) is the processing or production of nitric oxide. Nitric oxide is a reactive molecule (Goodrich, Paulat, Praneeth, & Lehnert, 2010; Heinrich et al., 2013) assuming multiple roles in cyanobacterial, algal, and plant cells (Bowman et al., 2011; Gupta, Hebelstrup, Mur, & Igamberdiev, 2011; Wang & Hargrove, 2013). It is an intra- and intercellular messenger implicated in signalling and defence mechanisms. A well-described source of intracellular nitric oxide in algae is NR acting on nitrite (Sakihama et al., 2002) Nitric oxide can also be produced from nitrite by the plant mitochondrion (Igamberdiev, Ratcliffe, & Gupta, 2014) and the mitochondrion of certain algae such as *Chlorella sorokiniana* (Tischner, Planchet, & Kaiser, 2004). Chloroplastic source may also contribute, principally through the action of nitrite reductase (Kumar, Castellano, Patti, Palumbo, & Buia, 2015). Some haemoglobins can also generate nitric oxide from nitrite (Section 4.2.3). Whether or not algae possess NOSs using L-arginine as a substrate is a matter of debate (Corpas & Barroso, 2015);

the green algae *O. tauri* and *O. lucimarinus*, however, were recently shown to produce an enzyme similar to the endothelial, inducible, and neuronal NOSs found in animals (Foresi et al., 2010).

4.2.2 Haemoglobins and Nitric Oxide

Unlike dioxygen and carbon monoxide, which have no affinity for the ferric state of haemoglobins, nitric oxide can combine with both the ferric and ferrous states with considerable affinities (Traylor & Sharma, 1992). Nitric oxide binding to ferrous haeme proteins can be very tight (dissociation constants in the pM range) and generally weakens the iron–histidine bond in a '*trans*' effect (Traylor, Duprat, & Sharma, 1993). Haeme modification can also be caused by nitric oxide (Yi & Richter-Addo, 2012). Haeme proteins have therefore evolved to adjust their susceptibility to and relative binding constants for diatomic ligands. For haeme protein sensors, a 'sliding scale rule' relates nitric oxide, carbon monoxide, and dioxygen affinities (Tsai, Berka, Martin, & Olson, 2012). In assessing the importance a various chemistries involving nitric oxide and dioxygen, the local concentrations, fluxes, and binding constants are essential.

4.2.3 Nitric Oxide Dioxygenation

Under oxic conditions, the nitric oxide dioxygenation reaction generates nitrate according to

$$\mathrm{Fe(II)} - \mathrm{O_2} + {}^{\bullet}\mathrm{NO} \rightarrow \mathrm{Fe(III)} + \mathrm{NO_3}^{-}.$$

As presented by Gardner (2012) who inspected in great detail the capacity of a haeme protein for dioxygenation of nitric oxide, the conditions for enzymatic nitric oxide dioxygenase (NOD) activity are (1) a high oxygen affinity, (2) a mechanism for decreasing nitric oxide binding to haeme, (3) a superoxide radical-like character of the bound oxygen, (4) a protected pocket for the peroxynitrite intermediate, (5) an O-atom isomerisation mechanism, (6) a mechanism for nitrate egress, and (7) a mechanism for univalent reduction. *In vitro* studies can verify that conditions 1 to 6 are satisfied. *In vivo* condition 7 amounts to the identification of a suitable reductase and has been problematic except for FHbs.

THB1 is capable of NOD activity *in vitro* (Johnson et al., 2014; Sanz-Luque et al., 2015). This by itself, however, is unremarkable, as most haemoglobins can perform the conversion of nitric oxide to nitrate (Gardner, 2005). The established control of NR and THB1 expression by the NIT2 transcription factor raises the attractive possibility that NR,

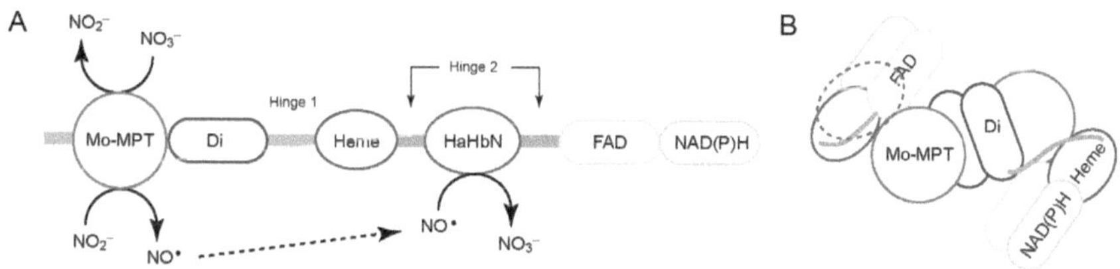

Figure 15 (A) The architecture of *Heterosigma akashiwo* nitrate reductase (Stewart & Coyne, 2011). Nitrate reduction takes place at the molydopterin (Mo-MPT) site. Di is a dimerisation domain; haeme indicates the active site of a *b* cytochrome subunit. FAD and NAD(P)H are in the C-terminal domain of the protein. A globin domain (HaHbN) is present between the FAD domain and the cytochrome domain. The Mo-MPT cofactor is thought to reduce nitrite, producing nitric oxide recycled to nitrate by HaHbN. (B) Possible tertiary structure of the nitrate reductase. HaHbN is outlined in dashed red (dark grey in the print version) lines.

which perform the two-electron reduction of nitrate to nitrite, is responsible for the recycling of THB1 (Sanz-Luque et al., 2015). In one instance described in Section 2.2.2.2, a TrHb1 domain is an integral part of an NR, suggesting that the electrons can indeed be diverted to the ferric globin in order to carry out multiple turnovers of nitric oxide dioxygenation. A representation of this scenario is shown in Fig. 15.

4.2.4 Nitrite Reduction

Under anoxic conditions, some globins are capable of reducing nitrite (Gladwin, Grubina, & Doyle, 2009; Igamberdiev, Bykova, Shah, & Hill, 2010; Sturms, DiSpirito, & Hargrove, 2011; Tejero & Gladwin, 2014). The process can be summarised as

$$\mathrm{Fe(II)}{-}\mathrm{NO_2}^- + \mathrm{H}^+ \rightarrow \mathrm{Fe(III)} + {}^{\bullet}\mathrm{NO} + \mathrm{OH}^-$$

In contrast to the production of nitric oxide by NOS (which is in question for most algae), haemoglobin reduction of nitrite occurs anaerobically and could take place in the dark. Thus, the *in vitro* production of nitric oxide by some THBs may have relevance to nitrogen metabolism. THB1, THB2, and THB4 have been shown to perform this reaction *in vitro* (Ciaccio et al., 2015).

4.3 Other Reactions

Hydroxylamine processing is a recent addition to the family of reactions considered for globins (Sturms, DiSpirito, Fulton, & Hargrove, 2011). The reaction produces ammonia:

$$2Fe(II) + NH_2OH + 2H^+ + 2e^- \rightarrow 2Fe(III) + NH_3 + H_2O$$

In *C. reinhardtii*, and presumably in other chlorophytes, hydroxylamine is an intermediate in the reduction of nitrite by the ferredoxin-dependent nitrite reductase (Hirasawa et al., 2010). If the nitrite reductase 'leaks' this intermediate under hypoxic conditions, it is conceivable that a globin could intervene. Cyanobacterial and plant haemoglobins, for chemical reasons that are yet to be elucidated, seem especially capable of catalysing this reaction. Of note is the requirement for the transfer of two electrons whereas monomeric globins can transfer only one. Because hydroxylamine reduction has been tested with a pair of globins from plant (rice) and cyanobacterial (*Synechocystis*), both of which are *bis*-histidine, it is possible that this particular ligation scheme favours consecutive electron transfers (Sturms, DiSpirito, Fulton, et al., 2011).

The various nitrogen oxides and derivatives have a rich chemistry that has not been entirely explored in algal systems. In addition to the species mentioned above, other transient forms may play an active role in cellular metabolism. They include HNO, NO^+, N_2O_3, and N_2O (Bowman et al., 2011; Igamberdiev et al., 2010).

4.4 Moving Forward with *C. reinhardtii*

4.4.1 Physiological Characterisation

The study of haemoglobin in algae is undergoing a surge of interest. In part, this is being driven by the recent appreciation that these proteins are versatile agents in the maintenance of ROS, RNS, and redox status of the cell, shadowing catalases, NOSs, peroxidases, and perhaps other dedicated enzymes. In part, it is also driven by the adoption of *C. reinhardtii* as an all-purpose model organism. One aspect specific to *C. reinhardtii* is the choice of 'wild-type' strain. CC124 and CC125 (sometimes referred to 137c) strains are often used to represent the wild-type *C. reinhardtii*. However, both of these strains were recognised early on to have a set of mutations within the nitrogen assimilation pathway, the *nit1−nit2−* mutation set, that prevents the cells from growing in medium using nitrate as a nitrogen source. The commonly used Tris–Acetate–Phosphate (TAP) medium contains ammonium chloride as the sole nitrogen source, negating the impact of the *nit1−nit2−* mutations in most experiments. Both *THB1* and *THB2* genes are under the regulatory control of the *NIT2* gene (Sanz-Luque et al., 2015) so that the use of a *nit2−* background prevents the full complement of haemoglobins to be expressed in these cells. Also, a strain that carries the *nit1−* mutation is by definition defective in the product of the *NIT1* gene, the nitrate reductase. Multiple researchers have now determined that NR is a major source, and under some conditions the only source, of endogenous nitric oxide for *C. reinhardtii* (Sakihama et al., 2002; Wei et al., 2014). Given the evidence pointing to the role of *C. reinhardtii* THBs in nitric oxide metabolism, a cell devoid of NR activity should have altered nitric oxide metabolism.

Finally, the TAP medium itself may impact *C. reinhardtii*'s behaviour with respect to haemoglobin function. The source of nitrogen in the growth medium is an intimate component of the metabolic landscape of the alga. The origins of nitric oxide in *C. reinhardtii* are still not well understood under different growth conditions, but it is generally agreed that endogenous nitric oxide is generated by very few mechanisms (Hemschemeier et al., 2013; Liping, Hongbo, Xiaohua, & Zhaopu, 2013; Wei et al., 2014; Yordanova et al., 2010). One of those is the reduction of nitrite by a variety of cellular means (including THBs potentially) or non-enzymatically by acidification below pH 5. Because TAP medium uses only ammonium as a nitrogen source and *C. reinhardtii* is not equipped for ammonium oxidation, it would appear that TAP-grown cells would contain little to no nitrite. Other pathways for nitric oxide production may exist, possibly involving an NOS as in other organisms. However, NOS-type nitric oxide production is an obligate aerobic reaction requiring one molecule of oxygen per nitric oxide molecule created. Nitric oxide production in *C. reinhardtii* has been linked to hypoxic stress (Hemschemeier et al., 2013); thus, the metabolic and catabolic path of nitric oxide remains an important question under both anaerobic and aerobic conditions.

4.4.2 Physical Characterisation

The study of recombinant haemoglobins can provide complementary insight into physiological properties. To be most useful, these proteins should be proven to mimic the globins as found in the cell. This could entail producing material devoid of tags and with proper post-translational modifications, a serious practical challenge. As pointed out, when an activity is detected *in vitro* (e.g. nitric oxide dioxygenation), it is not sufficient evidence to infer that this activity is relevant to the cell. However, the reverse may be significant provided that the conditions are reasonably close to physiological. For example, a *bis*-histidine protein incapable of exogenous ligand binding in the test tube is an unlikely candidate for dioxygenation reactions and perhaps a better one for electron transfer. At this time, additional information is required to formulate sound functional hypotheses for all THBs. Redox potential measurements, affinity for various metabolites in the ferric and ferrous states, kinetics of binding, and structural information are all essential components of the studies needed to obtain a clearer picture of the multiple roles of algal globins.

5. PERSPECTIVES

Most of our current knowledge about haemoglobins in algae comes from recent work in a single organism, *C. reinhardtii*. A possible lesson from this species is the prevalence of haemoglobins, with evidence for as many as 12 different types in use by the cell. It may be that we cannot expect a single function for haemoglobins, but rather a host of supporting roles, with each form of haemoglobin possessing unique attributes that limit it to perform one specific job under specific conditions.

Algal haemoglobin functions appear most closely related to nitric oxide chemistry. Externally, nitric oxide could be a sensitive messenger signalling the condition of an intimate symbiont or a weapon deployed during toxic red tides. Internally, it may be a by-product of normal metabolism during the assimilation of nutrients or a by-product of cellular machinery gone awry during cell stress. Unchecked, nitric oxide can nitrosylate indiscriminately or form highly reactive molecules like peroxinitrite, but the cell has evolved several control mechanisms, captured in Fig. 16, to manage nitrate, nitrite, and nitric oxide. Haemoglobins are potentially involved in these processes, which are excellent subjects for investigation.

It is interesting that each of the possibilities outlined in Fig. 16 could exist for most of the species of algae described within this review. The survey of algal haemoglobins has not presented any coherent order to the types, numbers, or distribution of haemoglobins within different groups of algae, nor even within different species of the same genus. But widely separated species of algae still contain haemoglobins with similar structure and perceived function, despite long evolutionary separation. If there are common physiologic needs for a haemoglobin in algae, why is there such a diversity of haemoglobin across species? Many species seem to have only one haemoglobin gene. Are these haemoglobins more versatile? Or are the cellular needs for a haemoglobin less stringent? Coordinated efforts from disciplines spanning genetics, biochemistry, chemistry, and biophysics will be required to provide the answers.

Algae are a collection of diverse species, a major force in driving ecosystems. They exist everyday under conditions of stress caused by nutrient

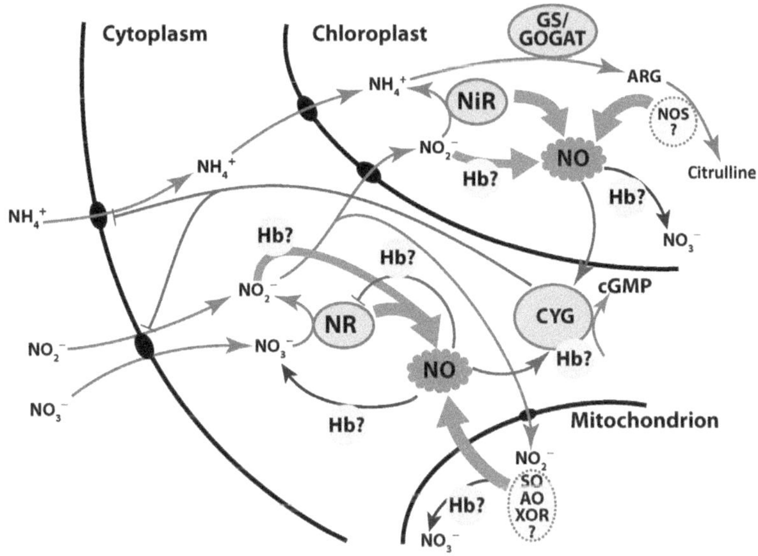

Figure 16 Possible roles for haemoglobins within algal cells (de Montaigu, Sanz-Luque, Galvan, & Fernandez, 2010; Wei et al., 2014). Hypothetical involvement of haemoglobins (yellow (light grey in the print version) circles) in relation to either nitric oxide (NO) metabolism or the nitrogen assimilation pathways is represented schematically. Grey lines follow the path of metabolites through various compartments of a hypothetical algal cell. Proteins possibly involved with algal haemoglobins are shown in blue (light grey in the print version), with hypothetical or unknown proteins shown with dotted outlines. Red (dark grey in the print version) lines denote possible regulatory influences; green (dark grey in the print version) line show possible routes for nitric oxide synthesis; and blue (dark grey in the print version) lines possible detoxification activity of haemoglobins within the cell. NR, nitrate reductase; NiR, nitrite reductase; NOS, hypothetical nitric oxide synthase; CYG, guanylate cyclase; GS/GOGAT, glutamine synthetase/glutamine oxoglutarate aminotransferase; SO, sulphite oxidase; AO, aldehyde oxidase; XOR, xanthine oxidase-dehydrogenase.

deficiencies, temperature, and light intensity fluctuations. The metabolism of nitric oxide is not one of the primary metabolic pathways normally considered, but it is this balanced system, of which haemoglobins are almost surely a part, that allows algae to function effectively in the uncertain environments they populate. Our understanding of haemoglobins helps define not just how these organisms function, but how they survive in their daily melee.

ACKNOWLEDGEMENT

This work was supported in part by the National Science Foundation, grant MCB 1330488.

REFERENCES

Andersen, R. A. (2004). Biology and systematics of heterokont and haptophyte algae. *American Journal of Botany*, *91*, 1508–1522.

Archibald, J. M., & Keeling, P. J. (2002). Recycled plastids: A 'green movement' in eukaryotic evolution. *Trends in Genetics*, *18*, 577–584.

Bhattacharya, D., Price, D. C., Chan, C. X., Qiu, H., Rose, N., Ball, S., et al. (2013). Genome of the red alga *Porphyridium purpureum*. *Nature Communications*, *4*, 1941.

Biasini, M., Bienert, S., Waterhouse, A., Arnold, K., Studer, G., Schmidt, T., et al. (2014). SWISS-MODEL: Modelling protein tertiary and quaternary structure using evolutionary information. *Nucleic Acids Research*, *42*, W252–W258.

Blanc, G., Duncan, G., Agarkova, I., Borodovsky, M., Gurnon, J., Kuo, A., et al. (2010). The *Chlorella variabilis* NC64A genome reveals adaptation to photosymbiosis, coevolution with viruses, and cryptic sex. *Plant Cell*, *22*, 2943–2955.

Bock, R. (2015). Engineering plastid genomes: Methods, tools, and applications in basic research and biotechnology. *Annual Review of Plant Biology*, *66*, 211–241.

Bolognesi, M., Bordo, D., Rizzi, M., Tarricone, C., & Ascenzi, P. (1997). Nonvertebrate hemoglobins: Structural bases for reactivity. *Progress in Biophysics and Molecular Biology*, *68*, 29–68.

Bonamore, A., Ilari, A., Giangiacomo, L., Bellelli, A., Morea, V., & Boffi, A. (2005). A novel thermostable hemoglobin from the actinobacterium *Thermobifida fusca*. *FEBS Journal*, *272*, 4189–4201.

Bowman, L. A. H., McLean, S., Poole, R. K., & Fukuto, J. M. (2011). The diversity of microbial responses to nitric oxide and agents of nitrosative stress: Close cousins but not identical twins. *Advances in Microbial Physiology*, *59*, 135–219.

Boxer, S. G., & Wright, K. A. (1979). Preparation and properties of a chlorophyllide-apomyoglobin complex. *Journal of the American Chemical Society*, *101*, 6791–6794.

Brantley, R. E., Jr., Smerdon, S. J., Wilkinson, A. J., Singleton, E. W., & Olson, J. S. (1993). The mechanism of autooxidation of myoglobin. *Journal of Biological Chemistry*, *268*, 6995–7010.

Brown, R. M., Larson, D. A., & Bold, H. C. (1964). Airborne algae: Their abundance and heterogeneity. *Science*, *143*, 583–585.

Bryant, D. A., & Frigaard, N. U. (2006). Prokaryotic photosynthesis and phototrophy illuminated. *Trends in Microbiology*, *14*, 488–496.

Ciaccio, C., Ocaña-Calahorro, F., Droghetti, E., Tundo, G. R., Sanz-Luque, E., Polticelli, F., et al. (2015). Functional and spectroscopic characterization of *Chlamydomonas reinhardtii* truncated hemoglobin. *PLoS One*, *10*, e0125005.

Corpas, F. J., & Barroso, J. B. (2015). Nitric oxide from a "green" perspective. *Nitric Oxide*, *45*, 15–19.

Couture, M., Chamberland, H., St-Pierre, B., Lafontaine, J., & Guertin, M. (1994). Nuclear genes encoding chloroplast hemoglobins in the unicellular green alga *Chlamydomonas eugametos*. *Molecular Genetics and Genomics*, *243*, 185–197.

Couture, M., Das, T. K., Lee, H. C., Peisach, J., Rousseau, D. L., Wittenberg, B. A., et al. (1999). *Chlamydomonas* chloroplast ferrous hemoglobin. Heme pocket structure and reactions with ligands. *Journal of Biological Chemistry*, *274*, 6898–6910.

Couture, M., Das, T. K., Savard, P. Y., Ouellet, Y., Wittenberg, J. B., Wittenberg, B. A., et al. (2000). Structural investigations of the hemoglobin of the cyanobacterium *Synechocystis* PCC 6803 reveal a unique distal heme pocket. *European Journal of Biochemistry*, *267*, 4770–4780.

Couture, M., & Guertin, M. (1996). Purification and spectroscopic characterization of a recombinant chloroplastic hemoglobin from the green unicellular alga *Chlamydomonas eugametos*. *European Journal of Biochemistry*, *242*, 779–787.

Cross, F. R., & Umen, J. G. (2015). The *Chlamydomonas* cell cycle. *The Plant Journal, 82,* 370–392.

Curtis, B. A., Tanifuji, G., Burki, F., Gruber, A., Irimia, M., Maruyama, S., et al. (2012). Algal genomes reveal evolutionary mosaicism and the fate of nucleomorphs. *Nature, 492,* 59–65.

Das, T. K., Couture, M., Lee, H. C., Peisach, J., Rousseau, D. L., Wittenberg, B. A., et al. (1999). Identification of the ligands to the ferric heme of *Chlamydomonas* chloroplast hemoglobin: Evidence for ligation of tyrosine-63(B10) to the heme. *Biochemistry, 38,* 15360–15368.

Davidson, K., Gowen, R., Tett, P., Bresnan, E., McKinney, A., Harrison, P., et al. (2012). Harmful algal blooms: How strong is the evidence that nutrient ratios and forms influence their occurrence? *Estuarine, Coastal and Shelf Science, 115,* 399–413.

Davy, S. K., Allemand, D., & Weis, V. M. (2012). Cell biology of cnidarian-dinoflagellate symbiosis. *Microbiology and Molecular Biology Reviews, 76,* 229–261.

Delwiche, C. F. (1999). Tracing the thread of plastid diversity through the tapestry of life. *The American Naturalist, 154,* S164–S177.

de Montaigu, A., Sanz-Luque, E., Galvan, A., & Fernandez, E. (2010). A soluble guanylate cyclase mediates negative signaling by ammonium on expression of nitrate reductase in *Chlamydomonas*. *Plant Cell, 22,* 1532–1548.

Dordas, C. (2009). Nonsymbiotic hemoglobins and stress tolerance in plants. *Plant Science, 176,* 433–440.

Dosztanyi, Z., Csizmok, V., Tompa, P., & Simon, I. (2005). IUPred: Web server for the prediction of intrinsically unstructured regions of proteins based on estimated energy content. *Bioinformatics, 21,* 3433–3434.

Erickson, E., Wakao, S., & Niyogi, K. K. (2015). Light stress and photoprotection in *Chlamydomonas reinhardtii*. *The Plant Journal, 82,* 449–465.

Foresi, N., Correa-Aragunde, N., Parisi, G., Calo, G., Salerno, G., & Lamattina, L. (2010). Characterization of a nitric oxide synthase from the plant kingdom: NO generation from the green alga *Ostreococcus tauri* is light irradiance and growth phase dependent. *Plant Cell, 22,* 3816–3830.

Fujita, Y., Tsujimoto, R., & Aoki, R. (2015). Evolutionary aspects and regulation of tetrapyrrole biosynthesis in cyanobacteria under aerobic and anaerobic environments. *Life (Basel), 5,* 1172–1203.

Gagné, G., & Guertin, M. (1992). The early genetic response to light in the green unicellular alga *Chlamydomonas eugametos* grown under light/dark cycles involves genes that represent direct responses to light and photosynthesis. *Plant Molecular Biology, 18,* 429–445.

Gardner, P. R. (2005). Nitric oxide dioxygenase function and mechanism of flavohemoglobin, hemoglobin, myoglobin and their associated reductases. *Journal of Inorganic Biochemistry, 99,* 247–266.

Gardner, P. R. (2012). Hemoglobin, a nitric-oxide dioxygenase. *Scientifica, 2012,* 683729.

Giordano, D., Pesce, A., Boechi, L., Bustamante, J. P., Caldelli, E., Howes, B. D., et al. (2015). Structural flexibility of the heme cavity in the cold-adapted truncated hemoglobin from the Antarctic marine bacterium *Pseudoalteromonas haloplanktis* TAC125. *FEBS Journal, 282,* 2948–2965.

Gladwin, M. T., Grubina, R., & Doyle, M. P. (2009). The new chemical biology of nitrite reactions with hemoglobin: R-state catalysis, oxidative denitrosylation, and nitrite reductase/anhydrase. *Accounts of Chemical Research, 42,* 157–167.

Goodrich, L. E., Paulat, F., Praneeth, V. K., & Lehnert, N. (2010). Electronic structure of heme-nitrosyls and its significance for nitric oxide reactivity, sensing, transport, and toxicity in biological systems. *Inorganic Chemistry, 49,* 6293–6316.

Gopalasubramaniam, S. K., Kovacs, F., Violante-Mota, F., Twigg, P., Arredondo-Peter, R., & Sarath, G. (2008). Cloning and characterization of a caesalpinoid (*Chamaecrista fasciculata*) hemoglobin: The structural transition from a nonsymbiotic hemoglobin to a leghemoglobin. *Proteins: Structure, Function, and Bioinformatics*, *72*, 252–260.

Gruber, H., Goetinck, S. D., Kirk, D. L., & Schmitt, R. (1992). The nitrate reductase-encoding gene of *Volvox carteri*: Map location, sequence and induction kinetics. *Gene*, *120*, 75–83.

Gupta, K. J., Hebelstrup, K. H., Mur, L. A. J., & Igamberdiev, A. U. (2011). Plant hemoglobins: Important players at the crossroads between oxygen and nitric oxide. *FEBS Letters*, *585*, 3843–3849.

Hallmann, A. (2011). Evolution of reproductive development in the volvocine algae. *Sexual Plant Reproduction*, *24*, 97–112.

Harris, E. H. (2001). *Chlamydomonas* as a model organism. *Annual Review of Plant Physiology and Plant Molecular Biology*, *52*, 363–406.

Harris, E. H. (2009). Introduction to *Chlamydomonas* and its laboratory uses. In *The Chlamydomonas sourcebook*; *Vol. 1*. New York: Academic Press.

Heinrich, T. A., da Silva, R. S., Miranda, K. M., Switzer, C. H., Wink, D. A., & Fukuto, J. M. (2013). Biological nitric oxide signalling: Chemistry and terminology. *British Journal of Pharmacology*, *169*, 1417–1429.

Hemschemeier, A., Duner, M., Casero, D., Merchant, S. S., Winkler, M., & Happe, T. (2013). Hypoxic survival requires a 2-on-2 hemoglobin in a process involving nitric oxide. *Proceedings of the National Academy of Sciences of the United States of America*, *110*, 10854–10859.

Herron, M. D., Hackett, J. D., Aylward, F. O., & Michod, R. E. (2009). Triassic origin and early radiation of multicellular volvocine algae. *Proceedings of the National Academy of Sciences of the United States of America*, *106*, 3254–3258.

Hill, R. D. (2012). Non-symbiotic haemoglobins—What's happening beyond nitric oxide scavenging? *AoB Plants*, *2012*, pls004.

Hirasawa, M., Tripathy, J. N., Sommer, F., Somasundaram, R., Chung, J.-S., Nestander, M., et al. (2010). Enzymatic properties of the ferredoxin-dependent nitrite reductase from *Chlamydomonas reinhardtii*. Evidence for hydroxylamine as a late intermediate in ammonia production. *Photosynthesis Research*, *103*, 67–77.

Holland, H. D. (2006). The oxygenation of the atmosphere and oceans. *Philosophical Transactions of the Royal Society, B: Biological Sciences*, *361*, 903–915.

Hoy, J. A., & Hargrove, M. S. (2008). The structure and function of plant hemoglobins. *Plant Physiology and Biochemistry*, *46*, 371–379.

Hoy, J. A., Kundu, S., Trent, J. T., 3rd., Ramaswamy, S., & Hargrove, M. S. (2004). The crystal structure of *Synechocystis* hemoglobin with a covalent heme linkage. *Journal of Biological Chemistry*, *279*, 16535–16542.

Hughes, M. N. (2008). Chemistry of nitric oxide and related species. In K. P. Robert (Ed.), *Methods in Enzymology: Vol. 436.* (pp. 3–19). USA: Academic Press.

Huwald, D., Schrapers, P., Kositzki, R., Haumann, M., & Hemschemeier, A. (2015). Characterization of unusual truncated hemoglobins of *Chlamydomonas reinhardtii* suggests specialized functions. *Planta*, *242*, 167–185.

Igamberdiev, A. U., Bykova, N. V., Shah, J. K., & Hill, R. D. (2010). Anoxic nitric oxide cycling in plants: Participating reactions and possible mechanisms. *Physiologia Plantarum*, *138*, 393–404.

Igamberdiev, A. U., Ratcliffe, R. G., & Gupta, K. J. (2014). Plant mitochondria: Source and target for nitric oxide. *Mitochondrion*, *19*(Pt. B), 329–333.

Igarashi, J., Kobayashi, K., & Matsuoka, A. (2011). A hydrogen-bonding network formed by the B10-E7-E11 residues of a truncated hemoglobin from *Tetrahymena pyriformis* is critical for stability of bound oxygen and nitric oxide detoxification. *Journal of Biological Inorganic Chemistry*, *16*, 599–609.

Ilari, A., Bonamore, A., Farina, A., Johnson, K. A., & Boffi, A. (2002). The X-ray structure of ferric *Escherichia coli* flavohemoglobin reveals an unexpected geometry of the distal heme pocket. *Journal of Biological Chemistry, 277*, 23725–23732.

Isogai, Y., & Ishida, M. (2009). Design of a novel heme protein with a non-heme globin scaffold. *Biochemistry, 48*, 8136–8142.

Jeong, H. J. (2011). Mixotrophy in red tide algae raphidophytes. *Journal of Eukaryotic Microbiology, 58*, 215–222.

Jinkerson, R. E., & Jonikas, M. C. (2015). Molecular techniques to interrogate and edit the *Chlamydomonas* nuclear genome. *The Plant Journal, 82*, 393–412.

Johnson, E. A., & Lecomte, J. T. (2013). The globins of cyanobacteria and algae. *Advances in Microbial Physiology, 63*, 195–272.

Johnson, E. A., & Lecomte, J. T. J. (2014). Characterization of the truncated hemoglobin THB1 from protein extracts of *Chlamydomonas reinhardtii*. *F1000Research, 3*, 294.

Johnson, E. A., Rice, S. L., Preimesberger, M. R., Nye, D. B., Gilevicius, L., Wenke, B. B., et al. (2014). Characterization of THB1, a *Chlamydomonas reinhardtii* truncated hemoglobin: Linkage to nitrogen metabolism and identification of lysine as the distal heme ligand. *Biochemistry, 53*, 4573–4589.

Kabsch, W., & Sander, C. (1983). Dictionary of protein secondary structure: Pattern recognition of hydrogen-bonded and geometrical features. *Biopolymers, 22*, 2577–2637.

Kachalova, G. S., Popov, A. N., & Bartunik, H. D. (1999). A steric mechanism for inhibition of CO binding to heme proteins. *Science, 284*, 473–476.

Kakar, S., Hoffman, F. G., Storz, J. F., Fabian, M., & Hargrove, M. S. (2010). Structure and reactivity of hexacoordinate hemoglobins. *Biophysical Journal, 152*, 1–14.

Keilin, D., & Ryley, J. F. (1953). Haemoglobin in protozoa. *Nature, 172*, 451–452.

Kiger, L., Tilleman, L., Geuens, E., Hoogewijs, D., Lechauve, C., Moens, L., et al. (2011). Electron transfer function versus oxygen delivery: A comparative study for several hexacoordinated globins across the animal kingdom. *PLoS One, 6*, e20478.

Kim, D., Kang, Y. S., Lee, Y., Yamaguchi, K., Matsuoka, K., Lee, K.-W., et al. (2008). Detection of nitric oxide (NO) in marine phytoplankters. *Journal of Bioscience and Bioengineering, 105*, 414–417.

Kornas, A., Kuźniak, E., Ślesak, I., & Miszalski, Z. (2010). The key role of the redox status in regulation of metabolism in photosynthesizing organisms. *Acta Biochimica Polonica, 57*, 143–151.

Kumar, A., Castellano, I., Patti, F. P., Palumbo, A., & Buia, M. C. (2015). Nitric oxide in marine photosynthetic organisms. *Nitric Oxide, 47*, 34–39.

Landfried, D. A., Vuletich, D. A., Pond, M. P., & Lecomte, J. T. J. (2007). Structural and thermodynamic consequences of *b* heme binding for monomeric apoglobins and other apoproteins. *Gene, 398*, 12–28.

Layer, G., Reichelt, J., Jahn, D., & Heinz, D. W. (2010). Structure and function of enzymes in heme biosynthesis. *Protein Science, 19*, 1137–1161.

Lechtreck, K. F., Johnson, E. C., Sakai, T., Cochran, D., Ballif, B. A., Rush, J., et al. (2009). The *Chlamydomonas reinhardtii* BBSome is an IFT cargo required for export of specific signaling proteins from flagella. *Journal of Cell Biology, 187*, 1117–1132.

Lesk, A. M., & Chothia, C. (1980). How different amino acid sequences determine similar protein structures: The structure and evolutionary dynamics of the globins. *Journal of Molecular Biology, 136*, 225–270.

Letunic, I., & Bork, P. (2007). Interactive Tree Of Life (iTOL): An online tool for phylogenetic tree display and annotation. *Bioinformatics, 23*, 127–128.

Letunic, I., & Bork, P. (2011). Interactive Tree Of Life v2: Online annotation and display of phylogenetic trees made easy. *Nucleic Acids Research, 39*, W475–W478.

Linding, R., Jensen, L. J., Diella, F., Bork, P., Gibson, T. J., & Russell, R. B. (2003). Protein disorder prediction: Implications for structural proteomics. *Structure, 11*, 1453–1459.

Liping, Z., Hongbo, S., Xiaohua, L., & Zhaopu, L. (2013). Gene regulation of iron-deficiency responses is associated with carbon monoxide and heme oxydase 1 in *Chlamydomonas reinhardtii*. *PLoS One*, *8*, e53835.

Lórenz-Fonfría, V. A., & Heberle, J. (2014). Channelrhodopsin unchained: Structure and mechanism of a light-gated cation channel. *Biochimica et Biophysica Acta—Bioenergetics*, *1837*, 626–642.

Ludwig, M., & Bryant, D. A. (2012). Acclimation of the global transcriptome of the cyanobacterium *Synechococcus* sp. strain PCC 7002 to nutrient limitations and different nitrogen sources. *Frontiers in Microbiology*, *3*, 145.

Marti, M. A., Crespo, A., Capece, L., Boechi, L., Bikiel, D. E., Scherlis, D. A., et al. (2006). Dioxygen affinity in heme proteins investigated by computer simulation. *Journal of Inorganic Biochemistry*, *100*, 761–770.

Merchant, S. S., Prochnik, S. E., Vallon, O., Harris, E. H., Karpowicz, S. J., Witman, G. B., et al. (2007). The *Chlamydomonas* genome reveals the evolution of key animal and plant functions. *Science*, *318*, 245–250.

Milani, M., Pesce, A., Ouellet, Y., Dewilde, S., Friedman, J., Ascenzi, P., et al. (2004). Heme-ligand tunneling in group I truncated hemoglobins. *Journal of Biological Chemistry*, *279*, 21520–25125.

Moore, C. E., & Archibald, J. M. (2009). Nucleomorph genomes. *Annual Review of Genetics*, *43*, 251–264.

Mukhi, N., Dhindwal, S., Uppal, S., Kumar, P., Kaur, J., & Kundu, S. (2013). X-ray-crystallographic structural characteristics of *Arabidopsis* hemoglobin I and their functional implications. *Biochimica et Biophysica Acta—Proteins and Proteomics*, *1834*, 1944–1956.

Murray, J. W., Delumeau, O., & Lewis, R. J. (2005). Structure of a nonheme globin in environmental stress signaling. *Proceedings of the National Academy of Sciences of the United States of America*, *102*, 17320–17325.

Muscatine, L., Goiran, C., Land, L., Jaubert, J., Cuif, J. P., & Allemand, D. (2005). Stable isotopes ($\delta^{13}C$ and $\delta^{15}N$) of organic matrix from coral skeleton. *Proceedings of the National Academy of Sciences of the United States of America*, *102*, 1525–1530.

Nardini, M., Pesce, A., Milani, M., & Bolognesi, M. (2007). Protein fold and structure in the truncated (2/2) globin family. *Gene*, *398*, 2–11.

Nixon, P. J. (2000). Chlororespiration. *Philosophical Transactions B*, *355*, 1541–1547.

Noordally, Z. B., & Millar, A. J. (2015). Clocks in algae. *Biochemistry*, *54*, 171–183.

Nothnagel, H. J., Preimesberger, M. R., Pond, M. P., Winer, B. Y., Adney, E. M., & Lecomte, J. T. J. (2011). Chemical reactivity of *Synechococcus* sp. PCC 7002 and *Synechocystis* sp. PCC 6803 hemoglobins: Covalent heme attachment and bishistidine coordination. *Journal of Biological Inorganic Chemistry*, *16*, 539–552.

Obornik, M., & Green, B. R. (2005). Mosaic origin of the heme biosynthesis pathway in photosynthetic eukaryotes. *Molecular Biology and Evolution*, *22*, 2343–2353.

Oldfield, C. J., & Dunker, A. K. (2014). Intrinsically disordered proteins and intrinsically disordered protein regions. *Annual Review of Biochemistry*, *83*, 553–584.

Orsini, M., Costelli, C., Malavasi, V., Cusano, R., Concas, A., Angius, A., et al. (2015). Complete sequence and characterization of mitochondrial and chloroplast genome of *Chlorella variabilis* NC64A. *Mitochondrial DNA*, 1–3.

Ouellet, Y., Milani, M., Couture, M., Bolognesi, M., & Guertin, M. (2006). Ligand interactions in the distal heme pocket of *Mycobacterium tuberculosis* truncated hemoglobin N: Roles of TyrB10 and GlnE11 residues. *Biochemistry*, *45*, 8770–8781.

Ozaki, S.-i., Roach, M. P., Matsui, T., & Watanabe, Y. (2001). Investigations of the roles of the distal heme environment and the proximal heme iron ligand in peroxide activation by heme enzymes via molecular engineering of myoglobin. *Accounts of Chemical Research*, *34*, 818–825.

Paracer, S., & Ahmadjian, V. (2000). *Symbiosis: An introduction to biological associations*. New York: Oxford University Press.

Pastore, A., & Lesk, A. M. (1990). Comparison of the structures of globins and phycocyanins: Evidence for evolutionary relationship. *Proteins, 8*, 133–155.

Perez, S., & Weis, V. (2006). Nitric oxide and cnidarian bleaching: An eviction notice mediates breakdown of a symbiosis. *Journal of Experimental Biology, 209*, 2804–2810.

Perutz, M. F. (1979). Regulation of oxygen affinity of hemoglobin: Influence of structure of the globin on the heme iron. *Annual Review of Biochemistry, 48*, 327–386.

Pesce, A., Couture, M., Dewilde, S., Guertin, M., Yamauchi, K., Ascenzi, P., et al. (2000). A novel two-over-two α-helical sandwich fold is characteristic of the truncated hemoglobin family. *EMBO Journal, 19*, 2424–2434.

Pettersen, E. F., Goddard, T. D., Huang, C. C., Couch, G. S., Greenblatt, D. M., Meng, E. C., et al. (2004). UCSF Chimera—A visualization system for exploratory research and analysis. *Journal of Computational Chemistry, 25*, 1605–1612.

Pietri, R., León, R. G., Kiger, L., Marden, M. C., Granell, L. B., Cadilla, C. L., et al. (2006). Hemoglobin I from *Lucina pectinata*: A model for distal heme-ligand control. *Biochimica et Biophysica Acta—Proteins and Proteomics, 1764*, 758–765.

Preimesberger, M. R., Pond, M. P., Majumdar, A., & Lecomte, J. T. J. (2012). Electron self exchange and self-amplified posttranslational modification in the hemoglobins from *Synechocystis* sp. PCC 6803 and *Synechococcus* sp. PCC 7002. *Journal of Biological Inorganic Chemistry, 17*, 599–609.

Prochnik, S. E., Umen, J., Nedelcu, A. M., Hallmann, A., Miller, S. M., Nishii, I., et al. (2010). Genomic analysis of organismal complexity in the multicellular green alga *Volvox carteri*. *Science, 329*, 223–226.

Pröll, S., Wilhelm, B., Robert, B., & Scheer, H. (2006). Myoglobin with modified tetrapyrrole chromophores: Binding specificity and photochemistry. *Biochimica et Biophysica Acta—Bioenergetics, 1757*, 750–763.

Read, B. A., Kegel, J., Klute, M. J., Kuo, A., Lefebvre, S. C., Maumus, F., et al. (2013). Pan genome of the phytoplankton *Emiliania* underpins its global distribution. *Nature, 499*, 209–213.

Reeder, B. J., & Hough, M. A. (2014). The structure of a class 3 nonsymbiotic plant haemoglobin from *Arabidopsis thaliana* reveals a novel N-terminal helical extension. *Acta Crystallographica D Biological Crystallography, 70*, 1411–1418.

Rice, S. L., Boucher, L. E., Schlessman, J. L., Preimesberger, M. R., Bosch, J., & Lecomte, J. T. J. (2015). Structure of *Chlamydomonas reinhardtii* THB1, a group 1 truncated hemoglobin with a rare histidine-lysine heme ligation. *Acta Crystallographica F Structural Biology Communications, F71*, 718–725.

Richards, T. A., & van der Giezen, M. (2006). Evolution of the Isd11-IscS complex reveals a single α-proteobacterial endosymbiosis for all eukaryotes. *Molecular Biology and Evolution, 23*, 1341–1344.

Rosic, N. N., Leggat, W., Kaniewska, P., Dove, S., & Hoegh-Guldberg, O. (2013). New-old hemoglobin-like proteins of symbiotic dinoflagellates. *Ecology and Evolution, 3*, 822–834.

Safavi-Hemami, H., Young, N. D., Doyle, J., Llewellyn, L., & Klueter, A. (2010). Characterisation of nitric oxide synthase in three cnidarian-dinoflagellate symbioses. *PLoS One, 5*, e10379.

Sakihama, Y., Nakamura, S., & Yamasaki, H. (2002). Nitric oxide production mediated by nitrate reductase in the green alga *Chlamydomonas reinhardtii*: An alternative NO production pathway in photosynthetic organisms. *Plant and Cell Physiology, 43*, 290–297.

Sanz-Luque, E., Ocaña-Calahorro, F., de Montaigu, A., Chamizo-Ampudia, A., Llamas, A., Galvan, A., et al. (2015). THB1, a truncated hemoglobin, modulates nitric oxide levels and nitrate reductase activity. *The Plant Journal, 81*, 467–479.

Savard, P. Y., Daigle, R., Morin, S., Sebilo, A., Meindre, F., Lagüe, P., et al. (2011). Structure and dynamics of *Mycobacterium tuberculosis* truncated hemoglobin N: Insights from NMR spectroscopy and molecular dynamics simulations. *Biochemistry*, *50*, 11121–11130.

Scaife, M. A., Nguyen, G. T. D. T., Rico, J., Lambert, D., Helliwell, K. E., & Smith, A. G. (2015). Establishing *Chlamydomonas reinhardtii* as an industrial biotechnology host. *The Plant Journal*, *82*, 532–546.

Scott, N. L., Falzone, C. J., Vuletich, D. A., Zhao, J., Bryant, D. A., & Lecomte, J. T. J. (2002). The hemoglobin of the cyanobacterium *Synechococcus* sp. PCC 7002: Evidence for hexacoordination and covalent adduct formation in the ferric recombinant protein. *Biochemistry*, *41*, 6902–6910.

Scott, N. L., & Lecomte, J. T. J. (2000). Cloning, expression, purification, and preliminary characterization of a putative hemoglobin from the cyanobacterium *Synechocystis* sp. PCC 6803. *Protein Science*, *9*, 587–597.

Scranton, M. A., Ostrand, J. T., Fields, F. J., & Mayfield, S. P. (2015). *Chlamydomonas* as a model for biofuels and bio-products production. *The Plant Journal*, *82*, 523–531.

Shen, M., Xu, J., Chiang, M. W., & Au, D. W. (2011). Unravelling the pathway of respiratory toxicity in goldlined seabream (*Rhabdosargus sarba*) induced by the harmful alga *Chattonella marina*. *Aquatic Toxicology*, *104*, 185–191.

Smith, D. R., Vinogradov, S. N., & Hoogewij, D. (2014). Hemoglobins in the genome of the cryptomonad *Guillardia theta*. *Biology Direct*, *9*, 7.

Stewart, J. J., & Coyne, K. J. (2011). Analysis of raphidophyte assimilatory nitrate reductase reveals unique domain architecture incorporating a 2/2 hemoglobin. *Plant Molecular Biology*, 77, 565–575.

Sturms, R., DiSpirito, A. A., Fulton, D. B., & Hargrove, M. S. (2011). Hydroxylamine reduction to ammonium by plant and cyanobacterial hemoglobins. *Biochemistry*, *50*, 10829–10835.

Sturms, R., DiSpirito, A. A., & Hargrove, M. S. (2011). Plant and cyanobacterial hemoglobins reduce nitrite to nitric oxide under anoxic conditions. *Biochemistry*, *50*, 3873–3878.

Tardif, M., Atteia, A., Specht, M., Cogne, G., Rolland, N., Brugiere, S., et al. (2012). PredAlgo: A new subcellular localization prediction tool dedicated to green algae. *Molecular Biology and Evolution*, *29*, 3625–3639.

Teh, A.-H., Saito, J. A., Baharuddin, A., Tuckerman, J. R., Newhouse, J. S., Kanbe, M., et al. (2011). Hell's Gate globin I: An acid and thermostable bacterial hemoglobin resembling mammalian neuroglobin. *FEBS Letters*, *585*, 3250–3258.

Tejero, J., & Gladwin, M. T. (2014). The globin superfamily: Functions in nitric oxide formation and decay. *Biological Chemistry*, *395*, 631–639.

Tilleman, L., De Henau, S., Pauwels, M., Nagy, N., Pintelon, I., Braeckman, B. P., et al. (2012). An N-myristoylated globin with a redox-sensing function that regulates the defecation cycle in *Caenorhabditis elegans*. *PLoS One*, 7, e48768.

Tischner, R., Planchet, E., & Kaiser, W. M. (2004). Mitochondrial electron transport as a source for nitric oxide in the unicellular green alga *Chlorella sorokiniana*. *FEBS Letters*, *576*, 151–155.

Traylor, T. G., Duprat, A. F., & Sharma, V. S. (1993). Nitric oxide-triggered heme-mediated hydrolysis—a possible model for biological reactions of NO. *Journal of the American Chemical Society*, *115*, 810–811.

Traylor, T. G., & Sharma, V. S. (1992). Why NO? *Biochemistry*, *31*, 2847–2849.

Trent, J. T., 3rd., Kundu, S., Hoy, J. A., & Hargrove, M. S. (2004). Crystallographic analysis of synechocystis cyanoglobin reveals the structural changes accompanying ligand binding in a hexacoordinate hemoglobin. *Journal of Molecular Biology*, *341*, 1097–1108.

Tsai, A. L., Berka, V., Martin, E., & Olson, J. S. (2012). A "sliding scale rule" for selectivity among NO, CO, and O_2 by heme protein sensors. *Biochemistry*, *51*, 172–186.

Urayama, P., Phillips, G. N., Jr., & Gruner, S. M. (2002). Probing substates in sperm whale myoglobin using high-pressure crystallography. *Structure*, *10*, 51–60.
Vázquez-Limón, C., Hoogewijs, D., Vinogradov, S. N., & Arredondo-Peter, R. (2012). The evolution of land plant hemoglobins. *Plant Science*, *191–192*, 71–81.
Vinogradov, S. N., Bailly, X., Smith, D. R., Tinajero-Trejo, M., Poole, R. K., & Hoogewijs, D. (2013). Microbial eukaryote globins. *Advances in Microbial Physiology*, *63*, 391–446.
Vinogradov, S. N., Fernandez, I., Hoogewijs, D., & Arredondo-Peter, R. (2011). Phylogenetic relationships of 3/3 and 2/2 hemoglobins in Archaeplastida genomes to bacterial and other eukaryote hemoglobins. *Molecular Plant*, *4*, 42–58.
Vinogradov, S. N., Hoogewijs, D., Bailly, X., Arredondo-Peter, R., Gough, J., Dewilde, S., et al. (2006). A phylogenomic profile of globins. *BMC Evolutionary Biology*, *6*, 31.
Vinogradov, S. N., Hoogewijs, D., Bailly, X., Mizuguchi, K., Dewilde, S., Moens, L., et al. (2007). A model of globin evolution. *Gene*, *398*, 132–142.
Vinogradov, S. N., & Moens, L. (2008). Diversity of globin function: Enzymatic, transport, storage, and sensing. *Journal of Biological Chemistry*, *283*, 8773–8777.
Vinogradov, S. N., Tinajero-Trejo, M., Poole, R. K., & Hoogewijs, D. (2013). Bacterial and archaeal globins—a revised perspective. *Biochimica et Biophysica Acta—Proteins and Proteomics*, *1834*, 1789–1800.
Vothknecht, U. C., & Westhoff, P. (2001). Biogenesis and origin of thylakoid membranes. *Biochimica et Biophysica Acta—Molecular and Cell Biology*, *1541*, 91–101.
Vuletich, D. A., & Lecomte, J. T. J. (2006). A phylogenetic and structural analysis of truncated hemoglobins. *Journal of Molecular Evolution*, *62*, 196–210.
Wang, X., & Hargrove, M. S. (2013). Nitric oxide in plants: The roles of ascorbate and hemoglobin. *PLoS One*, *8*, e82611.
Wei, L., Derrien, B., Gautier, A., Houille-Vernes, L., Boulouis, A., Saint-Marcoux, D., et al. (2014). Nitric oxide-triggered remodeling of chloroplast bioenergetics and thylakoid proteins upon nitrogen starvation in *Chlamydomonas reinhardtii*. *Plant Cell*, *26*, 353–372.
Wenke, B. B., Lecomte, J. T. J., Héroux, A., & Schlessman, J. L. (2014). The 2/2 hemoglobin from the cyanobacterium *Synechococcus* sp. PCC 7002 with covalently attached heme: Comparison of X-ray and NMR structures. *Proteins*, *82*, 528–534.
Winger, J., Derbyshire, E., Lamers, M., Marletta, M., & Kuriyan, J. (2008). The crystal structure of the catalytic domain of a eukaryotic guanylate cyclase. *BMC Structural Biology*, *8*, 42.
Wittenberg, J. B., Bolognesi, M., Wittenberg, B. A., & Guertin, M. (2002). Truncated hemoglobins: A new family of hemoglobins widely distributed in bacteria, unicellular eukaryotes and plants. *Journal of Biological Chemistry*, *277*, 871–874.
Wittenberg, J. B., Wittenberg, B. A., Peisach, J., & Blumberg, W. E. (1970). On the state of the iron and the nature of the ligand in oxyhemoglobin. *Proceedings of the National Academy of Sciences of the United States of America*, *67*, 1846–1853.
Wood, C. R., & Rosenbaum, J. L. (2015). Ciliary ectosomes: Transmissions from the cell's antenna. *Trends in Cell Biology*, *25*, 276–285.
Wright, K. A., & Boxer, S. G. (1981). Solution properties of synthetic chlorophyllide—and bacteriochlorophyllide—apomyoglobin complexes. *Biochemistry*, *20*, 7546–7556.
Yang, J., Yan, R., Roy, A., Xu, D., Poisson, J., & Zhang, Y. (2015). The I-TASSER Suite: Protein structure and function prediction. *Nature Methods*, *12*, 7–8.
Yi, J., & Richter-Addo, G. B. (2012). Unveiling the three-dimensional structure of the green pigment of nitrite-cured meat. *Chemical Communications*, *48*, 4172–4174.
Yoon, H. S., Hackett, J. D., Ciniglia, C., Pinto, G., & Bhattacharya, D. (2004). A molecular timeline for the origin of photosynthetic eukaryotes. *Molecular Biology and Evolution*, *21*, 809–818.

Yordanova, Z. P., Iakimova, E. T., Cristescu, S. M., Harren, F. J., Kapchina-Toteva, V. M., & Woltering, E. J. (2010). Involvement of ethylene and nitric oxide in cell death in mastoparan-treated unicellular alga *Chlamydomonas reinhardtii*. *Cell Biology International*, *9*, 301–308.

Yruela, I. (2013). Transition metals in plant photosynthesis. *Metallomics*, *5*, 1090–1109.

CHAPTER FIVE

The Challenging World of Biofilm Physiology

Joke Donné, Sylvia Dewilde[1]

Protein Chemistry, Proteomics and Epigenetic Signalling (PPES), Department of Biomedical Sciences, University of Antwerp, Antwerp, Belgium

[1]Corresponding author: e-mail address: sylvia.dewilde@uantwerpen.be

Contents

Abstract

Worldwide, infectious diseases are one of the leading causes of death among children. At least 65% of all infections are caused by the biofilm mode of bacterial growth. Bacteria colonise surfaces and grow as multicellular biofilm communities surrounded by a polymeric matrix as a common survival strategy. These sessile communities endow bacteria with high tolerance to antimicrobial agents and hence cause persistent and chronic bacterial infections, such as dental caries, periodontitis, otitis media, cystic fibrosis and pneumonia. The highly complex nature and the rapid adaptability of the biofilm population impede our understanding of the process of biofilm formation, but an important role for oxygen-binding proteins herein is clear. Much research on this bacterial lifestyle is already performed, from genome/proteome analysis to *in vivo* antibiotic susceptibility testing, but without significant progress in biofilm treatment or eradication. This review will present the multiple challenges of biofilm research and discuss possibilities to cross these barriers in future experimental studies.

Advances in Microbial Physiology, Volume 67
ISSN 0065-2911
http://dx.doi.org/10.1016/bs.ampbs.2015.09.003

ABBREVIATION

AB antibiotic
AHL acyl-homoserine lactone
AI auto-inducer
A-site active site
BAI biofilm-associated infections
BHI brain heart infusion
c-di-GMP cyclic di-guanosine monophosphate
CE capillary electrophoresis
CFU colony forming units
CLSM confocal laser scanner microscopy
CV crystal violet
CVCs central venous catheters
DGC diguanylate cyclase
DMMB dimethylmethylene blue
EAL motif Glutamic acid-Alanine-Leucine motif
EPS extracellular polymeric substance
FDA fluorescein diacetate
FISH fluorescence *in situ* hybridisation
GCS globin-coupled sensor
GGDEF motif Glycine-Glycine-Asparticacid-Glutamicacid-Phenylalanine motif
GGEEF motif Glycine-Glycine-Glutamicacid-Glutamicacid-Phenylalanine motif
GMP guanosine monophosphate
ICAT isotope-coded affinity tag
I-site inhibitory binding site
iTRAQ isobaric tag for relative and absolute quantitation
LB Luria–Bertani
MOPS 3-(*N*-morpholino)propanesulphonic acid
MRD modified Robbins device
MS mass spectrometry
PDE phosphodiesterase
PGA poly-β-1,6-*N*-acetyl-glucosamine
PMA propidium monoazide
ppGpp guanosine tetraphosphate
qPCR quantitative real-time PCR
QS quorum sensing
TA toxin–antitoxin pair
TOF-MS time-of-flight mass spectrometry
TSB tryptic soy broth
XTT 2,3-bis-(2-methoxy-4-nitro-5-sulphophenyl)-2*H*-tetrazolium-5-carboxanilide

1. BIOFILM STRUCTURE AND FORMATION

The investigation of natural microbial habitats verifies that most of the bacteria, perhaps 99%, are sessile and attached as biofilms to a surface. In such

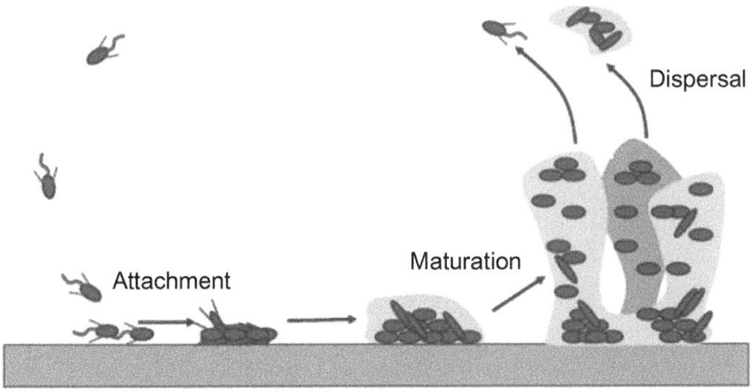

Figure 1 Steps of biofilm formation. *Aadapted from Abed, Ibnsouda, Latrache, and Hamadi (2012).*

a biofilm, the bacteria behave very differently than in a free-floating planktonic state. Furthermore, according to the U.S. National Institutes of Health, more than 80% of all microbial infections are associated with biofilm growth. The development of such a microbial biofilm is a complex and dynamic process involving several steps: approaching a surface, attachment, maturation and dispersal. All these steps are tightly regulated by complex signalling networks connecting a variety of physiological pathways such as motility, adhesion, communication by quorum sensing (QS), stress responses, virulence, matrix production and responses to antibiotic treatment. Furthermore, the ability of the biofilm to adapt continuously to changing environments and conditions, such as oxygen supply, is of utmost importance. In every stage of the biofilm formation (Fig. 1), the bacteria need to change their behaviour: from swimming to swarming, from attaching to growing and from defending to detaching.

1.1 Biofilm Structure

The three basic components of a biofilm culture are (a) microorganisms, (b) an extracellular matrix and (c) a surface. It must be noted that because of the wide range of environments where biofilms are found, it is very hard to generalise the structure and characteristics of biofilms. Many factors constantly affect the architecture of the biofilm population, from external influences such as the chemical composition of the environment, to internal factors such as the genetic profile of the microorganisms. Most species act differently in different environments, which makes biofilm formation difficult to cure and reproduce in a research setting.

1.1.1 Microorganisms

A biofilm can be formed by a single bacterial species, but in most environments, they are composed of a rich variety of species. Given the right conditions, almost all bacteria are able to form a biofilm, with different growth and stability characteristics. Also fungi, algae, yeasts and protozoa have already been observed in a biofilm mode (Cos, Tote, Horemans, & Maes, 2010). For example, most nosocomial *Candida albicans* bloodstream infections are caused by polymicrobial biofilms. Bacteria such as *Staphylococcus aureus*, *Staphylococcus epidermidis* and *Enterococcus* species are commonly co-isolated with *C. albicans*. The fungal hyphae enhance *S. epidermidis* growth, and the presence of *C. albicans* increases the vancomycin antibiotic (AB) resistance of *S. epidermidis* and *S. aureus* (Harriott & Noverr, 2011). In a polymicrobial biofilm, there is competition for nutrients and oxygen from the environment, effecting mostly the cells in the deeper layers, that are in turn longer protected from toxic agents. Nevertheless, metabolised products by one coloniser may stimulate the growth of the others present. These metabolites can even be ligands allowing the attachment of other species (Cos et al., 2010; Dunne, 2002). The variance in multispecies biofilms is discussed further in Section 4.5.

1.1.2 Matrix

The maintenance of the biofilm architecture depends mainly on the presence of the extracellular matrix. Biofilms are actually composed of more matrix (80–85%) than microorganisms (15–20%) (Dufour, Leung, & Lévesque, 2012). This extracellular slime layer surrounds and protects the microbial cells against harmful factors in the surrounding environment. The production of the polymeric matrix is a social activity wherein each bacterium of the biofilm community participates to create a protective biofilm society. There is a wide variety in the composition of the matrix, which depends especially on the bacterial species present and the surface (Pamp, Gjermansen, & Tolker-Nielsen, 2007). The biofilm extracellular polymeric substance (EPS) is a complex of secreted polysaccharides, proteins and nucleic acids from lysed cells and absorbed nutrients and ions from the surrounding area. The production of the polysaccharides and proteins of the extracellular matrix, which determine the viscosity, occurs by the bacteria themselves and is species dependent. Table 1 shows the basic matrix components of an *Escherichia coli* biofilm population (Beloin, Roux, & Ghigo, 2008) as an example. To conclude, the polymeric matrix is a very important feature in biofilm development and, for that reason, an interesting research topic. Nevertheless, its very sticky characteristic makes it difficult to analyse and separate from the microbial cells.

Table 1 *E. coli* Biofilm Matrix Components

1. Polysaccharides	
	Cellulose
	Colonic acid
	Poly-β-1,6-*N*-acetyl-glucosamine(PGA)
	Lipopolysaccharides
2. Proteins	
Multimeric cell appendages	Flagellin protein (flagellum)
	Curlin protein (curli)
	Fimbrin protein (fimbriae)
Surface proteins	Ag43 adhesin
	AidA and TibA proteins
3. Extracellular DNA	
4. Signalling molecules	Acyl-homoserine lactones (AHL)
	Others
5. Water, nutrients, antimicrobial molecules, ions, etc., from growth environment	

1.1.3 Surfaces

Biofilms can form on nearly every surface (Fig. 2), on natural surfaces like fruit and vegetables, on rocks and in soil. Biofilms also appear in the medical field where they grow on medical devices such as catheters, tubes and valves, contact lenses, in wounds or on teeth. Besides health care, biofilms also create problems in industry. They can cause food and water contamination, metal surface corrosion and clogging. Moreover, biofilm formation on for instance pipelines and fishing nets decreases the efficiency of this equipment which gives problems in every industrial setting. Consequently, biofilm development has an economical effect (Dufour et al., 2012).

1.2 Biofilm Formation

1.2.1 Different Steps in Biofilm Formation and Maturation

When free-floating planktonic bacteria experience a surface, they are able to swim towards it and attach to it. Besides passive movements due to Brownian and gravitational forces, they are also able to migrate in the direction of favourable conditions. This kind of active motility is dependent on flagellum machinery. With this active swimming mechanism, the bacterial

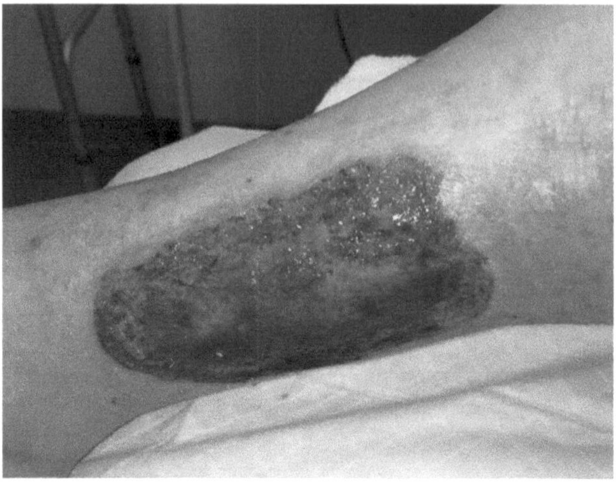

Figure 2 Biofilm accumulation in wound. An infected wound on a human leg, with a grey, slimy biofilm. *Photograph is a kindly gift from Jennifer Hurlow.*

cells are able to overcome repulsive electrostatic and hydrodynamic forces when approaching surfaces, which makes them able to interact and attach (Beloin et al., 2008). Thus, flagella are not only required for the propeller-based swimming capacity of the bacteria but also play a surface-sensing role in the process of attachment (Belas, 2014). This process involves (i) a decrease in the rotation speed of the flagella which will make the bacteria swim slower but keep them sufficiently energetic to exceed the repulsive forces of the surface, and (ii) an induction of the synthesis of cell appendages or adhesins which are necessary for adherence. Several additional factors will affect the efficiency of attachment: (i) the physicochemical properties of the substratum; (ii) the presence or absence of a conditioning film that can be formed as a coating on the substratum; (iii) the hydrodynamics of the medium which is defined by the flow velocity; (iv) the physicochemical characteristics of the medium like pH, nutrient concentration, temperature, ionic strength and viscosity; and (v) the surface properties (hydrophobicity, the presence and number of appendages, the composition of the outer membrane layer) of the bacterial cell that wants to attach (Donlan, 2002).

There are three types of cell-surface components expressed by bacteria that play a role in strengthening the bacteria-to-surface interactions: fimbriae, curli and conjugative pili. Fimbriae and curli are filamentous proteinaceous tubular structures which are present in large numbers (e.g. between 100 and 500 fimbriae) at the bacterial surface (Beloin et al., 2008). Mutants in both *fimA* and *fimH*, the genes encoding the major fimbriae subunit and the adhesin at the tip of the filament, respectively, have been reported to reduce initial attachment of *E. coli* to abiotic surfaces (Beloin et al., 2004). The conjugative pili are mainly responsible for cell–cell interactions that stabilise the structure of the biofilm. Furthermore, they are also involved in horizontal transfer of DNA and plasmids between interacting bacteria. It is suggested that these plasmids can carry determinants for biofilm initiation and possibly also for maturation, virulence and resistance (Roberts & Kreth, 2014). The attached bacteria will subsequently proliferate, grow and produce a slimy matrix. The resulting 3D biofilm community colonises the surface and forms what is assumed to be a mature dense mushroom-like population which will resist harmful conditions. The mechanisms that are involved in this protective lifestyle are discussed in more detail in Section 2. The process of maturation encompasses several pathways such as cell-to-cell adhesion, secretion of toxins and signal molecules, sensing microenvironments and environmental factors, quorum-sensing signalling, regulation of metabolism, stress responses, virulence and resistance. Finally, biofilms can spread through the detachment of cells, which can attach to downstream sites at the surface. Dispersal is a result of ageing and environmental shifts, when for instance the population density is too high, nutrient and oxygen limitations appear or when there is an increased temperature variation in the surrounding (Romeo, 2006). It is clear that this biofilm life cycle is a highly regulated process based on a complex regulatory network in order to tightly control every step of biofilm formation. The nucleotide-based second messenger, cyclic di-guanosine monophosphate (c-di-GMP), is a key molecule in this signalling network. Its concentration is controlled by oxygen-binding proteins, among others.

1.2.2 Role of c-di-GMP and Oxygen-Binding Proteins

c-di-GMP interferes in the entire spectrum of cellular processes such as adhesion, motility, virulence, QS and persistence and plays a key role in the transition between a motile single-cell planktonic lifestyle and a sessile multicellular biofilm lifestyle. The cellular c-di-GMP levels are controlled by two groups of enzymes, diguanylate cyclases (DGCs) and phosphodiesterases (PDEs). c-di-GMP is synthesised from two molecules GTP by DGCs and requires dimerisation of two intact GGDEF (Gly-Gly-Asp-Glu-Phe) or GGEEF (Gly-Gly-Glu-Glu-Phe) motifs in the active site (A-site) of the functional domain. An inhibitory binding site (I-site) in close proximity to the A-site is responsible for allosteric product inhibition in these DGCs (Kalia et al., 2012; Pesavento & Hengge, 2009) (Fig. 3A). This I-site is

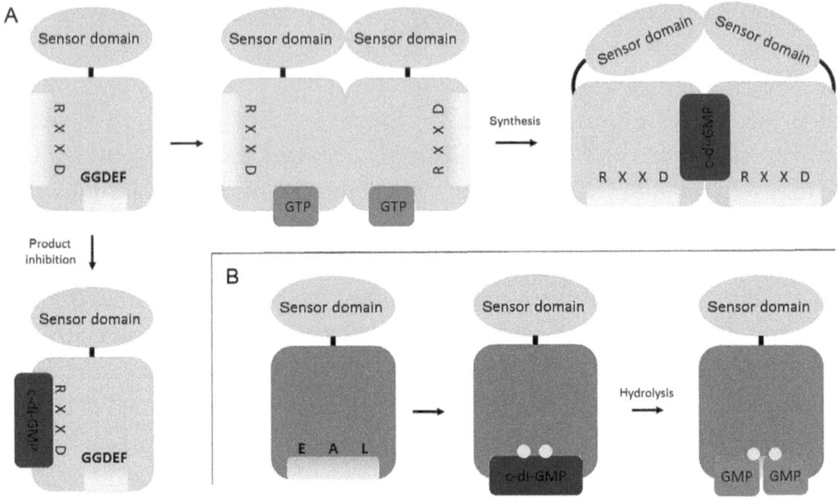

Figure 3 Structures of the c-di-GMP-metabolising enzymes, DGCs and PDEs. (A) Synthesis of c-di-GMP (red; dark grey in the print version) by dimerisation of DGC (light blue; light grey in the print version) and product inhibition via binding in the I-site(RXXD motif); (B) c-di-GMP hydrolysis by PDE (blue; grey in the print version) requires two Mg^{2+} ions (yellow; white in the print version).

characterised by a conserved RXXD motif. Notably, not all DGCs have an I-site for allosteric feedback control. The degradation of c-di-GMP is controlled by PDEs. Two classes of PDEs exist. One class contains an EAL (Glu-Ala-Leu) domain which cleaves c-di-GMP into a linear pGpG which can be broken into two GMP molecules, while others have a HD-GYP domain which hydrolyses c-di-GMP directly into GMP. Both classes require Mg^{2+} or Mn^{2+} ions for the phosphodiester hydrolysis. A detailed mechanism of the involvement of the Mg^{2+} or Mn^{2+} cations in the active site of PDEs has not been unravelled, but it is assumed that they play a role in the Lewis acid activation of the phosphate centre of c-di-GMP as well as activating the bound water molecule, which are both necessary for cleavage (Kalia et al., 2012) (Fig. 3B). Other divalent cations (Ca^{2+}, Zn^{2+}, etc.) will inhibit their activity.

Many bacteria contain several genes that code for these c-di-GMP-metabolising proteins. *E. coli* genes encode 29 c-di-GMP-metabolising enzymes, 12 DGCs (containing a GGDEF domain), 10 PDEs (containing an EAL domain) and 7 hybrid proteins containing both EAL and GGDEF domains. Remarkably, no PDE with a HD-GYP domain is expressed in *E. coli*. c-di-GMP signalling displays a high diversity concerning the

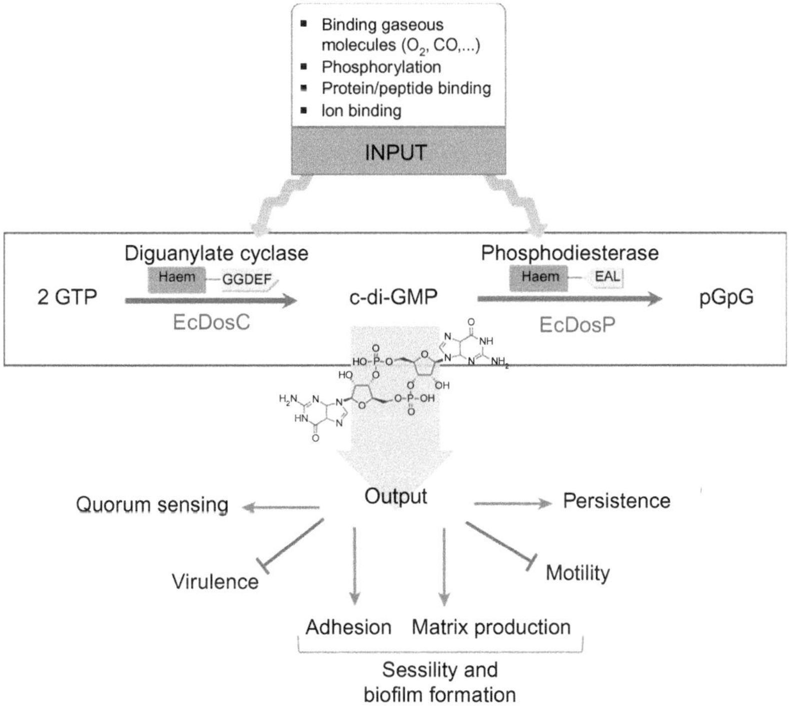

Figure 4 c-di-GMP metabolism and targeted pathways.

N-terminal sensory input domains fused to the majority of GGDEF, EAL and HD-GYP domains. Two c-di-GMP-metabolising enzymes from *E. coli*, EcDosC and EcDosP, possess a haem-based sensor domain and are characterised as globin-coupled sensors (GCSs) (Fig. 4).

The DGC EcDosC is transcribed together with the PDE EcDosP, controlling together the intracellular c-di-GMP concentration in an oxygen-dependent way. When the haem group is in its oxy (Fe(II)-O_2) or oxidised (Fe(III)) form, the EcDosC is active, and when the unligated (Fe(II)) haem is present, the protein is inactive. The oxygen-binding affinity for EcDosC ($K=0.07$ μM) is extremely low compared to Sperm whale Mb ($K=1.27$ μM) and other GCSs, e.g. the GCS from *Bordetella pertussis* ($K=1.8$ μM) and the haem-based aerotaxis transducer from *Bacillus subtilis* ($K=1.39$ μM) (Nakajima et al., 2012). Together with the relative fast autoxidation rate of the EcDosC (Kitanishi et al., 2010), this DGC enzyme is regulated in a very narrow oxygen range, living in equilibrium between its active oxygenated or oxidised form and the inactive ferrous Fe(II) form, which results in a slow c-di-GMP synthesis. A reducing system that regenerates the reduced (Fe(II)) form is however not reported yet. Under anaerobic conditions, the synthesis will stop and switch to a controlled EcDosP-dependentc-di-GMP hydrolysis (Kitanishi et al., 2010). Indeed, the deoxy form of the PDE EcDosP is an active form, although a very slow hydrolysis of c-di-GMP (17-fold slower) is also observed when the sensor domain is saturated with oxygen. However, EcDosP requires high oxygen concentrations (75–256 μM oxygen) for activation (30–100% saturation). Moreover, CO saturation enhanced the EcDosP activity in the same way as oxygen saturation (Tuckerman et al., 2009).

There is less information known about the c-di-GMP effectors. Mainly based on sequence studies, there appear to be the following types of c-di-GMP effectors: (a) PilZ-domain effector proteins that bind c-di-GMP to regulate other proteins/enzymes via protein–protein interactions, (b) proteins with inactive GGDEF or EAL domains to modulate the neighbouring catalytic or regulatory domains, (c) proteins with an intact I-site to regulate directly their activity, (d) transcription factors that bind c-di-GMP to regulate gene expression and (e) riboswitches that regulate transcription or translation upon c-di-GMP binding (Kalia et al., 2012; Römling, Galperin, & Gomelsky, 2013; Tamayo, Pratt, & Camilli, 2007). An example of c-di-GMP effectors in *E. coli* are BcsA and YcgR, the only two PilZ-domain proteins in *E. coli*. They are the key components in the motility-to-sessility transition and biofilm formation processes. The switch from a motile to an adhesive lifestyle when entering the stationary growth phase requires not only an accumulation of curli fimbriae but also a precisely timed shutdown of flagella expression. Hence, two inversely coordinated signalling cascades, dependent on separate c-di-GMP modules, are controlling this lifestyle switch. At the top of the protein network, there is competition for the RNA polymerase between sigma factors RpoD (housekeeping sigma factor), RpoS (the starvation/stationary phase sigma factor) and FliA (flagellar sigma factor). When cells enter the stationary growth phase, flagella expression stops because the RpoS-controlled pathway is induced, which stimulates DGCs YegE and YedQ. The resulting c-di-GMP production stimulates on one hand the interference of the YciR effector protein with the flagellar complex that slows down the flagellar rotation and on the other hand the transcription of the adhesive curli fimbriae activator gene *csgD*. Also, DGC/PDE pairs YdaM/YciR and YaiC/YoaD are induced, which do not contribute to motility. YdaM/YciR strongly and specifically control *csgD* transcription that results in curli fimbriae expression, necessary for attachment. The induced YaiC/YoaD pair regulates matrix production via the activity of the cellulose synthase BcsA. Moreover, also the RpoS-induced EcDosC/EcDosP pair regulates curli fimbriae expression, but in a CsgD-independent way. It also promotes the production of the matrix component PGA (Tagliabue, Antoniani, et al., 2010; Tagliabue, Maclg, Antoniani, & Landini, 2010). A brief overview of this transition network is presented in Fig. 5.

2. BIOFILM RESISTANCE

2.1 An Advantageous Way of Life

Since biofilms are the bacterial dominant lifestyle, there has to be significant advantages when living in sessile communities compared to free-floating single planktonic cells. Bacteria use several mechanisms to create a protected environment in the biofilm population to tolerate harmful external conditions and host defences (Fig. 6). Their high tolerance to AB is their most exceptional characteristic. The AB resistance of planktonic bacteria involves usually (i) inactivation or destruction of the drug by enzymes like β-lactamases; (ii) target modification via mutations; and (iii) exclusion of the drug through efflux pumps. These mechanisms require specific regulations of gene expression. However, a biofilm population not only uses the above factors to be tolerant to antimicrobial agents; a main plank in their defence is the structural nature of the biofilm itself (Jolivet-Gougeon & Bonnaure-Mallet, 2014; Lewis, 2013, Paraje, 2011). Since the biofilm architecture itself creates such a protective environment, it can be seen as an 'innate' resistance, which will be discussed below in more detail.

2.1.1 Biofilm Matrix as Diffusion Barrier

The biofilm matrix can act as a physical diffusion barrier that prevents antimicrobial agents from reaching their target. The slime layer slows down the diffusion rate which results in an increased resistance. The high viscosity and the size exclusion of the polymeric matrix prevent obtaining effective concentration of AB being reached in the deeper layers of the biofilm community. Bacteria in the outer layer will die, while bacteria buried in the deeper layers may react with adaptive responses. Furthermore, penetration through the biofilm can also be inhibited by the formation of ionic interactions between the negatively charged exopolysaccharide matrix components, such as colanic acid, and positively charged ABs like aminoglycosides. Finally, enzymes such as catalases and β-lactamases within the slime layer can inactivate the AB before they can even reach the bacterial population

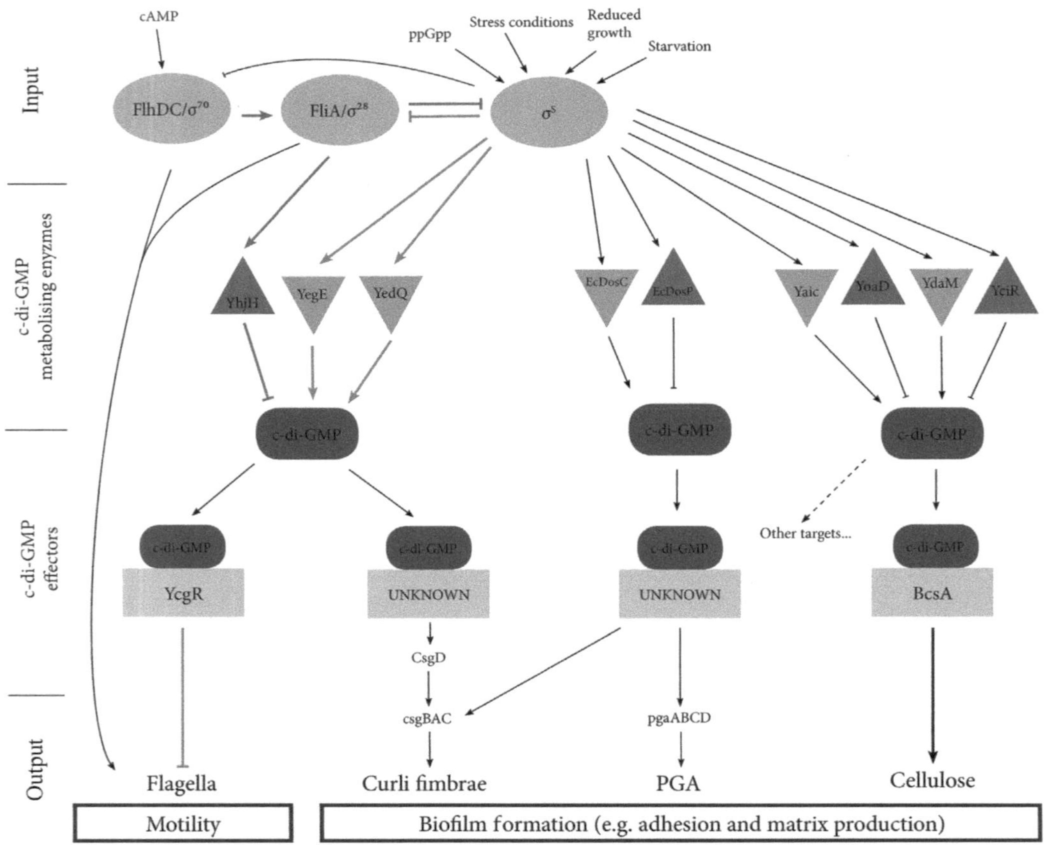

Figure 5 Schematic overview of the signalling network which regulates the switch from a motile planktonic to a sessile biofilm lifestyle. Red (dark grey in the print version) arrows indicate the signalling network of free-floating planktonic cells, while the green (grey in the print version) arrows show the switch to a sessile lifestyle. Blue (grey in the print version): PDE, light blue (light grey in the print version): DGC, brown (light grey in the print version): sigma factors, yellow (light grey in the print version): c-di-GMP effectors.

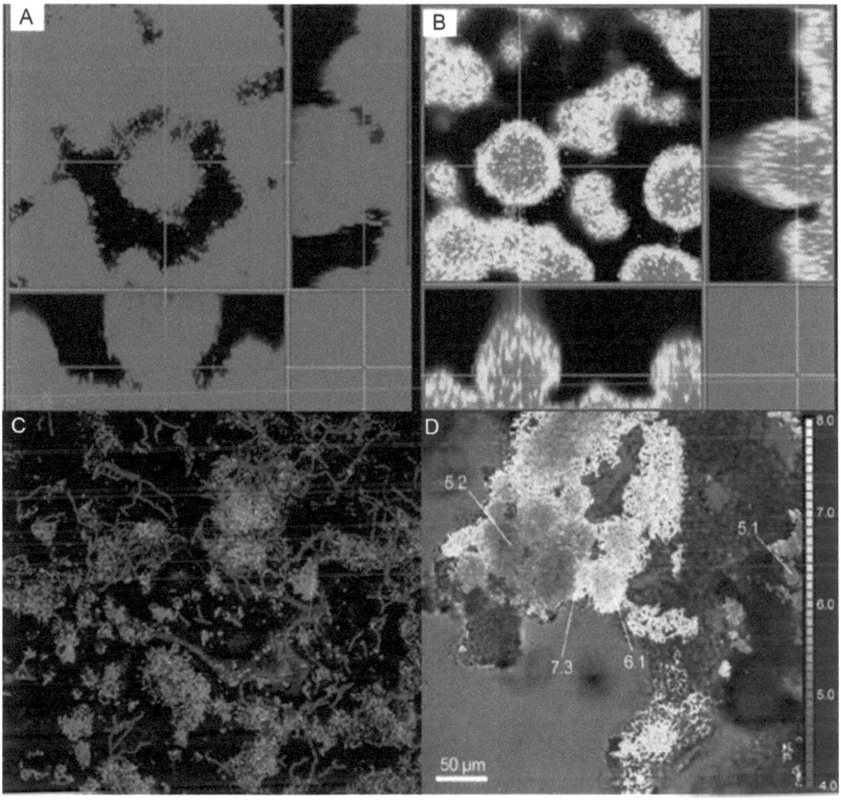

Figure 6 Biofilm physiology and the effect of the microenvironment. The physical, chemical and biological heterogeneity that can develop in a biofilm population is shown. (A) Wild-type *P. aeruginosa* biofilms, grown in *in vitro* flow cells, visualised by confocal laser scanning microscopy (Bjarnsholt et al., 2013). (B) The same wild-type *P. aeruginosa* biofilm as (A) treated with 100 mg/ml tobramycin for 24 h. Green bacteria are alive and yellow/red bacteria are dead. The AB cannot diffuse to the interior of the biofilm population because of mechanisms of tolerance (Bjarnsholt et al., 2013). (C) Micrograph of 18-h-old multispecies biofilm grown in flow cells. *Fusobacterium nucleatum* (red), *Aggregatibacter actinomycetemcomitans* (green), *Veillonella* sp. (dark blue) and *Streptococcus oralis* (light blue) (Häussler & Parsek, 2010). (D) *E. coli* PHL628 biofilm confocal and tomographic pH imaging, with the PH scale shown at the right. The biofilm population is composed of microenvironments of different pHs (Hidalgo et al., 2009). (See the colour plate.)

(Cos et al., 2010; Paraje, 2011). These β-lactamase genes are highly induced in response to AB sensed in the environment. However, this diffusion barrier is species dependent and antibiotic dependent and is not sufficient to explain the higher tolerance of biofilms against antimicrobial agents, compared to planktonic cultures. A low rate of vancomycin AB penetration was observed through the mucoid matrix of *S. aureus* biofilms using fluorescent-labelled vancomycin (Jefferson, Goldmann, & Pier, 2005). A study analysing the penetration of five different antibiotics in *S. aureus* and *S. epidermidis* biofilms, published by Singh, Ray, Das, and Sharma (2010), demonstrated a reduced penetration of AB oxacillin, cefotaxime and vancomycin, while the diffusion of amikacin and ciprofloxacin was unaffected (Singh et al., 2010). Moreover, Anderl and co-workers demonstrated that ciprofloxacin and chloride ion quickly penetrated the 48-h-old*Klebsiella pneumoniae* biofilm, while ampicillin was unable to penetrate when using 4-h treatment periods (Anderl, Franklin, & Stewart, 2000).

2.1.2 Microenvironments and Stress Responses

The spatial organisation and the physiological function of the whole biofilm community are vulnerable to environmental signals. The responses of the biofilm bacteria to switching environmental cues are not homogenous throughout the biofilm population. This can be explained by the spatial organisation in the stacked multilayered biofilm structure: the chemical and biological characteristics vary across the layers of the biofilm population. Cells in the deeper layers do not experience the same conditions as cells in the outer layers as mentioned before. Indeed, a mature biofilm consists of concentration gradients of several solutes, such as nutrients, oxygen and metabolites, creating different local microscale environments in the different layers of the population. These result from the diffusion process through the matrix and the extent of metabolic activity of the present bacteria (Stewart & Franklin, 2008). Oxygen is able to diffuse through the aqueous biofilm matrix, but it fails to penetrate to the deepest part of the biofilm population, creating hypoxic zones, because the bacterial cells in the upper layers actively consume oxygen. For example, a *Pseudomonas aeruginosa* biofilm, grown on a membrane filter on tryptic soy agar for 48 h, showed an oxygen-defined-two-layer pattern based on oxygen microelectrode measurements. This study combined oxygen diffusion experiments with protein synthetic activity assays based on GFP reporter gene constructs (Werner et al., 2004). They reported that the top layers of cells, where oxygen can diffuse in, exhibit active protein synthesis, while the cells in the thick bottom layer experience an anaerobic nature and appear to be metabolically inactive. They concluded that in the *P. aeruginosa* colony biofilm, the oxygen is consumed faster than it can diffuse into the deeper layers of the population (Serra & Hengge, 2014).

These oxygen gradients will also have an effect on the c-di-GMP concentrations within the entire biofilm because of the oxygen-dependent activity of the c-di-GMP-metabolising DGCs and PDEs, as discussed earlier in Section 1.2.2. In contrast, metabolites and waste products are available in higher concentration in the interior compared to the outside of the biofilm. The methane production of a sewage outlet biofilm of 3.5 mm thick was analysed, showing high concentrations of methane, building up to the depth of the biofilm structure (Damgaard, Nielsen, & Revsbech, 2001). The methane concentration at the surface of the biofilm was approximately 10% of the concentration at 2 mm depth, which is explained by aerobic respiration, which consumes a small part of the produced methane.

Due to these microenvironments within the biofilm, bacteria change their gene expression and metabolic activity which leads to a physiological heterogeneity and phenotypic diversity within the community (Parker & Sperandio, 2009). Regions in the interior of the biofilm are limited in not only oxygen but also nutrients, creating a starved environment. During this period of starvation, the bacteria slow down metabolism and consequently lower their growth and multiplication rates as an induced stress response. Because ABs are more effective against fast growing cells, these slower growing bacteria are co-responsible for the high biofilm resistance. Furthermore, the diversity of the bacterial phenotypes within a biofilm is caused by regulation of gene expression due to adaptation to the local environment. Also random mutations, genetic rearrangements and horizontal gene transfer may generate variants that are favoured by natural selection which results in the creation of more heterogeneity (Fig. 7) (Stewart & Franklin, 2008).

Next to the already mentioned hypoxia and nutrient starvation, bacteria encounter a variety of stresses in their natural environments, such as exposure to reactive oxygen and nitrogen species, membrane damage, high temperatures and ribosome disruption (Poole, 2012). Different stresses can induce several distinct stress responses and the different stress responses are also correlated with each other (Table 2). The initiation of such bacterial stress responses positively impacts the antimicrobial resistance mechanisms of the pathogen in a direct or indirect way. When a stress-induced growth termination or dormancy occurs, the stress response has an indirect effect on the AB susceptibility. Direct effects of stress responses are for instance the stimulation of a biofilm growth mode, increase in antimicrobial efflux pumps, changes to antimicrobial targets, modifications to the membrane barrier functions and the induction of resistance mutations (Poole, 2012).

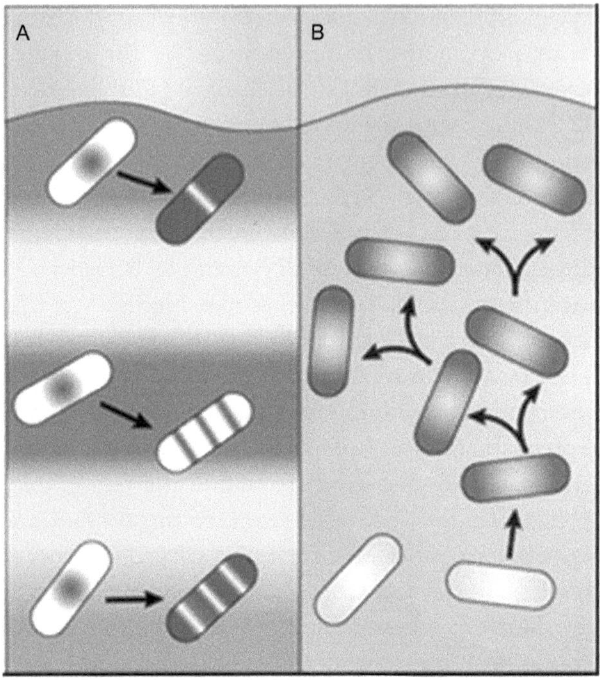

Figure 7 Biofilm heterogeneity. (A) Physiological heterogeneity. Cells respond to local microenvironmental conditions by turning on or off certain genes, (B) genotypic variation and natural selection. Random mutations or chromosomal rearrangements result in a variant (purple; dark grey in the print version) that multiplies according to its fitness in the biofilm. *Adapted from Stewart and Franklin (2008).*

Table 2 Bacterial Stress Responses as Influencing Factors of Antimicrobial Resistance (Poole, 2012b)

Types of Stress Responses	References
Stringent response	Chatterji and Kumar Ojha (2001)
SOS response	Baharoglu and Mazel (2014) and Simmons, Foti, Cohen, and Walker (2008)
Stress-induced mutagenesis	Joly et al. (2010)
General stress response	Hengge-Aronis (2002) and Battesti, Majdalani, and Gottesman (2011)
Induction of a persister phenotype	Boehm et al. (2009)

References are mostly based on *E. coli.*

The general stress response is controlled by the sigma factor S, encoded by the *rpoS* gene, which regulates many pathways. RpoS was shown to be required for a heat shock-promoted increase in carbapenem resistance in *P. aeruginosa*, which clearly demonstrates a stress response-induced antimicrobial resistance (Murakami et al., 2005). In *E. coli*, the general stress response is activated by oxygen and nutrient limitation, hyperosmolality, pH decrease, non-optimal temperatures, reduced growth rate and high cell density (Hengge-Aronis, 2002). This response allows cells to become more resistant not only to the stress that they are first exposed to but also to other stressful conditions. Such a cross-protection phenomenon is typical of the general stress response and contrasts with other specific stress responses that deal only with consequences of the inducing stress (Battesti et al., 2011). The SOS response is a set of physiological responses induced by DNA damage or replication fork blockages which can be caused by, e.g., oxidative stress and exposure to certain AB (Jolivet-Gougeon & Bonnaure-Mallet, 2014; Simmons et al., 2008). The produced DNA damage and mutations activate a set of SOS response-induced genes which express high fidelity DNA repair proteins and proteins that arrest the cell cycle of the bacteria. This mechanism allows bacteria to repair DNA before damaged DNA is segregated to daughter cells. Other tolerated mutations may lead to phenotypic variants within the microbial colony and be favourable allowing better adaptation to the host environment (Simmons et al., 2008). Consequently, increased mutation frequencies can result in AB resistance and latent pathogens (Jolivet-Gougeon & Bonnaure-Mallet, 2014). Moreover, the SOS response also contributes to AB resistance development by enhancing horizontal gene transfer and promoting the capture and expression of integrons, which are mobile genetic elements that carry multiple AB resistance genes and integrases that catalyse recombination (Poole, 2012). An increased mutation rate enlarges the chance to adaptation to a specific stress signal, but stress-induced mutations are rather random.

When bacteria experience nutritional limitations, or starvation, they adapt by changing to a slow-growing state which is characterised by an accumulation of guanosine tetraphosphates (ppGpps), called the stringent response. ppGpp negatively affects nucleic acid synthesis and positively regulates RpoS-dependent genes (Chatterji & Kumar Ojha, 2001). This positive regulation of RpoS leads to a broad effect of ppGpp because stationary genes and also genes controlling biofilm formation are RpoS dependent (Fig. 5). In *E. coli*, ribosome damage, for example, results in a high biofilm-forming capacity as a consequence of a decrease in the ppGpp levels and an increase of the c-di-GMP concentrations (Boehm et al., 2009). Furthermore, there are some doubts whether stress can also promote a dormant cell population, called persister cells (see Section 2.1.4).

2.1.3 Quorum Sensing

Bacteria also communicate with each other to respond to the environment. Such a cell-to-cell communication system is called QS. It is a cell density-dependent signal transduction process, which regulates a variety of physiological behaviours in bacteria. The close proximity of the cells in a biofilm allows rapid coordination of the behaviour of the entire community on the QS signals, which is beneficial for the biofilm collective responsiveness to the environment. Necessary collective actions of the biofilm population include adapting to different levels of nutrients in the environment as mentioned earlier in Section 2.1.2, defending against other competitive microorganisms, avoiding toxic materials and invading or infecting a host (Galloway, Hodgkinson, Bowden, Welch, & Spring, 2011).

During QS, bacteria secrete small signal molecules, called autoinducers(AIs), in the extracellular environment, which can be recognised by specific receptors on the surface or even in the cytoplasm of neighbours. When the concentration of the signal molecules reaches a threshold level, a signal transduction cascade activates the expression of certain genes of the responding cells, influencing many regulatory pathways among which biofilm formation and biofilm-related processes. Gram-negative bacteria possess three different cell-to-cell communication systems (Parker & Sperandio, 2009). The main QS mechanism used by these bacteria is based on acyl-homoserine lactone (AHL) molecules, commonly referred to as AI-1 molecules. AHL is synthesised by AHL synthase (LuxI and its homologues) and sensed by LuxR-like receptor proteins. *E. coli* however lack AHL synthase encoding genes but are able to respond to such AI-1 signals because they do express SdiA, a LuxR-type AHL receptor (Van Houdt, Aertsen, Moons, Vanoirbeek, & Michiels, 2006). Thus, AHL-dependent QS signalling is a communication system between only Gram-negative bacteria, in response to secreted AHL molecules (Van Houdt et al., 2006; Whitehead, Barnard, Slater, Simpson, & Salmond, 2001). Another QS pathway, which is found in both Gram-positive and Gram-negative bacteria, is based on AI-2 molecules that are derived from 4,5-dihydroxy-2,3-pentadione. In *E. coli*, AI-2 molecules are synthesised by LuxS and are bound to the ABC-type transporter Lsr. However, there are several different bacterial receptors for the detection of QS system 2 molecules, which allows communication among all bacterial species. Since these receptors can vary in structure and function, AI-2 signalling can induce different responses (Rezzonico, Smits, & Duffy, 2012). The third QS circuit used by *E. coli* uses the receptor QseC, which has a histidine kinase activity. Upon binding of AI-3 signals, auto-phosphorylation takes place in order to activate the response regulator QseB. Less is known about this QS pathway, but Kendall et al. showed that LuxS is also involved in the production of AI-3 (Kendall, Rasko, & Sperandio, 2007; Winzer & Williams, 2003).

2.1.4 Differentiation into Persister Cells

In the deepest interior of the biofilm, a specific non-growing biofilm phenotype, called persister cells, exists. These cells are significantly different from other bacteria in the biofilm population because they tolerate excessive antibiotic concentrations. Persister cells gain their resistance property by entering a state of dormancy. A specific and extraordinary characteristic of these dormant persister cells is recalcitrance, which means that after antibiotic treatment, the survived persister cells can relocate and form a new biofilm again with the same sensitivity as the original population. This unique property explains why persister cells are the major concern in biofilm treatment (Lewis, 2007).

Several mechanisms are hypothesised to explain the formation of this persistent subpopulation of a biofilm (Fig. 8) (Williams & Hergenrother, 2012). It is reported that dormant cells arise when the population enters the mid-exponential growth phase and they reach a maximum in the stationary phase (Lewis, 2007). It is assumed that a phenotypic switch takes place due to stochastic gene expression changes.

Another hypothesis states that persister formation can be induced by stresses such as exposure to AB. A specific level of SOS response induction is required for persister formation based on a study where all actively growing cells were exposed to ciprofloxacin, but where not all cells turned into persister cells, suggesting that there are different mechanisms of persister formation to different antibiotics (Dörr, Lewis, & Vulić, 2009).

A transcriptome study showed that genes coding for toxin–antitoxin(TA) module proteins were highly expressed in *E. coli* persisters, compared to cells in the stationary phase (Shah et al., 2006). This indicates a role for TA systems in the formation of persister cells. TA systems typically consist of a stable toxin protein that disturbs an essential cellular process like, e.g., translation, and an antitoxin (RNA or a protein) that prevents the toxin from being poisonous.

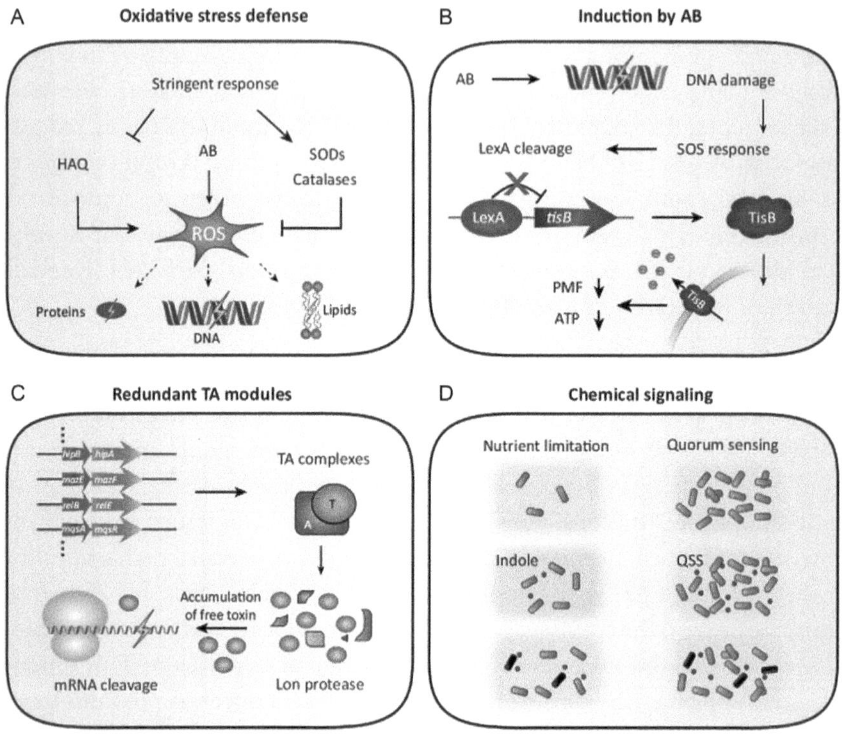

Figure 8 Four different models (A–D) to explain persister formation. *Copied from Kint, Verstraeten, Fauvart, and Michiels (2012).*

Type I antitoxins are RNA antitoxins that inhibit translation by acting as antisense RNA. When an RNA antitoxin binds and inhibits the toxin protein, it is known as a type III antitoxin. Type II TA systems are based on protein–protein interactions between the toxin and the antitoxin, which hinder toxin activity. Type IV protein antitoxins prevent the toxin from binding its target instead of inhibiting the toxin directly, and type V protein antitoxins cleave specifically the toxin mRNA (Wood, Knabel, & Kwan, 2013). The *E. coli* chromosome encodes several TA systems such as the *relBE, mazEF, chpBIK, hipBA, dinJ–yafQ, tisAB–istR* and *yefM–yoeB* systems (Fiedoruk, Daniluk, Swiecicka, Sciepuk, & Leszczynska, 2014; Karimi, Ghafourian, Kalani, & Jalilian, 2015; Tsilibaris, Maenhaut-Michel, Mine, & Van Melderen, 2007). In *E. coli*, the SOS response probably induces persister cell formation via the type I TA pair tisAB–istR. An SOS-activated cleavage of LexA stimulates the transcription of the *tisB* gene, which encodes a membrane-acting toxin. As a consequence, free TisB inserts in the membrane and disturbs the proton motive force which leads to decreased ATP concentrations and a state of dormancy and persistence (Cavalcanti et al., 2015; Dörr et al., 2009; Kint et al., 2012).

Although environmental changes seem to positively influence the persister phenotype, they can also result in the degradation of an antitoxin which is highly unstable (Cos et al., 2010; Paraje, 2011). Released toxin proteins can now inhibit cellular processes such as replication, translation, cell division and ATP synthesis. Additionally, several antitoxins are degraded by Lon proteases, whose cleavage activity is controlled by environmental factors. Both facts lead to the hypothesis that the amount of persister cells is dependent of the surrounded conditions (Unterholzner, Poppenberger, & Rozhon, 2013).

3. BIOFILM-ASSOCIATED INFECTIONS

The same microorganism can cause an acute or a chronic infection. The nature of the pathogenesis is dependent on environmental signals, such as the route of entry for infection in the host, the immune and nutrition status of the host, the tissue integrity and the nutrient and oxygen availability in the surrounding. Additionally, the type of infection is also correlated with the growth mode of the bacteria when entering the patient: growing and proliferating rapidly in the host is mostly associated with an acute infection and will involve planktonic bacteria, while settling in slow-growing communities, such as biofilms, mostly gives rise to chronic infections (Bjarnsholt et al., 2013; Furukawa, Kuchma, & O'Toole, 2006). Some common biofilm-associated infections (BAI) are shown in Table 3.

During an acute infection, planktonic bacteria cause a short-time illness, and ABs and vaccines are effective treatments of such infectious diseases. On the contrary, in patients with a chronic infection, the infection lasts longer but is mostly associated with mild symptoms, sometimes going unnoticed for a few months to sometimes years. These 'silent' infections rarely lead to indications in the bloodstream which gives difficulties for early diagnosis. Therefore, some criteria are formulated to evaluate several BAI characteristics (Hall-Stoodley & Stoodley, 2009; Hall-Stoodley et al., 2012; Parsek & Singh, 2003) (Table 4). Routine microbial examinations include sample collection, cultivation, identification and antibiotic susceptibility testing. The first step in this diagnostic protocol is not only the most important one but also the most challenging one. A correct sample collection is essential for a reliable diagnosis of the BAI but is associated with invasive procedures since the microorganisms from the biofilm-attached surfaces have to be

Table 3 Examples of Biofilm-Associated Infections in Humans

Infection or Disease	**Biofilm Species**
Tissue-related infections	
Dental carries	Acidogenic Gram-positive cocci (e.g. *S. mutans*)
Periodontitis	Gram-negative anaerobic oral bacteria (e.g. *P. gingivalis*)
Otitis media	*Haemophilus influenzae*
Osteomyelitis	Various bacterial and fungal species (e.g. *S. aureus*)
Cystic fibrosis pneumonia	*P. aeruginosa* and *Burkholderia cenocepacia*
Infective endocarditis	*Streptococci*, *Enterococci*, *Staphylococci*
Bacterial prostatitis	*E. coli* and other Gram-negative bacteria
Chronic wounds	*S. aureus*, *P. aeruginosa* and several others
Urinary tract infections	*E. coli*
Device-related infections	
Central venous catheters	*S. aureus*, *S. epidermidis*, *P. aeruginosa*, *C. albicans*, *Enterococcus faecalis*
Urinary catheters	*S. epidermidis*, *E. coli*, *E. faecalis*, *K. pneumoniae*, *Proteus mirabilis*
Heart valves	*Streptococci*, *Enterococci*
Orthopaedic implants	*P. aeruginosa*, *E. coli*, *Enterococci*, *Streptococci*
Endotracheal tubes	Various bacterial and fungal species
Contact lenses	*P. aeruginosa* and Gram-positive cocci

Adapted from Aparna and Yadav (2008), Lebeaux, Chauhan, Rendueles, and Beloin (2013) and Cos et al. (2010).

isolated (Wu, Moser, Wang, Høiby, & Song, 2014). When the isolated colonised surface is removed, microscopic analysis can already determine the biofilm phenotype based on diagnostic criteria 1 and 2 (Table 4). Furthermore, it has to be noted that the presence of biofilms is not necessarily associated with a positive blood culture neither when dealing with intravascular device-related biofilm infections. This is illustrated in criteria 3 and 5 (Table 4). Indeed, peripheral blood cultures reflect individual bacteria floating around in the bloodstream, but they do not detect biofilm-attached cells which can be present. The method used for detection is as such crucial. Post et al. studied diagnostic assays to detect bacterial pathogens in chronic otitis media patients (Post et al., 1995). They were able to detect a pathogen using

Table 4 Diagnostic Criteria for Biofilm-Associated Infections

1.	Pathogenic bacteria are associated with a surface.
2.	Direct examination of infected tissue or materials demonstrates aggregated cells encapsulated in a matrix, which may be of bacterial and host origin.
3.	Infection is confined to a particular site in the host.
4.	Recalcitrance to antibiotic treatment in spite of a demonstrated standard susceptibility testing of the specific bacterium (since standard tests are performed on planktonic single cells).
5.	Culture-negative result in spite of clinically documented high suspicion of infection (since localised bacteria in a biofilm may be missed in a conventional blood sample or aspirate).
6.	Evidence of ineffective host clearance with bacterial aggregates (microcolonies) demonstrated by the co-localisation of host inflammatory cells with discrete areas of the host tissue.

Adapted from Hall-Stoodley et al. (2012) and Hall-Stoodley and Stoodley (2009).

standard cultural methods in only 25–30% of the patients, while in an acute otitis media pathogenesis 90% of the patients report a positive culturing result. On the contrary, a PCR-based detection system and/or fluorescence *in situ* hybridisation (FISH) combined with confocal laser scanner microscopy (CLSM) identified in 80–100% of the chronic patients a harmful microorganism (Post et al., 1995; Wilkins, Hall-Stoodley, Allan, & Faust, 2014). Additionally, as mentioned above, biofilms can withstand both antibiotics and host defences, since they are persistent (10–100 times more resistant than their planktonic life form), which explains criteria 4 and 6 for the diagnosis of BAI (Hall-Stoodley et al., 2012).

There is an increasing concern about the role played by microbial biofilms in infection. Recent surveys clearly indicate that bacteraemia from catheter-related infections is by far the leading cause of nosocomial infections in intensive care units. Today, the common hospital-used-strategy to treat patients with device-related biofilm infections is the removal of the colonised indwelling device. Medical devices such as central venous catheters (CVCs), endotracheal tubes and urinary catheters are easy to remove and can be replaced by new ones, but for pacemakers or prostheses, it is more difficult and a surgical operation is required (Amalaradjou & Venkitanarayanan, 2013). In addition, ABs are administered prior to catheterisation in order to kill the incoming pathogens that are at that point still planktonic.

To try to avoid colonisation of the medical devices, a lot of research has been made on the biofilm susceptibility of the material the devices are made of and the possible pretreatment of this material. The surface properties, such as the hydrophobicity, charge and shape of the surface, can be changed in such a way that attachment will be inefficient. Biocompatibility of the catheter material, and the ability of the device to perform its function without provoking undesirable side effects, is of crucial importance and also dependent on the physical and chemical properties of the material and on the body's reaction to it. The most modern tubes used today are silicone-based or polyurethane catheters. They are non-allergenic, show a high flexibility and have low surface tension, diminishing the chance of biofilm formation (Hawser & Douglas, 1994). However, silicones are also hydrophobic which makes in turn the bacteria eager to adhere (Cohen et al., 2011; Cox, 1990; Curtis & Colas, 2004). In addition, a conditional film will be formed on, for example, CVC tubing. Components of the blood flow will form a natural coating on the surface area. The adsorption of blood proteins, like fibrinogen and fibronectin, or other organic molecules will affect the adhesion capacity of bacteria in a positive way (Murga, Miller, & Donlan, 2001; Teughels, Assche, Sliepen, & Quirynen, 2006). That is why despite the fact that silicone and polyurethane catheters are showing the highest biocompatibility, additional coatings will be necessary to reduce the susceptibility to biofilm formation. The search for new synthetic films which are coated directly on the surface material (with, e.g., metals or antimicrobials) and which should prevent or diminish natural conditioning and/or bacterial attachment, is progressing, but to date not entirely successful. That is why a combinational approach is necessary where not only a coated catheter is used but where the patient is treated with bacterial attachment inhibitors. Several anti-adhesion therapies have been reviewed such as the use of adhesin analogues (synthetic or recombinant adhesin fragments) and anti-adhesin antibodies. However, such therapies are adhesin specific and are not effective enough to prevent bacterial attachment. Another strategy is the administration of receptor-like carbohydrates, which interact with adhesins before they can attach to the glycoprotein/glycopeptide receptors on host cells. This is a more attractive therapy because of its broad-spectrum anti-adhesion activity. The anti-adhesion effects of dietary inhibitors are questioned. The most studied dietary inhibitor in *E. coli* adhesion in urinary tract infections is cranberry juice (Ofek, Hasty, & Sharon, 2003).

If such procedures are not feasible, as for instance when biofilms are already formed on tissue surfaces or implanted medical devices, other approaches must be implicated. A possibility to manage such biofilm infections is to drain biofilm components out of the body. This will not be sufficient to remove tightly attached cells and surgically removing the biofilm population will be the last option. However, some studies demonstrate an enhanced antimicrobial effect when using different AB dosing approaches, dependent on the bacterial species and the used antimicrobial agent. A pulse-dosing method, once-daily or thrice-daily, for example, can result in a different response dependent on the pathogen and the type of ABs (Ibrahim, Gunderson, Hermsen, Hovde, & Rotschafer, 2004). Grant and Bott published in 2003 a study investigating the effects of dose concentration, duration and frequency of dosing of biocides on *Pseudomonas fluorescens* biofilms attached to glass tubes (Grant & Bott, 2003). However, more research about evaluating different dosing approaches is required to achieve an optimum dosing method for each biofilm. Such a strategy could have promising beneficial effects in future treatments of biofilm-related infections, but complete eradication cannot be achieved with AB alone. There is a need for specific anti-biofilm agents that focus on the manipulation of important parameters in the process of biofilm formation such as attachment, motility, matrix production and QS (Table 5).

An interesting target for therapy is the bacterial QS communication system (see Section 2.1.3). There are four different ways to attenuate this system and consequently reduce biofilm virulence: (i) inhibition of the enzymes responsible for AI synthesis; (ii) targeting the AI receptors; (iii) preventing AI-binding on receptor by, e.g., AI antibodies; and (iv) inhibition of downstream proteins that link the QS receptor to the activation of QS-regulated genes (Sintim, Smith, Wang, Nakayama, & Yan, 2010). Because AI-1 molecules are produced by species-specific enzymes, it is hard to discover broad-spectrumAI-1 synthase antagonists. On the contrary, more than 70 bacterial species synthesise AI-2 molecules via a LuxS enzyme, which makes LuxS inhibitors superior therapeutic agents. Some attempts in finding LuxS substrate analogues were published (Shen, Rajan, Zhu, Bell, & Pei, 2006; Wnuk et al., 2009). Furthermore, up till now, there are no QS receptors found in humans, which means anti-AI-receptor molecules will show low cytotoxicity. Mostly, these molecules are structurally similar to the natural AI and thus are called AI analogues. Using AI-binding macromolecules that prevent the signal molecule binding to its receptor is another principle for biofilm research. A protective role for monoclonal antibodies that recognise AHL compounds against *P. aeruginosa* infections has been published (Palliyil, Downham,

Table 5 Anti-Biofilm Therapy

Point of Interference	Specific Inhibitory Effect
Motility	Stimulation of flagella expression or activity
EPS biofilm matrix	Matrix-degrading enzymes (e.g. dispersin B, cellulase, DNase)
	Inhibition of matrix production
Attachment	Anti-adhesion coatings
QS communication system	Inhibition AI synthesis
	AI antibodies
	Competitive AI receptor-binding molecules
c-di-GMP signalling	DGC inhibition
	PDE stimulation
	Interference with c-di-GMP effectors (e.g. c-di-GMP analogues)
AB tolerance	Inhibitors of efflux pumps
	Antimicrobial lock solutions (e.g. chelators, anticoagulants)
	Membrane-acting AB to kill also persisters
	Antimicrobial peptides
	Probiotics
	Stimulation of cell death (e.g. TA activation)
	Bacteriophage therapy

Different points of interference for biofilm therapy are demonstrated, each with some examples of possible inhibitory approaches.

Broadbent, Charlton, & Porter, 2014). Also the use of AI-degrading enzymes, such as lactonases and acylases, is a promising approach. Acylase enzymes, for example, remove a side chain from a ring-like molecule by hydrolysing the connecting amide bond (Sio et al., 2006). The enzymes were listed that have the ability to degrade AHL signal molecules that are characterised to date (Fetzner, 2014).

Another interesting point of interference is c-di-GMP signalling (see Section 1.2.2). Controlling the intracellular c-di-GMP concentrations may suppress biofilm formation. A decreased DGC activity or an increased PDE activity will lead to lower levels of c-di-GMP and result in the disturbance of the biofilm signalling network. A screen for small molecules that inhibit the c-di-GMP synthesising DGC WspR of *P. aeruginosa* identified ebselen as an I-site binding compound (Lieberman, Orr, Wang, & Lee, 2014). Also c-di-GMP analogues have potential to treat BAI because they can occupy the active site of c-di-GMP receptors. The two previously discussed possible therapeutic approaches (QS and c-di-GMP signalling as targets) appear more beneficial than conventional ABs because they are not directly involved in bacterial growth and may be less sensitive to the development of resistance.

Finally, compounds that are capable of inducing cell death have potential as antibacterial agents. Chromosomal TA systems are identified in almost all bacterial pathogens and are involved in several cellular pathways, including stress response, biofilm formation, QS, virulence and persistence, making them an interesting therapeutic target. A direct activation approach is based on TA disruptors that, e.g., inhibit the antitoxin and prevent it from complex formation with the toxin, or that bind to and activate the toxin. An indirect toxin activation can take place due to the modulation of TA expression via interfering with the promotor regions of the TA genes or due to the enhancement of the proteolytic susceptibility of the antitoxin (Williams & Hergenrother, 2012). Overproduction of the toxin of the yefM–yoeB TA system leads to reduced cell growth and cell viability in *E. coli* (Nieto et al., 2007).

To conclude, all above discussed therapies could be very powerful antibiofilm agents because they also target the dormant persister cells of the population. Additionally, membrane-acting strategies can also be successful in killing the whole biofilm community.

4. BIOFILM RESEARCH AND ITS CHALLENGES

4.1 *In Vitro* and *In Vivo* Biofilm Models

To get a better understanding of the underlying mechanisms of biofilm formation and its resistant characteristic, a first essential step is to develop appropriate biofilm model systems to produce, quantify and evaluate the biofilm phenotype. Several different *in vitro* and *in vivo* models have been developed over the last decade, aiming at a general model that approximates the natural biofilm condition the most (Coenye & Nelis, 2010; Cos et al., 2010). Both static and dynamic *in vitro* biofilm production model systems are available to grow adherent biofilm populations. Static test systems contain diverse flat surfaces such as microtitre plates, petri dishes or plastic tubes. The most popular static biofilm production model is a 96-well plate, where the bacterial suspension is added to every well of the plate and incubated. Such static systems cannot mimic the reality of a natural biofilm environment, where there is a free flow of nutrients and waste products and where there are influences of, e.g., the host immune system. A dynamic *in vitro* system is developed to create biofilms under a continuous flow of nutrients. Centre for Disease Control biofilm reactors and modified Robbins devices (MRDs) (Fig. 9) are commercially available flow displacement model systems (Coenye & Nelis, 2010; Lebeaux et al., 2013). The MRD is a linear tank reactor with several ports that accept plug-holding discs where biofilms can be formed. After an adhesion phase where a microorganism suspension is flowed through the device, a controlled medium flow can be applied to mimic the *in vivo* biofilm formation process as in, for instance, water pipelines or medical tubing. Numerous custom-made dynamic models are appearing in the literature. A simple example is a catheter which is connected at one end to a pump which provides the medium and/or cell suspension flow and at the other end to a waste reservoir. Dynamic methods generally allow better control of the biofilm growth environment (e.g. shear forces, nutrient concentrations, waste product drainage) and for that reason they have been frequently used to study the physical and chemical resistance capacity of biofilm populations (Lebeaux et al., 2013).

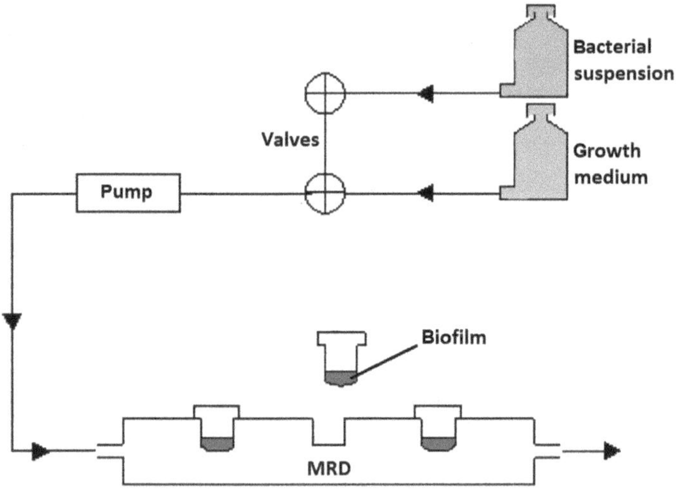

Figure 9 Schematic presentation of the modified Robbins device. Via a pump system, medium is pumped automatically through the device, containing several discs for biofilm formation. *Adapted from Oosterhof, Buijssen, Busscher, Van Der Laan, and Van Der Mei (2006).*

Both static and dynamic models are extensively used to characterise biofilm physiology in, for example, AB screens because of high throughput capability, and in genetic studies to investigate the roles of different genes in the process and regulation of biofilm formation. However, a major drawback of these *in vitro* models is that they do not take into account some important environmental factors, like host factors and other signalling molecules that are present in the surrounding of the adherent biofilm. These shortcomings can be partly addressed by using an *in vivo* model system. Many animal models have been developed to study BAI because they are better suited for observing the biological behaviour of both the host and the biofilm population. Most biofilm model systems use rats, mice or rabbits, not only for testing AB survival but also for studying factors important for biofilm formation. For example, rats are commonly used as an *in vivo* CVC model where catheters are inserted in the external jugular vein, followed by contamination with bacteria. This model gives the opportunity to study biofilm adhesion to CVC and evaluate the effect of antimicrobial agents, which can be injected. In foreign body infection models, a foreign body is implanted subcutaneously and allow biofilms to grow on the inserted material. The animals can be infected pre- or post-implantation. This model is well suited to compare the biofilm formation capacity on different substrates. Additionally, inserting foreign materials in the peritoneal cavity is performed to study chronic infections. Other model systems involve inserting polyethylene tubes in the bladder or endotracheal tube to develop, respectively, urinary tract and respiratory tract infections. For a more comprehensive overview of the used *in vivo* models, see Bjarnsholt et al. (2013) and Coenye and Nelis (2010). There are also *ex vivo* tissue/organ models such as teeth, kidneys and human cell lines, which may be used to study biofilm attachment to the tissues or cells. These models are, for example, used to study wound infections, where biofilms are formed on the wound surface (Rhoads, Wolcott, & Percival, 2008). Although these extracted tissue cells are placed in an artificial environment, they provide a good alternative to laboratory animals.

Because of the high biofilm diversity/complexity, today, there is no ideal laboratory biofilm model that is representative for all biofilms and suitable for biofilm susceptibility testing in order to develop anti-biofilm strategies. The choice of a suitable model system is mainly dependent on the objective of the biofilm study. Every model has strengths and weaknesses, which makes it suitable for different applications (McBain, 2009). For an extensive overview of all *in vitro* and *in vivo* biofilm models, discussing their advantages and disadvantages and their choice of application, see Lebeaux et al. (2013).

4.2 Biofilm Detection, Identification and Quantification Methods

To analyse biofilm growth in an *in vitro* or *in vivo* model system, it is necessary to measure the dimension/proportion of biofilm formation. Different approaches have been developed to visualise and quantify attached biofilm populations. They can be classified into biofilm biomass assays (quantification of matrix and both living and dead cells), viability assays (quantification of viable cells) and matrix assays (quantification of matrix). The most commonly used quantification techniques for microbial biofilms are listed in Table 6 (Coenye & Nelis, 2010; Peeters, Nelis, & Coenye, 2008). These staining methods are biofilm-destructive, but non-destructive visualisation of biofilms can be achieved by, for instance, CLSM. Besides microscopy, analytical techniques such as Raman spectroscopy, FTIR spectroscopy and magnetic resonance imaging have increased our knowledge of biofilm structure (Høiby et al., 2015).

The most popular biofilm biomass assay is crystal violet (CV) staining. CV is a basic cationic dye and will stain both the biofilm bacteria and the

Table 6 Mostly Used Quantification Assays for Microbial Biofilms

Quantification Technique	Type of Quantification	Based on
Crystal violet (CV) stain	Biomass	Binding of negatively charged surface molecules and polysaccharides in the extracellular matrix
Syto9 stain	Biomass	Binding of intracellular and extracellular DNA
XTT dye	Viability	Reduction to formazan by metabolically active cells
Resazurin	Viability	Reduction to pink, fluorescent resorufin by metabolically active cells
FDA		Reduction to yellow, fluorescent fluorescein by metabolically active cells
DMMB	Matrix	Binding of sulphated polysaccharides in matrix
Conventional plate count	Viability	Counting colony-forming units after plating biofilm cells
Life/dead PCR	Viability	Inhibition of PMA-modified DNA PCR amplification
Flow cytometry	Viability	Difference in fluorescence intensity using membrane-impermeable dyes

extracellular matrix. The protocol involves fixation of the biofilms grown in microtitre plates using methanol. After drying and washing, the stain is solubilised using 33% glacial acetic acid and absorbance is measured at 590 nm (Peeters et al., 2008). The most widely used methods to analyse the viability of a biofilm population are conventional plate counting and the resazurin assay. Resazurin is a blue non-fluorescent compound that can be reduced to pink resorufin, which is fluorescent. Resazurin is only reduced by metabolically active, viable cells. This compound is generally added to the biofilms in the wells, in combination with fresh medium, and incubated for 2–4 h before measuring the fluorescent signal (Peeters et al., 2008). The resazurin reduction kinetics are different for each strain, require a certain bacterial density and can be influenced by the cultured medium or testing compounds (Riss et al., 2004). Plate counting is a time-consuming,labour-intensive and subjective method that only allows determination of organisms that easily grow on solid medium and is as such not a good technique for high-throughput analysis (Kerstens et al., 2014). Flow cytometry, combined with the usage of membrane-impermeable dyes, is a powerful and faster alternative for cell counts. A TO-PRO-3iodide-based flow cytometry approach to isolate and quantitate viable and dead *S. aureus*, *E. coli* or *Bacillus subtilis* cells has been described (Kerstens et al., 2014). It is clear that parameters such as high-throughput ability, sensitivity and rapid performance need to be taken into account when choosing the assay to use.

Another alternative for viability assays is a PCR-based assay that discriminates between live and dead bacteria in a community. Because propidium monoazide (PMA), which can only enter cells with a damaged membrane, strongly intercalates with DNA and inhibits PCR amplification, only DNA from living cells will be amplified. Based on standard curves, made from analysis of serial dilutions of bacterial suspensions, the viability can be determined (Clais et al., 2015; Nocker, Richter Heitmann, Montijn, Schuren, & Kort, 2010). Sheridan and co-workers noted that mRNA has a half-life of only a few minutes, indicating that the detection of mRNA levels by quantitative RT-PCR could also be a powerful tool to determine the amount of living cells (Sheridan, Masters, Shallcross, & Mackey, 1998).

Before applying any of the above-mentioned biofilm analysis methods, there is the need for a good sample preparation. This includes washing and collecting the cells. The adherent multilayered biofilms are usually washed with PBS to remove non-adherent cells and to keep only biofilm cells for detection and phenotyping. In order to maintain the integrity of the biofilm structure, two aspects of the washing step are relevant to incorporate in every protocol: (i) the number of washings and (ii) the methodology used for washing (Stepanović et al., 2007). Two washing steps appear to be sufficient to get rid of the planktonic cells and minimise biofilm impairment (Toté, Vanden Berghe, Maes, & Cos, 2008). Concerning the techniques used for washing, a variety of methods are applied, but the most simple and effective technique is pipetting carefully the medium out of the wells, while slightly lifting the plate (Deighton, Capstick, Domalewski, & Van Nguyen, 2001; Stepanović et al., 2007). However, monitoring the biofilm integrity during washing is very important and in order to be accurate, the wells that lose visible clusters of biofilm cells should be excluded from further calculations (Deighton et al., 2001). Collecting biofilm cells from well plates can be done using cell scrapers or sonication (Bjerkan, Witsø, & Bergh, 2009). However, when biofilm bacteria are separated from the surface they were attached to, they change their physiology and become planktonic cells again when back in suspension (Kaplan, 2010; Kostakioti, Hadjifrangiskou, & Hultgren, 2013). As a consequence, analysis of the isolated biofilm cells has to be done immediately, before gene expression variations arise.

A last comment about biofilm analysis refers to several applied microscopic techniques, from electron microscopy to fluorescence microscopy, which are used for the detection and visualisation of biofilms. During such applications, it is important to standardise the position in the well where the structure and thickness of the biofilm population will be analysed since the biofilm morphology at the centre (mid-point) of the well will differ significantly from the edges. The flow characteristics in well plates were analysed in order to link hydrodynamics to the behaviour of cell cultures using computational simulations (Salek, Sattari, & Martinuzzi, 2012). They concluded that the fluid dynamics and wall shear stress distribution within a cylindrical well undergoing orbital shaking are controlled by the liquid medium volume, the orbital radius of gyration and angular speed. The influence of fluctuating shear stresses on the *S. aureus* biofilm morphology and tolerance to ABs, using a six-well plate agitated by an orbital shaker, has been reported (Kostenko, Salek, Sattari, & Martinuzzi, 2010). They observed that the biofilm cell density increases significantly and proportionally to the radial distance from the centre, when the microtitre plates were agitated at 100 rpm. Also the 3D structures of the biofilms became thicker at the exterior of the edges. Resistance to ABs was correlated with the morphology of the bacteria and was mostly higher in regions with high shear stress.

4.3 Comparative -Omics Studies

Numerous genomic and proteomic studies have been performed to unravel the molecular pathways involved in the different stages of the biofilm development. Important studies are comparative set-ups which evaluate the functional difference between a bacteria living in a biofilm and a planktonic bacterium. Such comparative approaches have taught us that the biofilm phenotype is extremely different from their planktonic counterparts (De Angelis, Siragusa, Campanella, Di Cagno, & Gobbetti, 2015; Mukherjee, Ow, Noirel, & Biggs, 2011; Post et al., 2014; Ren, Bedzyk, Thomas, Ye, & Wood, 2004; Tremoulet, Duche, Namane, Martinie, & Labadie, 2002).

Genomic studies are mostly based on DNA microarrays and have resulted in differentially expressed gene sets between biofilm and planktonic organisms. These cDNA-based studies revealed that there is no common gene expression pattern for 'biofilms' (Sauer, 2003). However, some general differences between the two bacterial lifestyles are revealed (e.g. higher cell density, lower oxygen levels and higher osmolarity in biofilms than in a planktonic lifestyle (Prigent-Combaret, Vidal, Dorel, & Lejeune, 1999)). Differential genes in mRNAs of planktonic and 6-h biofilms of *Pseudomonas putida* were identified. Some genes encode for structural components of flagella and type IV pili, and other differential genes are involved in polysaccharide biosynthesis. With immunoblot analysis of FliC, they validated the presence of flagella in planktonic cultures, but as expected flagella were not expressed in 12 or 24 h biofilms (Sauer & Camper, 2001). Other studies confirmed flagellum repression in biofilm populations (Pratt & Kolter, 1998; Prigent-Combaret et al., 1999). Furthermore, Beloin et al. made an interesting comparison between biofilm cells, cells in exponential planktonic phase and cells in stationary planktonic phase with the aim of identifying true biofilm-specific genes using DNA microarrays (Beloin et al., 2004). They demonstrated that among the 64 genes found to be the most induced ($\geq$twofold ratio) in biofilm versus exponential phase, 39 of them were not induced in the biofilm condition when compared with stationary phase. These 39 genes (for instance, genes involved in cell division and ion transport and genes coding for fimbriae components) are not biofilm specific but rather related to a stationary growth phenotype within the mature *E. coli* biofilm. The remaining 25 genes were also overexpressed in biofilm versus stationary growth phase, defining a set of biofilm-specific gene expressions. These genes encode mainly membrane proteins, regulators involved in envelope stress and carbohydrate transporters. Other studies confirm the fact that bacteria develop stress responses within biofilms (Ferrières & Clarke, 2003; Otto & Silhavy, 2002; Wang et al., 2011). To validate the differentially expressed genes functionally, analysing the biofilm phenotype of specific mutants is a widely used approach. Most genes overexpressed in mature *E. coli*K-12 biofilms, such as *recA*, *cpxAPR*, *pspF* and *msr*A, were not involved in the early steps of the biofilm formation process (Beloin & Ghigo, 2005). The mature biofilm-specific gene mutants were not able to form a mature biofilm, but their adhesion capacities could not be distinguished from the wild-type strain (Beloin et al., 2004). On the other hand, *pgaABCD* mutagenesis experiments in *E. coli* showed that synthesis of the polymeric substance PGA promotes adhesion and consequently biofilm formation (Wang, Preston, & Romeo, 2004).

Many studies comparing different ages of biofilms have been performed (Domka, Lee, Bansal, & Wood, 2007; Park et al., 2014; Resch et al., 2006). The gene expression in *E. coli*K-12 biofilms of different ages (4, 7, 15 and 24 h), grown in Luria–Bertani(LB) medium on glass wool was analysed (Domka et al., 2007). Colonic acid genes are only induced in mature 24-h-old biofilms, while fimbriae-related genes were induced in biofilms of all ages. These transient gene expression profiles can be considered as snapshots of the biofilm developmental process, which is a better method than analysing only the mature biofilms, but still less realistic than *in vivo* approaches. Besides transient alterations in the process of biofilm formation, there is also a spatial phenotypic variation within the biofilm population itself (discussed in Section 2.1.2). Studying localised expression could give more realistic insights into the composition of the multilayered community. Using fluorescent reporter genes or fluorescent staining methods coupled to fluorescence microscopy are appropriate techniques for visualising and mapping biofilm heterogeneity *in vivo*. An overview of some fluorescence-based methods to measure respiratory activity, membrane permeability, species diversity and gene-specific expressions within biofilms was published (Stewart & Franklin, 2008). As an example, Finelli et al. used *in vivo* expression technology to screen a reporter gene library and to select for those clones in which gene expression is upregulated in *P. aeruginosa* grown to a 5-day-old biofilm (Finelli, Gallant, Jarvi, Lori, & Burrows, 2003). They used promoterless gene-fusion constructs that complement an adenine auxotrophic mutation (*purEK* deletion) to identify promoters that are induced *in vivo* in the biofilm by plating out biofilm cells on adenine-containing medium (Angelichio & Camilli, 2002). An interesting local transcriptomics analysis (Williamson et al., 2012) reported a microarray and quantitative real-time PCR (qPCR) experiment on two separate layers of a 72-h-old*P. aeruginosa* biofilm. Through cryo-based-micro-dissection, they were able to compare a biofilm layer 30 μm from the top with a population that grows 30 μm from the bottom from a total biofilm of 350 μm thick. At the top of the biofilm, the cells were actively dividing and showed high mRNA levels for genes regulated by the hypoxia-induced regulator Anr. Slow-growing cells deep in the biofilm showed little expression of Anr-regulated genes and thus may have sensed anoxia.

Combined localised and transient gene expression analysis can be performed by using gene-targeted GFP fusions to follow live, via *in vivo* fluorescence microscopy, specific gene expressions in the subpopulations of the biofilm architecture (Beloin & Ghigo, 2005). A detailed study of the spatio-temporal expression of the transcriptional regulator *agr*, important in biofilm formation and virulence, in *Listeria monocytogenes* during non-motile growth in two different nutritional environments was reported (Rieu et al., 2008). They revealed that *agr* was expressed preferentially in cells located outside the microcolonies, but not in static growth circumstances.

As the high AB tolerance characteristic of biofilm populations is still a limiting factor, an underutilised approach in this research area is a drug-induced comparative biofilm set-up. For example, the transcriptional response of sessile *Burkholderia cenocepacia* cultures to treatment with the AB chlorhexidine was analysed (Coenye et al., 2011). Upregulated genes in the treated biofilms encoded membrane-related proteins, efflux pumps and chemotaxis- and motility-related proteins. This, together with the downregulation of an adhesin-encoding gene, they hypothesised that sessile cells try to escape the biofilm when treated with AB. Also other induction-based experiments can be advantageous in exposing specific regulatory networks in biofilms, for example, induction by AI, acidity, oxygen supply, reactive oxygen species or even components of the immune system like cytokines (Kanangat, Postlethwaite, Cholera, Williams, & Schaberg, 2007; McLaughlin & Hoogewerf, 2006; Villa et al., 2012; Welin et al., 2003).

Remarkably, the small overlap in up and downregulated genes between different studies, together with the very small fraction of the total bacterial genome that experience variations in expression, demonstrates that biofilm formation is a complex developmental process. In addition, because proteins are a better representation of the bacterial activity; proteomics is increasing in popularity. Table 7 lists all proteomic studies based on *E. coli* biofilms.

Table 7 Studies of *E. coli* Biofilm Proteome

Research Goal	Biofilm Growth Conditions	Techniques Employed	Result	References
Spatial proteome of biofilm	UPEC cystitis isolate UTI89, ITO-coated glass slides, YESCA broth, 48 h	Imaging MS	Type 1 pili localised to the air-exposed region, curli fibres to the air–liquid interface	Floyd et al. (2015)
Effect cell extract of *Bifidobacterium* spp. on biofilm proteome	EHEC O157:H7, microtitre plate, LB, 24 h	2D gel, Coomassie Blue staining, MALDI-TOF/MS	7 differential proteins including formation of iron–sulphur protein (NifU), thiol: disulphide interchange protein (DsbA) and flagellar P-ring protein (FlgI)	Kim, Lee, Kang, Oh, and Griffiths (2012)
Outer membrane proteins biofilm versus planktonic	*E. coli* MG1655, glass wool, LB medium, 24 h	1D gel, Coomassie Blue staining, LC–MS/MS	Unique to the biofilm growth are proteins involved in stress resistance, metabolism of carbohydrates and amino acids and ATP synthesis	Mukherjee, Karunakaran, and Biggs (2012)
Biofilm versus planktonic	*E. coli* MG1655, glass wool, LB medium, 24 h	Gel free, iTRAQ, LC–MS/MS	Differential proteins involved in acid resistance, DNA protection and binding and ABC transporters	Mukherjee et al. (2011)
Impact of *rpoS* deletion on proteome biofilms versus planktonic	*E. coli* ZK126, glass wool, MSM, 3 days	2D gel, silver staining, MALDI-TOF/MS and LC–MS/MS	35 *rpoS*-regulated proteins, some of them dependent on growth condition	Collet et al. (2008)

Table 7 Studies of *E. coli* Biofilm Proteome—cont'd

Research Goal	Biofilm Growth Conditions	Techniques Employed	Result	References
Impact indole on biofilm versus planktonic	*E. coli*S17-1,6-well plate, LB, 20 h	2D gel, silver staining, MALDI-TOF/MS,N-terminal sequencing	14 proteins overexpressed, 37 underexpressed in biofilms in the presence of indole compared to planktonic	Collet et al. (2007)
Intracellular proteome of planktonic versus biofilm	STEC O157:H7, glass fibre filter, M9 agar, 5 days	2D gel, silver staining, MALDI-TOF/MS	Role of periplasmic antioxidant systems (SodC and Tpx) in biofilm growth	Kim et al. (2006)
Biofilms on 3 types of hydrogel-coated petri dishes versus planktonic	*E. coli* XL1 blue and *E. coli* U125544, hydrogel-coated petri dish, YT, 16 h	2D gel, silver staining, LC-MS/MS	Upregulation of outer membrane protein A (OmpA) in biofilms compared to planktonic	Orme, Douglas, Rimmer, and Webb (2006)
Biofilm versus planktonic	*E. coli* O157:H7 NCTC 12900, glass fibre filter, M9 agar, 7 days	2D gel, Coomassie Blue staining, MALDI-TOF/MS	17 differential proteins, including general metabolism proteins, sugar and amino acid transporters and regulator proteins	Tremoulet et al. (2002)

Based on Pubmed key word searches 'Proteomics and *Escherichia coli* and Biofilm' and 'Proteome and *Escherichia coli* and Biofilm'.

Gel-based (e.g. two-dimensional difference gel electrophoresis) as well as gel-free (e.g. isobaric tag for relative and absolute quantitation (iTRAQ), stable isotope labelling of amino acids in cell culture, isotope-coded affinity tag (ICAT)) approaches are employed (Mukherjee et al., 2011; Petrova & Sauer, 2009; Phillips et al., 2012; Post et al., 2014; Rathsam, Eaton, Simpson, Browne, Valova, et al., 2005). To improve the protein coverage of proteomics in bacterial biofilms, pre-fractionation is a commonly applied technique. Proteins extracted from different cellular compartments are analysed separately, which gives a more detailed view of the protein distribution in the biofilm structure. A technical paper compares five different fractionation protocols for isolating outer membrane, cytoplasmic membrane, periplasmic and cytosolic proteins and evaluated the separation quality and the suitability of the samples for mass spectrometric analysis (Thein, Sauer, Paramasivam, Grin, & Linke, 2010). Washing steps with chaotropic reagents increased the purity and the amount of isolated outer membrane proteins (36 proteins) as would be expected. This method showed, for example, a contamination of the isolated outer membrane fraction of 37% with soluble cytoplasmic proteins, which were mostly ribosomal proteins. Other methods resulted in 42%, 43%, 55% and 57% cytoplasmic contaminants. Another study used a pre-fractionation of cell wall-,membrane- and cytoplasmic proteins of both planktonic and biofilm *Streptococcus mutans* cells to compare both phenotypes by gel-based mass spectrometry (MS) analysis. Their method could hardly separate the cytoplasmic fraction from the membrane fraction, resulting in very few cytoplasmic protein spots (<100), while between 200 and 600 membrane protein spots (Rathsam, Eaton, Simpson, Browne, Berg, et al., 2005).

Nonetheless, only a small proportion (<10%) (Mukherjee et al., 2011; Schembri, Kjaergaard, & Klemm, 2003) of the genome and proteome undergoes changes in expression levels when switching from non-biofilm mode of growth to biofilm mode of growth. It is assumed that the key proteins are regulated in quality instead of quantity. That is why post-translational modifications (phosphorylation, glycosylation and nitrosylation) of selected or all proteins should be enclosed in future research. Petrova and Sauer published in 2009 a phosphoproteomic analyses during *P. aeruginosa* biofilm development (Petrova & Sauer, 2009). Phosphorylated proteins of planktonic cells and of five different ages of biofilms were enriched by metal oxide affinity chromatography and detected by both 2D gel-based immunoblot analysis with anti-Phospho-(Ser/Thr)Phe antibodies and LC–MS/MS analysis with ICAT labelling. They identified three undescribed two-component systems (BfiSR harbouring an RpoD-like domain, an OmpR-like BfmSR and MifSR) which were sequentially phosphorylated during biofilm formation. Moreover, this phosphoproteomic study is also a good example of a transient biofilm analysis. When biofilm development proceeds, the local chemical environments and thus also the bacterial behaviours and/or genetic profiles change. Consequently, it is important to identify transient gene/protein expression profiles in order to completely unravel the process of biofilm formation.

Next to genomic, transcriptomic and proteomic analyses, metabolome analysis or metabolomics maps the final output of the biological function and is directly related to pathogenesis. Measuring the produced metabolites can tell us which pathways are truly active in the cellular population. Capillary electrophoresis (CE) combined with time-of-flight mass spectrometry (TOF-MS), or CE–MS, is excellent in separating ionised small molecules and has been proven suitable to divide, identify and quantify metabolic intermediates (Takahashi, Washio, & Mayanagi, 2012). This was proven by a study where the metabolites from the carbohydrate pathways (central carbon metabolism, EMP pathway, pentose phosphate pathway and Krebs cycle) from supragingival plaque microflora from teeth were examined by CE–MS (Takahashi, Washio, & Mayanagi, 2010). Junka and co-workers searched differences in intracellular metabolite patterns of *S. aureus* biofilm versus planktonic forms (Junka et al., 2013). They used nuclear magnetic resonance spectroscopy to measure four significantly different metabolites between these conditions, namely elevated levels of isoleucine, 2,3-butanediol and alanine in biofilm cells and a higher glycine-betaine concentration in planktonic cells. The higher levels of the anaerobic fermentation product 2,3-butanediol in biofilm cells for instance support the hypothesis that biofilms experience anoxic conditions. This is also confirmed by the study where the comparison of planktonic and biofilm cells from *Acinetobacter baumannii* (Yeom, Shin, Yang, Kim, & Hwang, 2013) revealed increased levels of intracellular metabolites acetate, pyruvate and succinate in the sessile fraction that may explain pyruvate fermentation in the mature biofilm population. Also UDP-glucose, AMP, glutamate and lysine were increased in the biofilm condition compared to planktonic cells. A recent review includes information on designing a metabolomics experiment (Zhang & Powers, 2012).

4.4 Experimental Design

It is difficult to standardise the process of biofilm formation in a laboratory set-up. Therefore, it is important to define the different parameters that affect the biofilm formation capacity in the models being used. The experimental set-up is crucial to the outcome of the biofilm experiment and its interpretation.

4.4.1 Growth Conditions

First of all, the growth conditions chosen to produce biofilms, such as the surface used to grow on, the growth medium and the type of bacterial/yeast strain(s), are crucial in the standardisation of the biofilm formation process in a laboratory set-up.

4.4.1.1 Surface

As discussed earlier, the properties of the surface, where the biofilm is attached to, play an important role. The chemical composition of the material, but also the surface area where the biofilm adheres, will determine the biofilm growth. Working with, for instance, a 96- or a 24-well plate as an *in vitro* model to produce biofilms in a laboratory environment determines the final biofilm phenotype, even when growth is started from identical inocula. Additionally, the crude material of the surface influences the biofilm-forming capacity as the roughness and hydrophobicity of the surface affect the bacterial attachment phase. A comparison of stainless steel (hydrophilic), glass (hydrophilic) and polyvinyl chloride (hydrophobic) surfaces to grow *Salmonella* spp. biofilms, concluded that glass was the best surface material to diminish biofilm production of that species (De Oliveira et al., 2014). Additionally, a surface exposed in a liquid medium will inevitably and rather immediately (within seconds) become conditioned or coated by components from that surrounded medium, which changes the physicochemical properties (roughness, hydrophobicity, chemical composition) of the surface, affecting bacterial attachment. Such conditioning films facilitate bacterial attachment due to their polar character, which is a stimulus for biofilm formation (Garrett, Bhakoo, & Zhang, 2008). Additional oxidations and hydrations due to cell appendage interactions even induce irreversible adhesion (Garrett et al., 2008). As an example, fibrinogen-coated polymethylmethacrylate coverslips promoted adherence for several clinical isolated *S. aureus* strains to a higher extent than laminin-coated surfaces (Herrmann et al., 1988). Addition of bovine serum albumin to periwinkle wilt culture medium will also enhance the formation of a conditioning film on a glass surface (Lorite et al., 2011).

4.4.1.2 Effect of Nutrients and Oxygen

Environmental conditions such as the available nutrients, oxygen and vitamins, the pH and the temperature of the surrounding in which the bacteria are living influence the efficacy of biofilm growth (Garrett et al., 2008). For example, our own experiments comparing 1640 and tryptic soy broth (TSB) indicated that *E. coli* ATCC10536 produce abundantly more biofilm in 1640 medium than in TSB medium (Fig. 10). We suppose that this is due to the

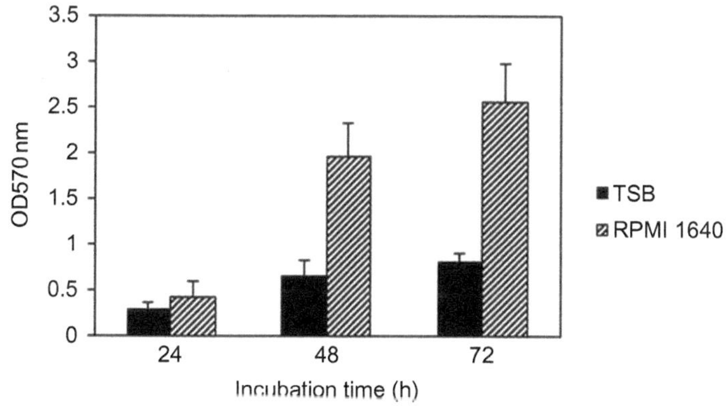

Figure 10 The effect of medium on biofilm formation. *E. coli* ATCC10536 biofilms were grown for 24, 48 and 72 h in 96-well plates ($N=24$) in two different media, TSB and 1640. They are inoculated with 10^5 CFU/ml. Error bars represent the standard deviation. Biofilm mass was determined using crystal violet staining (unpublished data).

presence of high vitamin concentrations and a reducing agent, glutathione, in the 1640 medium.

The effect of growth media was confirmed by Labrie et al. where *Actinobacillus pleuropneumoniae* biofilm formation, grown in five different culture media, was analysed by comparing the biofilm mass and the transcriptional profile (Labrie et al., 2010). They observed no biofilm formation for the *A. pleuropneumoniae* reference strain S4074 after 24 h of growth in a 96-well plate in LB medium, while there was a slight biofilm formation in wells with TSB, Mueller Hinton and brain heart infusion-A(BHI-A) medium. They tested BHI broth from two different suppliers, called BHI-A and BHI-B, with the latter showing a much higher biofilm formation than the former, which can be explained by the higher concentrations of zinc in BHI-A, which inhibits biofilm formation. Transcriptional profiles comparing *A. pleuropneumoniae* biofilms in these media demonstrated an upregulation of autotransporter adhesin genes and genes involved in matrix biosynthesis and zinc transport after growth in BHI-B. A study evaluating the effects of growth medium, temperature and incubation time on biofilm formation by *Enterobacter cloacae* strains using a microtitre plate model system (Nyenje, Green, & Ndip, 2013) showed stronger biofilm formation when incubated at 37 °C compared to 25 °C, which is expected due to the faster growth rate of bacteria at higher temperatures. However, for *Stenotrophomonas maltophilia*, a higher biofilm mass was measured at 32 °C compared to 37 °C (Di Bonaventura, Stepanović, Picciani,

Pompilio, & Piccolomini, 2007). It is postulated that the expression of bacterial surface appendages as pili and flagella that has an impact on biofilm formation is temperature dependent (Herald & Zottola, 1988). The attachment of *L. monocytogenes* isolates to stainless steel at three different temperatures, 10, 21 and 35 °C, resulted in cells with fibrils only at 21 °C (Herald & Zottola, 1988). Thus, depending on the strain and species, an optimum growing temperature will favour the bacterial growth efficiency and consequently biofilm formation (Garrett et al., 2008).

Protein expression and bacterial growth are also influenced by the intracellular and extracellular pH of the bacteria. pH changes disturb several metabolic processes, leading to an alteration of the biofilm development (Garrett et al., 2008). Many papers report the effect of temperature and pH on biofilm formation (Del Carpio-Perochena et al., 2015; Di Bonaventura et al., 2007; Hostacká, Ciznár, & Stefkovicová, 2010; Marsh, 2009; Nostro et al., 2012), but no general conclusion can be drawn, since the effect is again species dependent.

Changing the agitation velocity can also change the lifestyle of the bacteria, from free-living cells at a very high rotation speed to sticky biofilm cells when no or slow shaking is used. For producing mature biofilms in microtitre plates, we have demonstrated that a slow horizontal shaking at 25 rpm promotes more biofilm formation than no agitation at all (unpublished data) (Fig. 11). We suppose that biofilm formation will be enhanced by low-speed

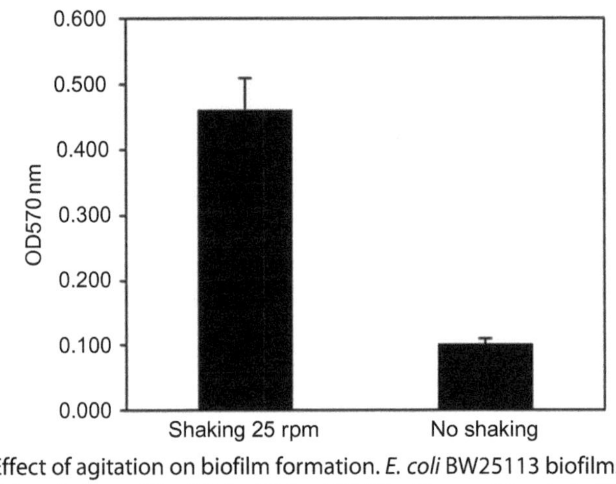

Figure 11 Effect of agitation on biofilm formation. *E. coli* BW25113 biofilms were grown in 1640, in 24-well plates ($N = 20$) for 72 h at 37 °C with or without shaking at 25 rpm. The biofilm mass is quantitated with 0.1% CV staining. The error bars represent the standard deviation (unpublished data).

shaking because it supports a uniform dispersion of nutrients and oxygen. However, some researchers showed that mechanical agitation at high speed can partly remove biofilm populations and thus enhances dispersal (Jiang, Pei, & Hu, 2010). The effect of agitation in *in vitro*set-ups can be extrapolated to the effect of flow rates in *in vivo* situations. The effects of these flow rates on the biofilm-forming capacity can be investigated in a laboratory system with dynamic models. It is reported that *E. coli* biofilm formation was favoured at low flow rates in a flow cell system. It is hypothesised that the shear stress effects were playing a more important role than mass transfer (Moreira et al., 2013).

4.4.1.3 Strain

The above discussed factors affecting biofilm development are species dependent. Every bacterial species exhibits a different biofilm phenotype, probably due to morphology and metabolism. Some microorganisms produce a more 'slimy' matrix than others (Silva et al., 2009), and some bacterial biofilms grow more in height, some more in width. But even within the same species group, there are differences. For example, the ATCC10536 *E. coli* show a much stronger biofilm phenotype than *E. coli* BW25113, based on its total biofilm mass (unpublished data), even when using the same inoculation size. In addition, the use of inocula derived from a culture medium, a nutrient agar or a cryogenic stock can influence the biofilm production. All tested *E. cloacae* isolates were able to produce a biofilm, cultured for 24 h in BHI medium in 96-well plates at 37 °C, when starting from a nutrient agar inoculum, while only 93% of the isolates from culture medium were able to form a biofilm under the same growth conditions (Nyenje et al., 2013). It is also important to inoculate single cells and thus avoid cell clusters to diminish false-positive results. Such bacterial clusters will experience a small advantage over single cells to form biofilms. Consequently, thoroughly mixing the cell suspension is necessary before inoculation (Stepanović et al., 2007).

4.4.2 Expression Profiles

The gene expression profiles are dependent on the experimental set-up. As a comprehensive example, two separate DNA microarray-based studies (Ren et al., 2004; Schembri et al., 2003), which both compare the gene expression profiles of *E. coli* as a biofilm phenotype and as planktonic cells, will be discussed. Ren and co-workers identified only 19 altered genes (induced and repressed) in *E. coli*K-12 biofilms, compared to planktonic cells, while Schembri et al. reported hundreds of differential genes. Remarkably, both researchers used the same statistical *t*-test analysis, reporting genes as significantly differential when fold changes were greater than 2.5 and *p*-values less than 0.05. There were less than 10 differentially expressed genes in common, based on their published data. Such incomparability is likely due to differences in growth conditions, the type of strain and the age of the biofilm. Ren et al. grew *E. coli* K-12 ATCC25404 biofilms in LB medium on metal plates in continuous reactors for 5 days, while Schembri et al. used a four-channel flow system to form *E. coli* K-12 MG1655 biofilms of 42 h old in 3-(*N*-morpholino)propanesulphonic acid (MOPS) minimal media supplemented with 0.2% glucose. The influence of the experimental design on the genetic character of the biofilm is also relevant in differential proteomic studies. Twelve identified and quantified proteins were determined in an iTRAQ study that were significantly increased or decreased in abundance between biofilm and planktonically grown *E. coli* cells (Mukherjee et al., 2011). There were only a few differential proteins in common with the experiments of Collet et al., which can be partly explained by their different biofilm growth conditions (*E. coli*S17-1 biofilms grown in LB in a six-well plate for 20 h versus *E. coli* MG1655 biofilms grown in LB on glass wool for 24 h). These two studies also differ in their used proteomic approach: a gel-based method versus a study based on peptide labelling (Table 7) (Collet et al., 2007; Mukherjee et al., 2011).

4.5 Multispecies Biofilms

Biofilms in nature consist of multiple species. This characteristic is a serious impediment in the progress of biofilm research. Limited experiments have been performed on multispecies biofilms since many methodologies are not able to distinguish between different types of cells within the biofilm population. It is crucial to identify and localise the different species in polymicrobial biofilms. Viable plate count on selective and differential agars (Table 8) can select for certain bacteria, but this is mostly a species-unspecific detection. The most popular methods to identify the diversity of microorganisms are based on ribosomal amplification, e.g., denaturing gradient gel electrophoresis of PCR-amplified 16S rRNA or qPCR analysis with a Taqman probe. FISH in combination with CLSM is another method to identify and visualise microbial species in the multispecies biofilms using species-dependent probes (Yang et al., 2011). A five-colour multiplex FISH analysis is an interesting example (Al-Ahmad et al., 2007). Furthermore, it was demonstrated that both qPCR levels and fluorescence microscopic

Table 8 Some Selective and Differential Agars

Medium	Selective and Differential Agents	Selected Organisms	Differentiation
Blood agar	Sheep blood cells	Most bacteria	Haemolytic bacteria, e.g. – *S. aureus*: yellow colonies – *S. epidermidis*: white pigment
Chocolate agar	Haemoglobin and supplements	*Haemophilus* and *Neisseria* species	
Eaton's agar	Penicillin	*Mycoplasma pneumoniae*	
Eosin methylene blue agar	Lactose, eosin and methylene blue	Gram-negative enteric species	– *Enterobacter aerogenes*: large, pink colonies with dark centre – *E. coli*: small, dark colonies with a green sheen – *Pseudomonas*, *Proteus*, *Salmonella* and *Shigella* sp.: colourless colonies
Hektoen Enteric agar	Lactose, sucrose, salicin, bile salts, sodium thiosulphate, ferric ammonium citrate, bromothymol blue, acid fuchsin	Gram-negative enteric species, especially *Salmonella* and *Shigella*	
MacConkey's agar	Lactose, bile salts, neutral red, crystal violet	Gram-negative bacteria	– Lactose fermenting bacteria (*E. coli*, *Enterobacter*, *Klebsiella*): red/pink colonies – Non-lactose fermenting bacteria (*P. aeruginosa*, *Salmonella*, *Shigella*, *Proteus* sp.): white/colourless colonies
MRS agar	Sodium acetate	*Lactobacillus* species	
Mannitol salt agar	7.5% NaCl, mannitol, phenol red	*Staphylococci*	– *S. aureus*: yellow colonies – *S. epidermidis*: red colonies
Phenylethyl alcohol agar	Phenylethyl alcohol	Gram-positive bacteria	

counts based on species-specific staining by FISH or immunofluorescence yielded in higher values than colony forming units (CFU) counts on selective agars when quantifying species in polymicrobial biofilms (Ammann, Bostanci, Belibasakis, & Thurnheer, 2013).

Interactions between different species (inter-species interactions) often change the physiology of individual bacteria as well as the whole biofilm community. Mostly these interactions are competitions for nutrients to stimulate their own growth (Yang et al., 2011). In multispecies communities, the different bacteria can stimulate or inhibit each other's biofilm-forming ability comparing to the biofilm production by any of the single species (Varposhti, Entezari, & Feizabadi, 2014; Wen, Yates, Ahn, & Burne, 2010). If the cooperation between the species is synergistic or non-synergistic depends mostly on their secreted molecules, which can be for instance signal molecules, metabolic products or toxins (see Sections 2.1.4). *Aggregatibacter actinomycetemcomitans* inhibits *C. albicans* biofilm formation by means of the signalling molecule AI-2 that was secreted during growth. *C. albicans* biofilm formation increased when they were co-cultured with the Gram-negative bacteria that were deficient in AI-2 production (Bachtiar et al., 2014). Besides the biofilm-forming capacity, other physiological functions can alter in polymicrobial biofilms, e.g., virulence and drug resistance. Some interesting examples are discussed below. A study showed an increased biofilm phenotype when growing *S. aureus* in the presence of *C. albicans*, compared to the mono-bacterial biofilm. The *S. aureus* bacteria seemed to become coated in the secreted matrix of the underlying yeast cells. In addition, a higher vancomycin resistance of *S. aureus* was demonstrated when living in a dual-species biofilm (Harriott & Noverr, 2009, 2010). Furthermore, a mixed-biofilm study of *S. mutans* and *Veillonella parvula* showed no significant difference in biofilm growth between single- and dual-species biofilms, grown for 72 h, based on the viable count levels (Kara, Luppens, & Ten Cate, 2006). However, a much higher lactic acid concentration was detected in the medium of 72-h-oldsingle-species biofilms of *S. mutans* compared to the dual-species community, although the pH of the medium was similar. This could be explained by the significant correlation between decreasing concentrations of lactic acid and increasing concentrations of acetic acid. Another study evaluated the phenazine toxin production of *P. aeruginosa* in the presence of *C. albicans*. The signalling molecule farnesol produced by *C. albicans* stimulates the production of phenazine toxins in *P. aeruginosa* clinical isolates. Even bacterial mutants that lack the QS regulator responsible for phenazine expression regained their toxin-producing ability when cultured with *C. albicans* (Cugini, Morales, & Hogan, 2010). To conclude, many dual-species biofilm research studies can be found in the literature, but multi-species biofilm experiments are less frequently reported (Cavalcanti et al., 2015; Da Silva Fernandes, Kabuki, & Kuaye, 2015; Kim & Izadjoo, 2015). It is important in biofilm research to have a better understanding of multispecies biofilms and find ways to manipulate the biofilm properties, especially its virulence. Effective biofilm inhibitory compounds, based on single-species experiments, are probably less antibacterial to multispecies biofilms because many more factors are involved. Some attempts to approach a more realistic scenario have been made by using spent growth medium during biofilm formation. In this way, secreted molecules from other bacteria or yeasts are incorporated in the biofilm development. For example, Kim et al. screened the spent media from 4104 separate *actinomycetes* strains in order to identify anti-biofilm strains against *P. aeruginosa* (Kim et al., 2012). The spent media of two strains, namely *Streptomyces* sp. BFI 230 and *Kribbella* sp. BFI 1562, inhibited *P. aeruginosa* biofilm formation by 90% without affecting the growth of planktonic *P. aeruginosa* cells. Transcriptome analysis revealed an interference with the iron acquisition by *P. aeruginosa*. Another study (Bandara, Cheung, Watt, Jin, & Samaranayake, 2013) investigated the effects of *E. coli* biofilms supernatants of different ages on the biofilm-forming capacity of *Candida* species and demonstrated a significant inhibitory effect due to the *E. coli* secretory elements. Another study reported future research ideas concerning spent medium analysis. Seventy-nine cell-free culture supernatants from a variety of single oral streptococci was screened to identify extracellular compounds that inhibit biofilm formation by the oral anaerobe *Porphyromonas gingivalis* strain 381 (Christopher, Arndt, Cugini, & Davey, 2010). These authors discovered a 48-kDa arginine deiminase protein that was responsible for inhibition. By using qPCR, it was concluded that *P. gingivalis* can sense this extracellular protein, produced by an oral *Streptococcus intermedius*, and respond by down-regulating the expression of cell-surface appendages.

Before producing multispecies biofilms in a laboratory for experimental research, we want to firstly note that the order in which the organisation of a multispecies biofilm happens in nature is unknown. That is why simultaneous inoculation of the several species in a lab model system is a good alternative. For example, *S. aureus* bacteria nestle in the secreted matrix and the hyphae of the underlying *C. albicans* yeast cells, when culturing them together (Harriott & Noverr, 2009, 2010). Secondly, distinguishing between the identified genes or proteins derived from the different species is extremely hard to accomplish, as well as discovering the cause of the differentially expressed genes/proteins by each species. An outstanding technical paper demonstrated a powerful technique to investigate the complexity of transcriptional changes in mixed-species biofilms, which is based on multiplex-labelled cDNA hybridisations with a NimbleGen DNA array (Redanz, Standar, Podbielski, & Kreikemeyer, 2011). With a 385 KTM chip, 385,000 gene probes could be tested allowing coverage of several bacterial genomes on one array.

To conclude, these concerns of species identification, inter-species interactions, species-specific expression profiling and inoculation order hinder researchers to venture to competitive polymicrobial biofilm research.

REFERENCES

Abed, S. E., Ibnsouda, S. K., Latrache, H., & Hamadi, F. (2012). Scanning electron microscopy (SEM) and environmental SEM: Suitable tools for study of adhesion stage and biofilm formation. In V. Kazmiruk (Ed.), *Scanning electron microscopy* (pp. 717–730). INTECH Open Access Publisher.

Al-Ahmad, A., Wunder, A., Auschill, T. M., Follo, M., Braun, G., Hellwig, E., et al. (2007). The in vivo dynamics of *Streptococcus* spp., *Actinomyces naeslundii*, *Fusobacterium nucleatum* and *Veillonella* spp. in dental plaque biofilm as analysed by five-colour multiplex fluorescence in situ hybridization. *Journal of Medical Microbiology*, *56*, 681–687.

Amalaradjou, M., & Venkitanarayanan, K. (2013). Role of bacterial biofilms in catheter-associated urinary tract infections (CAUTI) and strategies for their control. In T. Nelius (Ed.), *Recent advances in the field of urinary tract infections* (pp. 1–15). INTECH Open Access Publisher.

Ammann, T. W., Bostanci, N., Belibasakis, G. N., & Thurnheer, T. (2013). Validation of a quantitative real-time PCR assay and comparison with fluorescence microscopy and selective agar plate counting for species-specific quantification of an in vitro subgingival biofilm model. *Journal of Periodontal Research*, *48*, 517–526.

Anderl, J. N., Franklin, M. J., & Stewart, P. S. (2000). Role of antibiotic penetration limitation in *Klebsiella pneumoniae* biofilm resistance to ampicillin and ciprofloxacin. *Antimicrobial Agents and Chemotherapy*, *44*(7), 1818–1824.

Angelichio, M. J., & Camilli, A. (2002). In vivo expression technology. *Society*, *70*(12), 6518–6523.

Aparna, M. S., & Yadav, S. (2008). Biofilms: Microbes and disease. *The Brazilian Journal of Infectious Diseases: An Official Publication of the Brazilian Society of Infectious Diseases*, *12*, 526–530.

Bachtiar, E. W., Bachtiar, B. M., Jarosz, L. M., Amir, L. R., Sunarto, H., Ganin, H., et al. (2014). AI-2 of *Aggregatibacter actinomycetemcomitans* inhibits *Candida albicans* biofilm formation. *Frontiers in Cellular and Infection Microbiology*, *4*, 94.

Baharoglu, Z., & Mazel, D. (2014). SOS, the formidable strategy of bacteria against aggressions. *FEMS Microbiology Reviews*, *38*, 1126–1145.

Bandara, H. M. H. N., Cheung, B. P. K., Watt, R. M., Jin, L. J., & Samaranayake, L. P. (2013). Secretory products of *Escherichia coli* biofilm modulate Candida biofilm formation and hyphal development. *Journal of Investigative and Clinical Dentistry*, *4*, 186–199.

Battesti, A., Majdalani, N., & Gottesman, S. (2011). The RpoS-mediated general stress response in *Escherichia coli*. *Annual Review of Microbiology, 65*, 189–213.

Belas, R. (2014). Biofilms, flagella, and mechanosensing of surfaces by bacteria. *Trends in Microbiology, 22*, 517–527.

Beloin, C., & Ghigo, J. M. (2005). Finding gene-expression patterns in bacterial biofilms. *Trends in Microbiology, 13*(1), 16–19.

Beloin, C., Roux, A., & Ghigo, J.-M. (2008). *Escherichia coli* biofilms. *Current Topics in Microbiology and Immunology, 322*, 249–289.

Beloin, C., Valle, J., Latour-Lambert, P., Faure, P., Kzreminski, M., Balestrino, D., et al. (2004). Global impact of mature biofilm lifestyle on *Escherichia coli*K-12 gene expression. *Molecular Microbiology, 51*, 659–674.

Bjarnsholt, T., Alhede, M., Alhede, M., Eickhardt-Sørensen, S. R., Moser, C., Kühl, M., et al. (2013). The in vivo biofilm. *Trends in Microbiology, 21*(9), 466–474.

Bjerkan, G., Witsø, E., & Bergh, K. (2009). Sonication is superior to scraping for retrieval of bacteria in biofilm on titanium and steel surfaces in vitro. *Acta Orthopaedica, 80*(2), 245–250.

Boehm, A., Steiner, S., Zaehringer, F., Casanova, A., Hamburger, F., Ritz, D., et al. (2009). Second messenger signalling governs *Escherichia coli* biofilm induction upon ribosomal stress. *Molecular Microbiology, 72*, 1500–1516.

Cavalcanti, Y. W., Morse, D. J., da Silva, W. J., Del-Bel-Cury, A. A., Wei, X., Wilson, M., et al. (2015). Virulence and pathogenicity of *Candida albicans* is enhanced in biofilms containing oral bacteria. *Biofouling, 31*, 27–38.

Chatterji, D., & Kumar Ojha, A. (2001). Revisiting the stringent response, ppGpp and starvation signaling. *Current Opinion in Microbiology, 4*, 160–165.

Christopher, A. B., Arndt, A., Cugini, C., & Davey, M. E. (2010). A streptococcal effector protein that inhibits *Porphyromonas gingivalis* biofilm development. *Microbiology, 156*, 3469–3477.

Clais, S., Boulet, G., Van kerckhoven, M., Lanckacker, E., Delputte, P., Maes, L., et al. (2015). Comparison of viable plate count, turbidity measurement and real-time PCR for quantification of *Porphyromonas gingivalis*. *Letters in Applied Microbiology, 60*, 79–84.

Coenye, T., & Nelis, H. J. (2010). In vitro and in vivo model systems to study microbial biofilm formation. *Journal of Microbiological Methods, 83*(2), 89–105.

Coenye, T., Van Acker, H., Peeters, E., Sass, A., Buroni, S., Riccardi, G., et al. (2011). Molecular mechanisms of chlorhexidine tolerance in *Burkholderia cenocepacia* biofilms. *Antimicrobial Agents and Chemotherapy, 55*(5), 1912–1919.

Cohen, A., Dagli, M., Stavropoulos, S. J., Mondschein, J., Soulen, M., Shlansky-Goldberg, R., et al. (2011). Silicone and polyurethane tunneled infusion catheters: A comparison of durability and breakage rates. *Journal of Vascular and Interventional Radiology, 22*(5), 638–641.

Collet, A., Cosette, P., Beloin, C., Ghigo, J. M., Rihouey, C., Lerouge, P., et al. (2008). Impact of rpoS deletion on the proteome of *Escherichia coli* grown planktonically and as biofilm. *Journal of Proteome Research, 7*, 4659–4669.

Collet, A., Vilain, S., Cosette, P., Junter, G. A., Jouenne, T., Phillips, R. S., et al. (2007). Protein expression in *Escherichia coli*S17 1 biofilms: Impact of indole. *Antonie van Leeuwenhoek International Journal of General and Molecular Microbiology, 91*, 71–85.

Cos, P., Tote, K., Horemans, T., & Maes, L. (2010). Biofilms: An extra hurdle for effective antimicrobial therapy. *Current Pharmaceutical Design, 16*(20), 2279–2295.

Cox, A. J. (1990). Comparison of catheter surface morphologies. *British Journal of Urology, 65*, 55–60.

Cugini, C., Morales, D. K., & Hogan, D. A. (2010). *Candida albicans*-produced farnesol stimulates Pseudomonas quinolone signal production in LasR-defective*Pseudomonas aeruginosa* strains. *Microbiology, 156*, 3096–3107.

Curtis, J., & Colas, A. (2004). Medical applications of silicones. In B. D. Ratner, A. S. Hoffman, F. J. Schoen, & J. E. Lemons (Eds.), *Biomaterials science: An introduction to materials in medicine* (2nd ed., pp. 697–707). Canada: Elsevier Academic Press.

Damgaard, L. R., Nielsen, L. P., & Revsbech, N. P. (2001). Methane microprofiles in a sewage biofilm determined with a microscale biosensor. *Water Research, 35*(6), 1379–1386.

Da Silva Fernandes, S., Kabuki, D. Y., & Kuaye, A. (2015). Behavior of *Listeria monocytogenes* in a multi-species bio film with *Enterococcus faecalis* and *Enterococcus faecium* and control through sanitation procedures. *International Journal of Food Microbiology, 200*, 5–12.

De Angelis, M., Siragusa, S., Campanella, D., Di Cagno, R., & Gobbetti, M. (2015). Comparative proteomic analysis of biofilm and planktonic cells of *Lactobacillus plantarum* DB200. *Proteomics*, (Epub ahead of print).

Deighton, M. A., Capstick, J., Domalewski, E., & Van Nguyen, T. (2001). Methods for studying biofilms produced by *Staphylococcus epidermidis*. *Methods in Enzymology, 336*(1996), 177–195.

Del Carpio-Perochena, A., Bramante, C. M., de Andrade, F. B., Maliza, A. G. A., Cavenago, B. C., Marciano, M. A., et al. (2015). Antibacterial and dissolution ability of sodium hypochlorite in different pHs on multi-species biofilms. *Clinical Oral Investigations,* (Epub ahead of print).

De Oliveira, D. C. V., Fernandes Júnior, A., Kaneno, R., Silva, M. G., Araújo Júnior, J. P., Silva, N. C. C., et al. (2014). Ability of *Salmonella* spp. to produce biofilm is dependent on temperature and surface material. *Foodborne Pathogens and Disease, 11*(6), 478–483.

Di Bonaventura, G., Stepanović, S., Picciani, C., Pompilio, A., & Piccolomini, R. (2007). Effect of environmental factors on biofilm formation by clinical *Stenotrophomonas maltophilia* isolates. *Folia Microbiologica, 52*(1), 86–90.

Domka, J., Lee, J., Bansal, T., & Wood, T. K. (2007). Temporal gene-expression in *Escherichia coli*K-12 biofilms. *Environmental Microbiology, 9*, 332–346.

Donlan, R. M. (2002). Biofilms: Microbial life on surfaces. *Emerging Infectious Diseases, 8*(9), 881–890.

Dörr, T., Lewis, K., & Vulić, M. (2009). SOS response induces persistence to fluoroquinolones in *Escherichia coli*. *PLoS Genetics, 5*(12), e1000760.

Dufour, D., Leung, V., & Lévesque, C. M. (2012). Bacterial biofilm: Structure, function, and antimicrobial resistance. *Endodontic Topics, 22*, 2–16.

Dunne, W. M. (2002). Bacterial adhesion: Seen any good biofilms lately? *Society, 15*(2), 155–166.

Ferrières, L., & Clarke, D. J. (2003). The RcsC sensor kinase is required for normal biofilm formation in *Escherichia coli*K-12 and controls the expression of a regulon in response to growth on a solid surface. *Molecular Microbiology, 50*, 1665–1682.

Fetzner, S. (2014). Quorum quenching enzymes. *Journal of Biotechnology, 201*, 2–14.

Fiedoruk, K., Daniluk, T., Swiecicka, I., Sciepuk, M., & Leszczynska, K. (2014). Type II toxin-antitoxin systems are unevenly distributed among *Escherichia coli* phylogroups. *Microbiology, 161*, 158–167.

Finelli, A., Gallant, C. V., Jarvi, K., Lori, L., & Burrows, L. L. (2003). Use of in-biofilm expression technology to identify genes involved in *Pseudomonas aeruginosa* biofilm development. *Journal of Bacteriology, 185*(9), 2700–2710.

Floyd, K. A., Moore, J. L., Eberly, A. R., Good, J. A. D., Shaffer, C. L., Zaver, H., et al. (2015). Adhesive fiber stratification in uropathogenic *Escherichia coli* biofilms unveils oxygen-mediated control of type 1 pili. *PLoS Pathogens, 11*, e1004697.

Furukawa, S., Kuchma, S. L., & O'Toole, G. A. (2006). Keeping their options open: Acute versus persistent infections. *Journal of Bacteriology, 188*(4), 1211–1217.

Galloway, W. R. J., Hodgkinson, J. T., Bowden, S. D., Welch, M., & Spring, D. R. (2011). Quorum sensing in Gram-negative bacteria: Small-molecule modulation of AHL and AI-2 quorum sensing pathways. *Chemical Reviews*, *111*, 28–67.
Garrett, T. R., Bhakoo, M., & Zhang, Z. (2008). Bacterial adhesion and biofilms on surfaces. *Progress in Natural Science*, *18*, 1049–1056.
Grant, D. M., & Bott, T. R. (2003). Biocide dosing strategies for biofilm control. In *Heat exchanger fouling and cleaning: Fundamentals and applications Engineering conference international 2003*.
Hall-Stoodley, L., & Stoodley, P. (2009). Evolving concepts in biofilm infections. *Cellular Microbiology*, *11*, 1034–1043.
Hall-Stoodley, L., Stoodley, P., Kathju, S., Høiby, N., Moser, C., William Costerton, J., et al. (2012). Towards diagnostic guidelines for biofilm-associated infections. *FEMS Immunology and Medical Microbiology*, *65*, 127–145.
Harriott, M. M., & Noverr, M. C. (2009). *Candida albicans* and *Staphylococcus aureus* form polymicrobial biofilms: Effects on antimicrobial resistance. *Antimicrobial Agents and Chemotherapy*, *53*(9), 3914–3922.
Harriott, M. M., & Noverr, M. C. (2010). Ability of *Candida albicans* mutants to induce *Staphylococcus aureus* vancomycin resistance during polymicrobial biofilm formation. *Antimicrobial Agents and Chemotherapy*, *54*(9), 3746–3755.
Harriott, M. M., & Noverr, M. C. (2011). Importance of Candida—Bacterial polymicrobial biofilms in disease. *Trends in Microbiology*, *19*(11), 557–563.
Häussler, S., & Parsek, M. R. (2010). Biofilms 2009: New perspectives at the heart of surface-associated microbial communities. *Journal of Bacteriology*, *192*(12), 2941–2949.
Hawser, S. P., & Douglas, L. J. (1994). Biofilm formation by Candida species on the surface of catheter materials in vitro. *Infection and Immunity*, *62*(3), 915–921.
Hengge-Aronis, R. (2002). Signal transduction and regulatory mechanisms involved in control of the sigma(S) (RpoS) subunit of RNA polymerase. *Microbiology and Molecular Biology Reviews*, *66*(3), 373–395. table of contents.
Herald, P. J., & Zottola, E. A. (1988). Attachment of *Listeria monocytogenes* to stainless steel surfaces at various temperatures and pH values. *Journal of Food Science*, *53*(5), 1549–1562.
Herrmann, M., Vaudaux, P. E., Pittet, D., Auckenthaler, R., Lew, P. D., Schumacher-Perdreau, F., et al. (1988). Fibronectin, fibrinogen, and laminin act as mediators of adherence of clinical Staphylococcal isolates to foreign material. *The Journal of Infectious Diseases*, *158*(4), 693–701.
Hidalgo, G., Burns, A., Herz, E., Hay, A. G., Houston, P. L., Wiesner, U., et al. (2009). Functional tomographic fluorescence imaging of pH microenvironments in microbial biofilms by use of silica nanoparticle sensors. *Applied and Environmental Microbiology*, *75*(23), 7126–7135.
Høiby, N., Bjarnsholt, T., Moser, C., Bassi, G. L., Coenye, T., Donelli, G., et al. (2015). ESCMID guideline for the diagnosis and treatment of biofilm infections 2014. *Clinical Microbiology and Infection*, *21*, S1–S25.
Hostacká, A., Ciznár, I., & Stefkovicová, M. (2010). Temperature and pH affect the production of bacterial biofilm. *Folia Microbiologica*, *55*(1), 75–78.
Ibrahim, K. H., Gunderson, B. W., Hermsen, E. D., Hovde, L. B., & Rotschafer, J. C. (2004). Pharmacodynamics of pulse dosing versus standard dosing: In vitro metronidazole activity against *Bacteroides fragilis* and *Bacteroides thetaiotaomicron*. *Antimicrobial Agents and Chemotherapy*, *48*(11), 4195–4199.
Jefferson, K. K., Goldmann, D., & Pier, G. B. (2005). Use of confocal microscopy to analyze the rate of vancomycin penetration through *Staphylococcus aureus* biofilms. *Antimicrobial Agents and Chemotherapy*, *49*(6), 2467–2473.

Jiang, B., Pei, H., & Hu, W. (2010). The effect of mechanical agitation on the stripping of bio-film from ceramic particles. In *2010 4th international conference on bioinformatics and biomedical engineering, iCBBE 2010* (pp. 10–13) (2007).

Jolivet-Gougeon, A., & Bonnaure-Mallet, M. (2014). Biofilms as a mechanism of bacterial resistance. *Drug Discovery Today. Technologies, 11*, 49–56.

Joly, N., Engl, C., Jovanovic, G., Huvet, M., Toni, T., Sheng, X., et al. (2010). Managing membrane stress: The phage shock protein (Psp) response, from molecular mechanisms to physiology. *FEMS Microbiology Reviews, 34*(1), 797–827.

Junka, A. F., Deja, S., Smutnicka, D., Szymczyk, P., Ziółkowski, G., Bartoszewicz, M., et al. (2013). Differences in metabolic profiles of planktonic and biofilm cells in *Staphylococcus aureus*—1H nuclear magnetic resonance search for candidate biomarkers. *Acta Biochimica Polonica, 60*(4), 701–706.

Kalia, D., Merey, G., Nakayama, S., Zheng, Y., Zhou, J., Luo, Y., et al. (2012). Nucleotide, c-di-GMP, c-di-AMP, cGMP, cAMP, (p)ppGpp signaling in bacteria and implications in pathogenesis. *Chemical Society Reviews, 42*, 305–341.

Kanangat, S., Postlethwaite, A., Cholera, S., Williams, L., & Schaberg, D. (2007). Modulation of virulence gene expression in *Staphylococcus aureus* by interleukin-1β: Novel implications in bacterial pathogenesis. *Microbes and Infection, 9*, 408–415.

Kaplan, J. B. (2010). Biofilm dispersal: Mechanisms, clinical implications, and potential therapeutic uses. *Journal of Dental Research, 89*(3), 205–218.

Kara, D., Luppens, S. B. I., & Ten Cate, J. M. (2006). Differences between single- and dual-species biofilms of *Streptococcus mutans* and *Veillonella parvula* in growth, acidogenicity and susceptibility to chlorhexidine. *European Journal of Oral Sciences, 114*(12), 58–63.

Karimi, S., Ghafourian, S., Kalani, M. T., & Jalilian, F. A. (2015). Association between toxin-antitoxin systems and biofilm formation. *Jundishapur Journal of Microbiology, 8*(1), 1–6.

Kendall, M. M., Rasko, D. A., & Sperandio, V. (2007). Global effects of the cell-to-cell signaling molecules autoinducer-2,autoinducer-3, and epinephrine in a luxS mutant of enterohemorrhagic *Escherichia coli*. *Infection and Immunity, 75*(10), 4875–4884.

Kerstens, M., Boulet, G., Tritsmans, C., Horemans, T., Hellings, M., Delputte, P., et al. (2014). Flow cytometric enumeration of bacteria using TO-PRO(R)-3 iodide as a single-stain viability dye. *Journal of Laboratory Automation, 19*, 555–561.

Kim, H., & Izadjoo, M. J. (2015). Antibiofilm efficacy evaluation of a bioelectric dressing in mono- and multi-species biofilms. *Journal of Wound Care, 24*(2), 10–15.

Kim, Y., Lee, J. W., Kang, S. G., Oh, S., & Griffiths, M. W. (2012a). *Bifidobacterium* spp. influences the production of autoinducer-2 and biofilm formation by *Escherichia coli* O157: H7. *Anaerobe, 18*, 539–545.

Kim, Y. G., Lee, J. H., Kim, C. J., Lee, J. C., Ju, Y. J., Cho, M. H., et al. (2012b). Antibiofilm activity of Streptomyces sp. BFI 230 and Kribbella sp. BFI 1562 against *Pseudomonas aeruginosa*. *Applied Microbiology and Biotechnology, 96*, 1607–1617.

Kim, Y. H., Lee, Y., Kim, S., Yeom, J., Yeom, S., Kim, B. S., et al. (2006). The role of periplasmic antioxidant enzymes (superoxide dismutase and thiol peroxidase) of the Shiga toxin-producing*Escherichia coli* O157:H7 in the formation of biofilms. *Proteomics, 6*, 6181–6193.

Kint, C. I., Verstraeten, N., Fauvart, M., & Michiels, J. (2012). New-found fundamentals of bacterial persistence. *Trends in Microbiology, 20*(12), 577–585.

Kitanishi, K., Kobayashi, K., Kawamura, Y., Ishigami, I., Ogura, T., Nakajima, K., et al. (2010). Important roles of Tyr43 at the putative heme distal side in the oxygen recognition and stability of the Fe(II)-O2 complex of YddV, a globin-coupledheme-based oxygen sensor diguanylate cyclase. *Biochemistry, 49*(49), 10381–10393.

Kostakioti, M., Hadjifrangiskou, M., & Hultgren, S. J. (2013). Bacterial biofilms: Development, dispersal, and therapeutic strategies in the dawn of the postantibiotic era. *Cold Spring Harbor Perspectives in Medicine, 3*, 1–24.

Kostenko, V., Salek, M. M., Sattari, P., & Martinuzzi, R. J. (2010). *Staphylococcus aureus* biofilm formation and tolerance to antibiotics in response to oscillatory shear stresses of physiological levels. *FEMS Immunology and Medical Microbiology*, *59*, 421–431.

Labrie, J., Pelletier-Jacques, G., Deslandes, V., Ramjeet, M., Auger, E., Nash, J. H. E., et al. (2010). Effects of growth conditions on biofilm formation by *Actinobacillus pleuropneumoniae*. *Veterinary Research*, *41*, 3.

Lebeaux, D., Chauhan, A., Rendueles, O., & Beloin, C. (2013). From in vitro to in vivo models of bacterial biofilm-related infections. *Pathogens*, *2*, 288–356.

Lewis, K. (2007). Persister cells, dormancy and infectious disease. *Nature Reviews. Microbiology*, *5*, 48–56.

Lewis, K. (2013). Platforms for antibiotic discovery. *Nature Reviews. Drug Discovery*, *12*(5), 371–387.

Lieberman, O. J., Orr, M. W., Wang, Y., & Lee, V. T. (2014). High-throughput screening using the differential radial capillary action of ligand assay identifies ebselen as an inhibitor of diguanylate cyclases. *ACS Chemical Biology*, *9*(1), 183–192.

Lorite, G. S., Rodrigues, C. M., de Souza, A. A., Kranz, C., Mizaikoff, B., & Cotta, M. A. (2011). The role of conditioning film formation and surface chemical changes on Xylella fastidiosa adhesion and biofilm evolution. *Journal of Colloid and Interface Science*, *359*(1), 289–295.

Marsh, P. D. (2009). Dental plaque as a biofilm: The significance of pH in health and caries. *Compendium of Continuing Education in Dentistry (Jamesburg, N.J.: 1995)*, *30*, 76–78.

McBain, A. J. (2009). In vitro biofilm models: An overview. In S. Sima & M. G. Geoffrey (Eds.), *Advances in applied microbiology: Vol. 69* (1st ed., pp. 99–132): Academic Press.

McLaughlin, R. A., & Hoogewerf, A. J. (2006). Interleukin-1-induced growth enhancement of *Staphylococcus aureus* occurs in biofilm but not planktonic cultures. *Microbial Pathogenesis*, *41*, 67–79.

Moreira, J. M. R., Teodósio, J. S., Silva, F. C., Simões, M., Melo, L. F., & Mergulhão, F. J. (2013). Influence of flow rate variation on the development of *Escherichia coli* biofilms. *Bioprocess and Biosystems Engineering*, *36*, 1787–1796.

Mukherjee, J., Karunakaran, E., & Biggs, C. A. (2012). Using a multi-faceted approach to determine the changes in bacterial cell surface properties influenced by a biofilm lifestyle. *Biofouling*, *28*, 1–14.

Mukherjee, J., Ow, S. Y., Noirel, J., & Biggs, C. A. (2011). Quantitative protein expression and cell surface characteristics of *Escherichia coli* MG1655 biofilms. *Proteomics*, *11*, 339–351.

Murakami, K., Ono, T., Viducic, D., Kayama, S., Mori, M., Hirota, K., et al. (2005). Role for rpoS gene of *Pseudomonas aeruginosa* in antibiotic tolerance. *FEMS Microbiology Letters*, *242*, 161–167.

Murga, R., Miller, J. M., & Donlan, R. M. (2001). Biofilm formation by Gram-negative bacteria on central venous catheter connectors: Effect of conditioning films in a laboratory model. *Journal of Clinical Microbiology*, *39*(6), 2294–2297.

Nakajima, K., Kitanishi, K., Kobayashi, K., Kobayashi, N., Igarashi, J., & Shimizu, T. (2012). Leu65 in the heme distal side is critical for the stability of the Fe(II)-O2 complex of YddV, a globin-coupled oxygen sensor diguanylate cyclase. *Journal of Inorganic Biochemistry*, *108*, 163–170.

Nieto, C., Cherny, I., Seok, K. K., De Lacoba, M. G., Wai, T. C., Chew, C. Y., et al. (2007). The yefM-yoeBtoxin-antitoxin systems of *Escherichia coli* and *Streptococcus pneumoniae*: Functional and structural correlation. *Journal of Bacteriology*, *189*(4), 1266–1278.

Nocker, A., Richter-Heitmann, T., Montijn, R., Schuren, F., & Kort, R. (2010). Discrimination between live and dead cells in bacterial communities from environmental water samples analyzed by 454 pyrosequencing. *International Microbiology*, *13*, 59–65.

Nostro, A., Cellini, L., Di Giulio, M., D'Arrigo, M., Marino, A., Blanco, A. R., et al. (2012). Effect of alkaline pH on staphylococcal biofilm formation. *APMIS*, *120*(6), 733–742.

Nyenje, M. E., Green, E., & Ndip, R. N. (2013). Evaluation of the effect of different growth media and temperature on the suitability of biofilm formation by *Enterobacter cloacae* strains isolated from food samples in South Africa. *Molecules (Basel, Switzerland)*, *18*, 9582–9593.

Ofek, I., Hasty, D. L., & Sharon, N. (2003). Anti-adhesion therapy of bacterial diseases: Prospects and problems. *FEMS Immunology and Medical Microbiology*, *38*, 181–191.

Oosterhof, J. J. H., Buijssen, K. J. D., Busscher, H. J., Van Der Laan, B. F., & Van Der Mei, H. C. (2006). Effects of quaternary ammonium silane coatings on mixed fungal and bacterial biofilms on tracheoesophageal shunt prostheses. *Applied and Environmental Microbiology*, *72*(5), 3673–3677.

Orme, R., Douglas, C. W. I., Rimmer, S., & Webb, M. (2006). Proteomic analysis of *Escherichia coli* biofilms reveals the overexpression of the outer membrane protein OmpA. *Proteomics*, *6*, 4269–4277.

Otto, K., & Silhavy, T. J. (2002). Surface sensing and adhesion of *Escherichia coli* controlled by the Cpx-signaling pathway. *Proceedings of the National Academy of Sciences of the United States of America*, *99*(4), 2287–2292.

Palliyil, S., Downham, C., Broadbent, I., Charlton, K., & Porter, A. J. (2014). High-sensitivity monoclonal antibodies specific for homoserine lactones protect mice from lethal *Pseudomonas aeruginosa* infections. *Applied and Environmental Microbiology*, *80*(2), 462–469.

Pamp, S. J., Gjermansen, M., & Tolker-Nielsen, T. (2007). The biofilm matrix: A sticky framework. In M. Givskov & S. Kjelleberg (Eds.), *The biofilm mode of life: mechanisms and adaptations (bacterial)*. Wymondham, UK: Horizon BioScience.

Paraje, M. G. (2011). Antimicrobial resistance in biofilms. In A. Méndez-Vilas (Ed.), *Science against microbial pathogens: Communicating current research and technological advances: Vol. 2* (pp. 736–744). Badajoz: Formatex Research Center.

Park, A. J., Murphy, K., Krieger, J. R., Brewer, D., Taylor, P., Habash, M., et al. (2014). A temporal examination of the planktonic and biofilm proteome of whole cell *Pseudomonas aeruginosa* PAO1 using quantitative mass spectrometry. *Molecular & Cellular Proteomics*, *13*, 1095–1105.

Parker, C. T., & Sperandio, V. (2009). Cell-to-cell signalling during pathogenesis. *Molecular Microbiology*, *11*(3), 363–369.

Parsek, M. R., & Singh, P. K. (2003). Bacterial biofilms: An emerging link to disease pathogenesis. *Annual Review of Microbiology*, *57*, 677–701.

Peeters, E., Nelis, H. J., & Coenye, T. (2008). Comparison of multiple methods for quantification of microbial biofilms grown in microtiter plates. *Journal of Microbiological Methods*, *72*, 157–165.

Pesavento, C., & Hengge, R. (2009). Bacterial nucleotide-based second messengers. *Current Opinion in Microbiology*, *12*, 170–176.

Petrova, O. E., & Sauer, K. (2009). A novel signaling network essential for regulating *Pseudomonas aeruginosa* biofilm development. *PLoS Pathogens*, *5*(11), e1000668.

Phillips, N. J., Steichen, C. T., Schilling, B., Post, D. M. B., Niles, R. K., Bair, T. B., et al. (2012). Proteomic analysis of Neisseria gonorrhoeae biofilms shows shift to anaerobic respiration and changes in nutrient transport and outermembrane proteins. *PLoS One*, *7*(6), e38303.

Poole, K. (2012). Bacterial stress responses as determinants of antimicrobial resistance. *Journal of Antimicrobial Chemotherapy*, *67*(9), 2069–2089.

Post, D., Held, J. M., Ketterer, M. R., Phillips, N. J., Sahu, A., Apicella, M. A., et al. (2014). Comparative analyses of proteins from Haemophilus influenzae biofilm and planktonic populations using metabolic labeling and mass spectrometry. *BMC Microbiology*, *14*, 1–16.

Post, J., Preston, R., Aul, J., Larkins-Pettigrew, M., Rydquist-White, J., Anderson, K., et al. (1995). Molecular analysis of bacterial pathogens in otitis media with effusion. *JAMA, the Journal of the American Medical Association*, *273*(20), 1598–1604.

Pratt, L. A., & Kolter, R. (1998). Genetic analysis of *Escherichia coli* biofilm formation: Roles of flagella, motility, chemotaxis and type I pili. *Molecular Microbiology*, *30*, 285–293.

Prigent-Combaret, C., Vidal, O., Dorel, C., & Lejeune, P. (1999). Abiotic surface sensing and biofilm-dependent regulation of gene expression in *Escherichia coli*. *Journal of Bacteriology*, *181*(19), 5993–6002.

Rathsam, C., Eaton, R. E., Simpson, C. L., Browne, G. V., Berg, T., Harty, D. W. S., et al. (2005a). Up-regulation of competence—But not stress-responsive proteins accompanies an altered metabolic phenotype in *Streptococcus mutans* biofilms. *Microbiology*, *151*, 1823–1837.

Rathsam, C., Eaton, R. E., Simpson, C. L., Browne, G. V., Valova, V. A., Harty, D. W. S., et al. (2005b). Two-dimensional fluorescence difference gel electrophoretic analysis of *Streptococcus mutans* biofilms. *Journal of Proteome Research*, *4*, 2161–2173.

Redanz, S., Standar, K., Podbielski, A., & Kreikemeyer, B. (2011). A five-species transcriptome array for oral mixed-biofilm studies. *PLoS One*, *6*(12), e27827.

Ren, D., Bedzyk, L. A., Thomas, S. M., Ye, R. W., & Wood, T. K. (2004). Gene expression in *Escherichia coli* biofilms. *Applied Microbiology and Biotechnology*, *64*, 515–524.

Resch, A., Leicht, S., Saric, M., Pásztor, L., Jakob, A., Götz, F., et al. (2006). Comparative proteome analysis of *Staphylococcus aureus* biofilm and planktonic cells and correlation with transcriptome profiling. *Proteomics*, *6*, 1867–1877.

Rezzonico, F., Smits, T. H. M., & Duffy, B. (2012). Detection of AI-2 receptors in genomes of Enterobacteriaceae suggests a role of type-2 quorum sensing in closed ecosystems. *Sensors (Basel, Switzerland)*, *12*, 6645–6665.

Rhoads, D. D., Wolcott, R. D., & Percival, S. L. (2008). Biofilms in wounds: Management strategies. *Journal of Wound Care*, *17*(11), 502–508.

Rieu, A., Briandet, R., Habimana, O., Garmyn, D., Guzzo, J., & Piveteau, P. (2008). *Listeria monocytogenes*EGD-e biofilms: No mushrooms but a network of knitted chains. *Applied and Environmental Microbiology*, *74*(14), 4491–4497.

Riss, T. L., Moravec, R. A., Niles, A. L., Benink, H. A., Worzella, T. J., Minor, L., et al. (2004). Cell viability assays. In G. S. Sittampalam, N. P. Coussens, H. Nelson, M. Arkin, D. Auld, & C. Austin, et al. *Assay guidance manual [Internet]*. Bethesda (MD): Eli Lilly & Company and the National Center for Advancing Translational Sciences. 2013 May 1 [updated 2015 Jun 29].

Roberts, A. P., & Kreth, J. (2014). The impact of horizontal gene transfer on the adaptive ability of the human oral microbiome. *Frontiers in Cellular and Infection Microbiology*, *4*, 1–9.

Romeo, T. (2006). When the party is over: A signal for dispersal of *Pseudomonas aeruginosa* biofilms. *Journal of Bacteriology*, *188*(21), 7325–7327.

Römling, U., Galperin, M. Y., & Gomelsky, M. (2013). Cyclic di-GMP: The first 25 years of a universal bacterial second messenger. *Microbiology and Molecular Biology Reviews*, 77(1), 1–52.

Salek, M. M., Sattari, P., & Martinuzzi, R. J. (2012). Analysis of fluid flow and wall shear stress patterns inside partially filled agitated culture well plates. *Annals of Biomedical Engineering*, *40*, 707–728.

Sauer, K. (2003). The genomics and proteomics of biofilm formation. *Genome Biology*, *4*, 219.

Sauer, K., & Camper, A. K. (2001). Characterization of phenotypic changes in *Pseudomonas putida* in response to surface-associated growth. *Society*, *183*(22), 6579–6589.

Schembri, M. A., Kjaergaard, K., & Klemm, P. (2003). Global gene expression in *Escherichia coli* biofilms. *Molecular Microbiology*, *48*, 253–267.

Serra, D. O., & Hengge, R. (2014). Stress responses go three dimensional—The spatial order of physiological differentiation in bacterial macrocolony biofilms. *Environmental Microbiology*, *16*, 1455–1471.

Shah, D., Zhang, Z., Khodursky, A., Kaldalu, N., Kurg, K., & Lewis, K. (2006). Persisters: A distinct physiological state of *E. coli*. *BMC Microbiology*, *6*, 53.

Shen, G., Rajan, R., Zhu, J., Bell, C. E., & Pei, D. (2006). Design and synthesis of substrate and intermediate analogue inhibitors of S-ribosylhomocysteinase. *Journal of Medicinal Chemistry*, *49*(10), 3003–3011.

Sheridan, G. E. C., Masters, C. I., Shallcross, J. A., & Mackey, B. M. (1998). Detection of mRNA by reverse transcription-PCR as indicator of viability in *Escherichia coli* cells. *Applied and Environmental Microbiology*, *64*(4), 1313–1318.

Silva, S., Henriques, M., Martins, A., Oliveira, R., Williams, D., & Azeredo, J. (2009). Biofilms of non-*Candida albicans* Candida species: Quantification, structure and matrix composition. *Medical Mycology*, *47*, 681–689.

Simmons, L. A., Foti, J. J., Cohen, S. E., & Walker, G. C. (2008). The SOS regulatory network. *EcoSal Plus*, 1–48. http://dx.doi.org/10.1128/ecosalplus.5.4.3.

Singh, R., Ray, P., Das, A., & Sharma, M. (2010). Penetration of antibiotics through *Staphylococcus aureus* and *Staphylococcus epidermidis* biofilms. *Journal of Antimicrobial Chemotherapy*, *65*, 1955–1958.

Sintim, H. O., Smith, J. A., Wang, J., Nakayama, S., & Yan, L. (2010). Paradigm shift in discovering next-generationanti-infective agents: Targeting quorum sensing, c-di-GMP signaling and biofilm formation in bacteria with small molecules. *Future Medicinal Chemistry*, *2*(6), 1005–1035.

Sio, C. F., Otten, L. G., Cool, R. H., Stephen, P., Braun, P. G., Bos, R., et al. (2006). Quorum quenching by an N-acyl-homoserine lactone acylase from *Pseudomonas aeruginosa* PAO1. *Infection and Immunity*, *74*(3), 1673–1682.

Stepanović, S., Vuković, D., Hola, V., Di Bonaventura, G., Djukić, S., Ćirković, I., et al. (2007). Quantification of biofilm in microtiter plates: Overview of testing conditions and practical recommendations for assessment of biofilm production by staphylococci. *APMIS*, *115*, 891–899.

Stewart, P. S., & Franklin, M. J. (2008). Physiological heterogeneity in biofilms. *Nature Reviews. Microbiology*, *6*, 199–210.

Tagliabue, L., Antoniani, D., Macia̧g, A., Bocci, P., Raffaelli, N., & Landini, P. (2010). The diguanylate cyclase YddV controls production of the exopolysaccharide poly-N-acetylglucosamine(PNAG) through regulation of the PNAG biosynthetic pgaABCD operon. *Microbiology*, *156*(10), 2901–2911.

Tagliabue, L., MacIg, A., Antoniani, D., & Landini, P. (2010). The yddV-dos operon controls biofilm formation through the regulation of genes encoding curli fibers' subunits in aerobically growing *Escherichia coli*. *FEMS Immunology and Medical Microbiology*, *59*(3), 477–484.

Takahashi, N., Washio, J., & Mayanagi, G. (2010). Metabolomics of supragingival plaque and oral bacteria. *Journal of Dental Research*, *89*, 1383–1388.

Takahashi, N., Washio, J., & Mayanagi, G. (2012). Metabolomic approach to oral biofilm characterization—A future direction of biofilm research. *Journal of Oral Biosciences*, *54*(3), 138–143.

Tamayo, R., Pratt, J. T., & Camilli, A. (2007). Roles of cyclic diguanylate in the regulation of bacterial pathogenesis. *Annual Review of Microbiology*, *61*, 131–148.

Teughels, W., Assche, N., Sliepen, I., & Quirynen, M. (2006). Effect of material characteristics and/or surface topography on biofilm development. *Clinical Oral Implants Research*, *17*(Suppl. 2), 68–81.

Thein, M., Sauer, G., Paramasivam, N., Grin, I., & Linke, D. (2010). Efficient subfractionation of Gram-negative bacteria for proteomics studies. *Journal of Proteome Research*, *9*, 6135–6147.

Toté, K., Vanden Berghe, D., Maes, L., & Cos, P. (2008). A new colorimetric microtitre model for the detection of *Staphylococcus aureus* biofilms. *Letters in Applied Microbiology*, *46*, 249–254.

Tremoulet, F., Duche, O., Namane, A., Martinie, B., & Labadie, J. (2002). A proteomic study of *Escherichia coli* O157: H7 NCTC 12900 cultivated in bio film or in planktonic growth mode. *FEMS Microbiology Letters*, *215*, 7–14.

Tsilibaris, V., Maenhaut-Michel, G., Mine, N., & Van Melderen, L. (2007). What is the benefit to *Escherichia coli* of having multiple toxin-antitoxin systems in its genome? *Journal of Bacteriology*, *189*(17), 6101–6108.

Tuckerman, J. R., Gonzalez, G., Sousa, E. H. S., Wan, X., Saito, J. A., Alam, M., et al. (2009). An oxygen-sensing diguanylate cyclase and phosphodiesterase couple for c-di-GMP control. *Biochemistry*, *48*(41), 9764–9774.

Unterholzner, S. J., Poppenberger, B., & Rozhon, W. (2013). Toxin-antitoxin systems: Biology, identification, and application. *Mobile Genetic Elements*, *3*, e26219.

Van Houdt, R., Aertsen, A., Moons, P., Vanoirbeek, K., & Michiels, C. W. (2006). N-Acyl-L-homoserine lactone signal interception by *Escherichia coli*. *FEMS Microbiology Letters*, *256*, 83–89.

Varposhti, M., Entezari, F., & Feizabadi, M. M. (2014). Synergistic interactions in mixed-species biofilms of pathogenic bacteria from the respiratory tract. *Revista da Sociedade Brasileira de Medicina Tropical*, *47*, 649–652.

Villa, F., Remelli, W., Forlani, F., Gambino, M., Landini, P., & Cappitelli, F. (2012). Effects of chronic sub-lethal oxidative stress on biofilm formation by Azotobacter vinelandii. *Biofouling*, *28*, 823–833.

Wang, X., Kim, Y., Hong, S. H., Ma, Q., Brown, B. L., Pu, M., et al. (2011). Antitoxin MqsA helps mediate the bacterial general stress response. *Nature Chemical Biology*, 7(6), 359–366.

Wang, X., Preston, J. F., & Romeo, T. (2004). The pgaABCD Locus of *Escherichia coli* promotes the synthesis of a polysaccharide adhesin required for biofilm formation. *Journal of Bacteriology*, *186*(9), 2724–2734.

Welin, J., Wilkins, J. C., Beighton, D., Wrzesinski, K., Fey, S. J., Mose-Larsen, P., et al. (2003). Effect of acid shock on protein expression by biofilm cells of *Streptococcus mutans*. *FEMS Microbiology Letters*, *227*, 287–293.

Wen, Z. T., Yates, D., Ahn, S.-J., & Burne, R. a. (2010). Biofilm formation and virulence expression by *Streptococcus mutans* are altered when grown in dual-species model. *BMC Microbiology*, *10*, 111.

Werner, E., Roe, F., Bugnicourt, A., Michael, J., Heydorn, A., Molin, S., et al. (2004). Stratified growth in *Pseudomonas aeruginosa* biofilms. *Applied and Environmental Microbiology*, *70*(10), 6188–6196.

Whitehead, N. A., Barnard, A. M. L., Slater, H., Simpson, N. J. L., & Salmond, G. P. C. (2001). Quorum-sensing in Gram-negative bacteria. *FEMS Microbiology Reviews*, *25*, 365–404.

Wilkins, M., Hall-Stoodley, L., Allan, R. N., & Faust, S. N. (2014). New approaches to the treatment of biofilm-related infections. *Journal of Infection*, *69*, S47–S52.

Williams, J. J., & Hergenrother, P. J. (2012). Artificial activation of toxin-antitoxin systems as an antibacterial strategy. *Trends in Microbiology*, *20*(6), 291–298.

Williamson, K. S., Richards, L. A., Perez-Osorio, A. C., Pitts, B., McInnerney, K., Stewart, P. S., et al. (2012). Heterogeneity in *Pseudomonas aeruginosa* biofilms includes expression of ribosome hibernation factors in the antibiotic-tolerant subpopulation and hypoxia-induced stress response in the metabolically active population. *Journal of Bacteriology*, *194*, 2062–2073.

Winzer, K., & Williams, P. (2003). *Escherichia coli* gets the message. *Nature Medicine*, *9*(9), 1118–1119.

Wnuk, S. F., Robert, J., Sobczak, A. J., Meyers, B. P., Malladi, V. L., Zhu, J., et al. (2009). Inhibition of S-ribosylhomocysteinase(LuxS) by substrate analogues modified at the ribosyl C-3 position. *Bioorganic and Medicinal Chemistry*, *17*(18), 6699–6706.

Wood, T. K., Knabel, S. J., & Kwan, B. W. (2013). Bacterial persister cell formation and dormancy. *Applied and Environmental Microbiology*, *79*, 7116–7121.

Wu, H., Moser, C., Wang, H.-Z., Høiby, N., & Song, Z.-J. (2014). Strategies for combating bacterial biofilm infections. *International Journal of Oral Science*, *7*, 1–7.

Yang, L., Liu, Y., Wu, H., Hóiby, N., Molin, S., & Song, Z. (2011). Current understanding of multi-species biofilms. *International Journal of Oral Science*, *3*, 74–81.

Yeom, J., Shin, J. H., Yang, J. Y., Kim, J., & Hwang, G. S. (2013). 1H NMR-based metabolite profiling of planktonic and biofilm cells in *Acinetobacter baumannii*1656-2. *PLoS One*, *8*(3), 1–7.

Zhang, B., & Powers, R. (2012). Analysis of bacterial biofilms using NMR-based metabolomics. *Future Medicinal Chemistry*, *4*(10), 1273–1306.

CHAPTER SIX

Avoid Excessive Oxygen Levels in Experiments with Organisms, Tissues and Cells

David Lloyd*^,1^, Catrin F. Williams*^,†^

*School of Biosciences, Cardiff University, Cardiff, Wales, United Kingdom
†School of Engineering, Cardiff University, Cardiff, Wales, United Kingdom
[1]Corresponding author: e-mail address: lloydd@cf.ac.uk

Contents

Abstract

O_2 levels encountered *in vivo* in cells and tissues are almost always at least an order of magnitude less than atmospheric pO_2 because of sensing, signalling and bioenergetic demand. Although deleterious reactions are minimized by protective mechanisms (residual toxic products scavenged and detoxified) ambient levels should be mimicked in experiments with whole organisms, their isolated organs, tissues or cells and also with cultures of cell lines. These are also important issues for microorganisms inhabiting low O_2 niches within higher organisms and their cells. Here, we highlight the importance of optimization of micro-aerobic conditions for experimentation and the deleterious consequences of not doing so.

Advances in Microbial Physiology, Volume 67
ISSN 0065-2911
http://dx.doi.org/10.1016/bs.ampbs.2015.09.001

1. INTRODUCTION

Aerobic living organisms harness the chemistry of dioxygen in a delicately controlled and uniquely advantageous way (Banerjee, Becker, Dickman, Gladyshev, & Ragsdale, 2008). Even those organisms that cannot exploit the huge energetic advantages of O_2 utilization cannot afford to be indifferent to it, and its presence even at extremely low concentrations produces major metabolic effects. If not utilized, it must be disposed of rapidly and safely. Respiration proceeds in parallel with oxygen scavenging, for the elimination of a potential toxicant. Partial reduction of O_2 produces oxygen-centred radicals ($O_2^{-\bullet}$, superoxide and $OH^{\bullet}$, hydroxyl) as well as H_2O_2 (hydrogen peroxide), $^1O_2$ (singlet oxygen), HOCl (hypochlorous acid) and O_3 (ozone). These reactive oxygen species (ROS) are potentially damaging at many intracellular sites and produce drastic consequences. These well-established phenomena include oxidation of protein cysteine thiols, lipid peroxidation and nucleic acid damage, leading to redox imbalance, membrane dysfunction and genetic effects: senescence and loss of viability of cell cultures, as well as apoptosis, and patho-physiological consequences *in vivo* (Wang, Yang, & Yi, 2011). However, low levels of ROS and macromolecular modulation events also serve in essential signalling roles, and so their balanced levels (Fig. 1) must be preserved for redox sensing and the maintenance and control of cellular functions (Aon et al., 2010).

To take cells from a low-O_2 niche within the organism for culture in a growth medium equilibrating with 95% air–5% CO_2 is a recipe for oxidative stress. Thus, understanding of O_2 effects are paramount for its considered use in experiments with whole organisms, their isolated organs, tissues or cells and also with cultures of cell lines. These are also important issues for microorganisms inhabiting regions of low O_2 within higher organisms and their cells (Lloyd, 2002). Here, we highlight the necessary considerations that must be made to match O_2 concentrations as nearly as possible to those of their natural environments and hence optimize aerobic or micro-aerobic conditions for experimentation.

1.1 Oxygen in the Atmosphere and Aqueous Solutions

The atmospheric gas mixing ratio at average sea level pressure (101,325 Pa) provides 20.92% O_2 (158 torr, i.e., mm Hg pressure or 20,900 Pa). O_2 is only sparingly soluble in liquids; factors influencing solubilities include

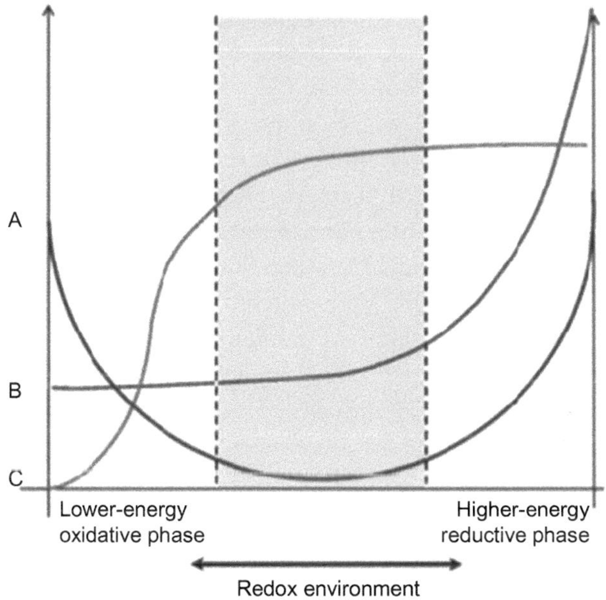

Figure 1 The redox-optimized reactive oxygen species (ROS) balance hypothesis. ROS levels (A), rate of ROS production by respiratory complexes (B) and the rate of ROS scavenging (C). The optimal conditions for physiological ROS signalling occur when ROS levels are minimal with redox potential being at an intermediate state (between the dashed lines of the plot). However, oxidative stress occurs at the extremes of redox potential, under highly reduced or highly oxidized intracellular or intramitochondrial environments. Under reducing conditions (right hand side of the plot), ROS accumulation exceeds scavenger capacity, whilst at oxidizing conditions (left hand side of plot), ROS levels increase as a result of scavenger depletion. *Reproduced from Aon, Cortassa, and O'Rourke (2010) with permission from Elsevier Science (USA).*

pressure, temperature, presence of other solutes (e.g. in pH- and osmotically active buffers) and properties of the liquid (hydrophobicity, dielectric properties, etc.). At 37 °C, O_2 at average sea level pressure water equilibrates in air to become a saturated solution at 278 μM (Wilhelm, Battino, & Woodcock, 1977). In normal saline at 37 °C, this value becomes 258 μM, and in some culture media becomes 220 μM (Gnaiger & Forstner, 1983).

The O_2-cascade through the human blood circulatory systems is exquisitely developed so that controlled and appropriate partial pressures of O_2 are delivered to all tissues. Although alveolar air in equilibrium with arterial blood contains the equivalent of 100 μM O_2, discharge from blood oxyhaemoglobin occurs within a range of 80–20 mm Hg≡60–30 μM O_2, depending upon CO_2 tension (Wittenberg & Wittenberg, 1989). The oxymyoglobin store of red muscle begins to unload at below 10 μM O_2. The working pO_2, there is about 3 mm Hg. Steep gradients of O_2 diminish this working range over which intracellular O_2 is poised, so that intramitochondrial O_2 in the vicinity of its reaction with cytochrome *c* oxidase may be as little as 100 nM (Tamura, Hazeki, Nioka, & Chance, 1989). The intracellular environment is characterized by steep gradients and is thus predominantly micro-aerobic. pO_2 values there are measured only with difficulty, although they may be derived from estimates obtained by *in situ* observation of the redox states of respiratory chain components using non-invasive fluorimetric determinations using NADH (Chance, Noka, Warren, & Yurtsever, 2005) and/or spectrophotometric, for myoglobin and cytochrome redox state measurements (Chance & Quistorff, 1978; Lloyd, Mellor, & Williams, 1983).

1.2 Partial Pressures of O_2 in Tissues

Highly reliable and reproducible measurements of *in vivo* O_2 tensions as determined under minimally perturbing procedures in body fluids, organs and tissues have been available for more than 40 years (Vanderkooi, Erecińska, & Silver, 1991), although rather little attention has been paid to the effects of anaesthesia. Table 1 shows typical values obtained for some mammalian species. It is notable that, despite the efficiency of O_2 delivery, high utilization rates dictate that steady state O_2 levels (Vanderkooi et al., 1991) (with few exceptions, e.g. the surface of the epidermis) (Stücker et al., 2002), are measured at values well below atmospheric partial pressure, and in some cases as little as 1% O_2 (Erecińska & Silver, 2001). Mimicking physiological conditions *in vitro* has rarely been reported.

1.3 Oxygen and the Growth, Differentiation, Senescence and Death of Cells in Culture

Experiments with cultured cells are routinely performed using a 95% air/5% CO_2 gas phase and this should not be accepted without question as it has been demonstrated that the resulting elevated exposure to O_2 can have multiple modulating and deleterious effects on cellular growth and differentiation, and lead to cellular senescence (Table 2). Similar reservations apply to values for activities of subcellular systems (e.g. suspensions of mitochondria) conducted in air-saturated buffer solutions. In the culture of neurons, for example, cells experience highly elevated O_2 tensions (21%) as compared

Table 1 Tissue PO_2 Values *In Situ* in Respired Air

	PO_2 (mm Hg)	Species	Remarks
Air	158	–	20.92% (258 μM) O_2 in normal saline buffer at 37 °C (Gnaiger & Förstner, 1983; Wilhelm et al., 1977)
Bloodstream (Semenza, 1998; Semenza, 2000; Tiede, Cook, Morsey, & Fox, 2011; Wenger, 2002)			
Capillary	10–60	H	E (Kessler, Lang, Sinagowitz, Rink, & Höper, 1973)
Macrophages at lung apices	150	M (D'Mello, Hill, & Poole, 1995)	
In tissues			
In venous blood	<10		
Periarterioles	40	R	E (Tsai, Johnson, & Intaglietta, 2003)
7–11 μm diam.			
28 μm diam.	31 ± 3		
41 μm diam.	46 ± 3	Ha	E (Lash & Bohlen, 1987)
60–100 μm diam.	50 ± 12		
Terminal arterioles	35 ± 4		
(10–20 μm diam.)	24 ± 3		
Capillaries			
Tissues	20 ± 3	H	E (Kessler et al., 1973)
Venular	8 ± 2		
	10–40		
Bone			
Marrow	<7.2–43		
Osteum	<7.2	H (Flamant et al., 2003)	

Continued

Table 1 Tissue PO_2 Values *In Situ* in Respired Air—cont'd

	PO_2 (mm Hg)	Species	Remarks
Brain			
Cortex	38.7 ± 0.93	C	E (1–5 μm) (Duling & Berne, 1970)
	24.5 ± 18	Rab	
Cortex—grey	19–40	R	E (<5 μm) (Nair, Whalen, & Buerk, 1975)
Cortex—white	6–16		E (1–5 μm) (Smith, Guilbeau, & Reneau, 1977)
Hippocampus	29–30		
Hypothalamus	6–16		
Pons fornix	1–3		
Vasculature arteriole	7.6	M	
Venule	44.4		P (Coogan et al., 2010)
Carotid body (0.15 μm from surface)	0–90	C	E (1–5 μm) (Cater, Garaltini, Marina, & Silver, 1961)
Eye			
Conjunctiva	79.5	Hu (Eliasson & Jonsson, 2010)	
Lens (gradient)	9–65	Rab	E (Acker, Lubbers, Purves, & Tan, 1980)
Retina (gradient)	30–50	R	P (Sakadzic et al., 2010)
	21.7–42.0	M	E (Barr & Roetman, 1974)
Gastrointestinal tract			
Duodenum and pancreas surface	32	D	E (10 μm) (Yu & Cringle, 2005)
Duodenum (jejunum at the villi)	<0.15–36 (Kessler et al., 1973)	S	Ms (Lloyd, 2002)
	40		

Table 1 Tissue PO_2 Values *In Situ* in Respired Air—cont'd

	PO_2 (mm Hg)	Species	Remarks
Small intestine surface	38	D	E (10 μm) (Yu & Cringle, 2005)
Stomach mucosa	15		E (1–2 μm) (Messmer et al., 1973)
Heart			
Endocardium	14.7–17.2	D	E (Bowen & Garg, 1977)
Left ventricle	27–31		
Subepicardium	6.2	C	Mb (Weiss & Winbury, 1972)
Joints			
Cartilage	10.8–43.2	Co, D, H, Ra (Weiss, Isenberg, McArthur, Del Signore, & McDonald, 2011)	
Synovial fluid	7.3–14	H	E (Honig & Gayeski, 1987)
Acute inflammation	20.8–49.0	Rab	E (Weiss et al., 2011)
Normal knee	21.6–68.4		
Osteoarthritis	40.0–61.9		
Rheumatoid arthritis	0–46	H	
Kidney			
Medulla/papilla	7.2	H	E (Edwards, Hallett, & Campbell, 1984)
Surface	40–100	D	E (Zhou, Cui, & Urban, 2004)
	10–50	H	E (Edwards et al., 1984)
Liver			
In tissue	23	R	

Continued

Table 1 Tissue PO_2 Values *In Situ* in Respired Air—cont'd

	PO_2 (mm Hg)	Species	Remarks
Surface	0–55	D (Zhou et al., 2004)	E (50–500 μm) (Brezis, Rosen, Silva, & Epstein, 1984)
Rumen	<0.15–2.0	S, Co, Go, Ms	E (Lloyd, 2002)
Skeletal muscle surface	18	D	E (Sato et al., 1983)
Gracilis at rest in tissue	5.4 ± 0.5	G	E (1–5 μm) (Sato et al., 1983)
Skin			
Dermis	37–50		
Epidermis	130–140	H	P (Stücker et al., 2002)
Tumour			
Subcutaneous			
Untreated	5–7	M	O (Lloyd et al., 2014)
After radio and chemotherapy	16–28		
Malignant glioma			
Untreated	5–7		
Cyclophosphamide for 10 days	19.7	M	O (Khan et al., 2009)
Vagina	8.0–35	H	E (Stücker et al., 2002)

C, cat; Co, cow; D, dog; G, guinea pig; Go, goat; Ha, hamster, H, human; M, mouse; Rab, rabbit; R, rat; E, electrode; P, phosphorescence quenching; Ph, photobacterium bioluminescence; Ms, mass spectrometry; Mb, myoglobin saturation; O, oximetry (electron paramagnetic resonance).

to the *in vivo* scenario. This can have drastic results, especially with regards to mitochondrial oxidative metabolism. A study by Tiede et al. (2011). showed that primary neurons which were cultured under physiological O_2 tensions were more highly polarized, had lower ROS production, greater mitochondrial networks, fractions and perimeters than cells cultured at higher O_2 tensions. Furthermore, when cells cultured at physiologically relevant O_2 tensions were challenged with human immunodeficiency virus virotoxins, the mitochondria underwent rapid depolarization and increased cell death, whereas cells cultured at higher O_2 tensions did not. Thus, reliable control of

Table 2 Role of Low pO_2 in Stem Cells and Consequences of Elevated Oxygenation (Mohyeldin, Garzón-Murdi, & Quiñones-Hinojosa, 2010; Toussaint, Weemaels, Debacq-Chainiaux, Scharffetter-Kochanek, & Wlaschek, 2011; Whalen & Nair, 1970)

Stem Cell Type	Role and Consequences of Elevated Oxygenation
Cancer stem cells	Tumour-initiating cells (leukaemic, breast, brain, melanoma, colon or pancreatic cancer precursors) especially in hypoxic microenvironments influence tumour initiation and progression. HIF2d regulation of Oct 4 and downstream targets. Nitric oxide and Notch signalling also implicated in the epithelial to mesenchyme transition whereby non-motile cells converted to promigratory and invasive phenotype
Embryonic stem cells	Modulation of cellular growth, pleuripotency and senescence. Homozygous knockouts of HIF1d, HIF-2a and HIFB subunits give lethal defects in vascularization of yolk sac, placenta, brachial arches, cranium and somites, neural-fold closure and blood vessel formation
Haematopoietic stem cells	Maintenance of slow primitive phenotype and (cell division) cycling in anaerobic bone marrow niches affords protection against oxidative stress mediated by HIF-1 signalling. O_2 exposure promotes cellular division and decreases capacity for normal haematopoiesis
Induced pleuripotent stem cells	Stemness induced and promoted by transcriptional activation at low pO_2, probably by regulation of HIF2α or HIF3α
Mesenchymal stem cells	Maintenance of undifferentiated state prior to osteogenic, chondrogenic and adipogenic differentiation *in situ*. Level of O_2 determines eventual differentiated state and expression of cell surface markers and adhesive properties
Neural stem cells	Support of the undifferentiated state *in vivo*, increase proliferative capacity when cultured *in vitro* and erythropoietin expression under "hypoxic" conditions. Elevated pO_2 results in mitotic arrest and promotes differentiation migration, and glial proliferation probably via phosphorylation of p53 and Notch signalling. Prolyl hydroxylation by EgIN hydroxylases regulate HIF and response: failure of EgIN3-mediated developmental apoptosis predisposes to sympathetic neuronal cancers

dissolved gases at is necessary if physiological conditions are to be achieved and maintained during experiments.

Detailed elucidation and optimization of growth factor requirements for cells in culture has not always been paralleled by concerns about the necessity for O_2 control. Accumulating evidence (summarized in Table 2) for the deleterious consequences of inappropriate oxygenation conditions stresses the need for careful poising of gas supply so as to mimic as far as possible the appropriate *in vivo* situation (examples of which are given in Table 1). The provision of excess O_2 over and above that experienced *in situ* leads to redox imbalance (Aon et al., 2010), diverse macromolecular modification and damage (Wang et al., 2011), and thereby also to metabolic and biosynthetic dysfunction (Table 2). Conversely, even transient O_2 starvation may lead to energy deficit, excessive hypoxic responses (Semenza, 1998, 2000; Wenger, 2002), and slowing or cessation of growth and differentiation. Signalling functions of redox-active metabolites must also be maintained within well-defined limits to minimize damage to all cellular constituents. Thus maintenance of redox states necessitates conservation of a balance (Fig. 1) between O_2 radical production during electron transport chain processes and protection against O_2 by scavenging processes (Aon et al., 2010).

Table 2 lists some well-established consequences of inappropriate O_2 supply to cultured cells (Toussaint et al., 2011). The immediate microenvironment of a cell *in vivo* is determined by distance from the vascular blood supply, the presence of matrix elements and the O_2 consumption rate of contiguous cells. As detailed above, the pO_2 may lie within the range of <1–8% O_2 in most tissues, with some examples exhibiting especially low oxygenation levels (in active striated muscle, kidney medullary cells, brain pons or fornix and regions of cartilage and bone marrow, Table 1). Cellular states in these low O_2 environments are under controls exerted by metabolic and transcriptional events exerted via "hypoxic" responses of O_2-sensitive ion channels, hypoxia-inducible transcription factors and the endoplasmic reticulum response (Semenza, 1998, 2000; Wenger, 2002).

Awareness of this hierarchy of oxygenation states through the circulatory system as well as in mammalian organs, tissues and cells (Vanderkooi et al., 1991) is essential both for the understanding of metabolic status and function *in vivo* and for the manipulation of cells *in vitro*. Inadequate regard for these principles casts doubt on the interpretation of many published data and may require re-evaluation by conducting further experiments under conditions resembling as far as possible appropriate physiological status.

2. METHODS FOR MEASURING O_2

2.1 Considerations for Choosing an Appropriate Method

The detection and quantitative measurement of O_2, either as a gas or as a dissolved in liquids is of such great importance to biological research that many of the most specific, sensitive and rapidly responding assays for O_2 have been elaborated as the result of this need (Table 3). The various aspects of biological demand (energetic performance, signalling) and of scavenging, as well as the heterogeneous localization of O_2 in cells, tissues and organs and in structured ecosystems (in the atmosphere, soil, sediments, marine and freshwater environments) place various criteria for selection of suitable methods of measurement of stoichiometries and rates of change. Fluctuations in O_2 levels, especially when rapid (e.g. during oscillatory intracellular reactions and functions) may often require instrumental response times of milliseconds or less. Spatial precision may demand use of microprobe methods.

The manometric methods introduced by Warburg and more recent automated innovations based on volumetric techniques for measuring gaseous exchange are still sometimes useful for medium (hour), or longer term experiments (days), but cannot be used for more rapid reactions or for

Table 3 Methods for Measurement of O_2

Method (Response Time)	Continuous Liquid Phase	Lower Limit
Haemoglobin/myoglobin (10 s)	–	10 nM (Barzu, 1984)
O_2 electrode, Clark membrane (30 s)	+	0.1 μM (Degn, Cox, & Toftlund, 1985)
Membrane inlet mass spec. (30 s)	+	0.1 μM (Lloyd, Bohátka, & Szilágyi, 1985)
Photobacterium suspension (1 s)	+	<0.1 pM (Lloyd, James, Williams, & Williams, 1981)
Photobacterium membrane probe (<8 s)	–	35 nM (Lloyd et al., 1981)
Phosphorescence quenching (1 s)	–	<0.1 μM (Coogan et al., 2010)

transient changes. Choice of a method thus depends on: (a) whether gas or liquid phase determinations are required, (b) sensitivity, (c) response time and (d) duration of measurement.

2.2 Oxygenation of Haemoglobins/Myoglobins

Mammalian oxyhaemoglobin (HbO_2) provides an efficient method of measuring oxygen at low, physiological tensions. The ideal O_2-buffering capacities of HbO_2 are attributed to the cooperative binding of O_2 to the four Hb subunits. Difference spectra of oxygenated and deoxygenated haemoglobin allow detection of O_2 at several wavelength ranges between 370 and 650 nm. The principle of the HbO_2 method is thus based on the ability of oxyhaemoglobin (or oxymyoglobin) to dissociate O_2 at decreasing oxygen tensions, by spectrophotometric measurement. At a concentration of 0.07 mM HbO_2, the OD after complete deoxygenation of the protein is 2.8 (path length 4 mm at 436 nm). This value is adjusted to zero and the OD increase as a result of subsequent oxygen consumption is recorded. Subsequent O_2 tensions are calculated according to a calibration curve of OD versus $[O_2]$. This method has been employed in a wide range of biochemical assays allowing determination of enzyme activity, mitochondrial respiration as well as O_2 consumption in isolated cells (Barzu, 1984).

2.3 Oxygen Electrode Methods

Oxygen electrode methods (Acker et al., 1980; Gnaiger & Förstner, 1983) use either polarographic sensors, in which the polarized electrodes produce a current in an external circuit, or the galvanic principle whereby a current is generated directly. These two devices, represented, respectively, by the Clark and Mackereth electrodes, especially when constructed as steam-sterilizable devices, are invaluable for measurements of O_2 in the liquid phase in cultures of cells or organisms growing in fermenters. These are reliable over long periods (up to many months) and through many cycles of sterilization.

An especially useful electrode which has provided the O_2 measuring device central to many of the applications described here, is that manufactured by Radiometer (Copenhagen). The advantages of this O_2 electrode include: (i) rapid response immersed in liquid ($t_{90\%} < 30$ s, depending on membrane material and thickness), (ii) high sensitivity (~0.1 μM, again depending on the membrane employed), (iii) stability (adequate for experiments over many hours or days) and (iv) low endogenous consumption of O_2 (due to the very small-area Pt cathode).

The electrode can be sterilized by using 70% ethanol; like all Ag–Pt-based Clark electrodes, it is affected by the presence of some other gases (H_2, CO_2, H_2S, Cl_2, Br_2 and I_2) and high intensity illumination. For measurements in tissues a variety of Pt or Au microelectrodes have been described, sometimes with a tip of 1–5 μm (Gnaiger & Förstner, 1983).

A closed O_2 electrode system, specially constructed to avoid the O_2-leakage problems inherent in many commercially available systems and capable of measuring O_2 to 0.1 μM has been developed and used extensively for studies of cells or mitochondria at low O_2 levels (and for the inhibitory effects of NO). This system uses Viton O-ring seals, a glass rather than perspex vessel and titanium stoppers. Glass-coated stirrer bars are better than Teflon-coated stirrers, as the latter acts as an O_2 buffer and O_2 leaks back at up to 30 pmol s^{-1} cm^{-3} when its concentration is lowered quickly.

2.4 Membrane Inlet Mass Spectrometry

Where other gases (or some membrane-permeable low molecular weight volatile organic compounds) are to be measured alongside O_2, membrane inlet mass spectrometry (MIMS) (Degn et al., 1985; Lloyd et al., 1985) provides an ideal monitoring technique. The specifications provided by a small quadrupole mass spectrometer (we have found ideal the ones that scan to mass 200 e.g. from Hiden analytical), fitted with a MIMS probe (a sealed stainless steel 0.16 mm tube with inlet orifice fitted with silicone rubber catheter sleeve) are adequate for many biomedical or environmental applications. Measurements of O_2 using this mass spectrometric method and the electrode method are similar in sensitivities, as it is the unstirred liquid layer on the gas-permeable membrane that limits sensitivities and response times for dissolved gas measurements for both techniques. However, the facility for parallel measurements of other gases of biological interest (CO_2, H_2, CH_4, H_2S, N_2, NO_x, etc.) as well as many low molecular weight volatiles (CH_3OH, CH_3COOH, C_2H_5OH, CH_3SH, etc.), provide many applications in biology.

2.5 Bacterial Bioluminescence

As a simple test for the presence of minute traces of O_2 in gas or liquid phase, bacterial bioluminescence (Lloyd et al., 1981, 1983; Tamura et al., 1989) has been used for more than a century and is still unsurpassed except by elaborate physical methods. Thus, studies by Beijerinck in 1889 were followed by many early reports describing the use of suspensions of luminous bacteria as a measure of the permeability of rubber membranes and films of various oils. Other experiments sought to detect the minimum concentration of O_2 detectable and values for just visible luminescence, by the dark-adapted eye, of O_2 at a gas pressure of about 7×10^{-4} mm Hg is possible. Thus, the photoemission of these marine species is O_2-limited at concentrations less than 1 nM and is proportional to concentrations over several orders of magnitude below this. The lower limit of sensitivity differs between species, depending on the affinity of the bacterial luciferase for O_2. Thus, monitoring of levels of <100 fM O_2 are feasible using those bacteria with the most avidly O_2-binding luciferase systems (those from the luminous organs of deep-sea fish).

There is a requirement for high salt (3% NaCl) by these marine bacteria, although buffers containing 0.5 M mannitol also serve well; pH values in the range 6–8 are also required. For five different strains, we have noted great stability of light output in a buffer containing 0.5 M mannitol, 0.17 M NaCl, 50 mM K-phosphate at pH 7.0.

A photobacterium probe in which a suspension of live bacteria (*Vibrio fischeri* MJ-1) was employed behind a gas-permeable membrane gives excellent performance characteristics and is threefold more sensitive to trace O_2 than a radiometer electrode (Table 3). Using thin (20 μm) silicone membrane, half-times of <8 s were recorded for both detectors. We have used this technique for many physiological determinations, e.g., the determination of the apparent O_2 affinities of intact aerobic and microaerophilic species of protists and for O_2 calibration of *in vivo* steady state redox levels of cytochromes, flavins and NAD(P)H of the respiratory chain (Lloyd et al., 1983).

2.6 Measurements of O_2 Using Phosphorescence Quenching

Phosphorescence quenching (Coogan et al., 2010; Sakadzic et al., 2010) is used successfully for measuring and for mapping O_2 gradients in tissues and cultured cells. Lifetime measurements have advantages over emission intensities as the former are less susceptible to interference by local environmental influences. In suspensions of cells and mitochondria, the pO_2 dependence of O_2 consumption used phosphorescence quenching of albumin-bound Pd-porphyrins in which intersystem crossing is sufficiently rapid to convert singlet into triplet excited state (Vanderkooi et al., 1991). Thereby these compounds show little or no fluorescence and the quantum efficiency of phosphorescence is high. Lifetimes can be determined using cameras or photomultipliers. Local O_2 concentrations in tissues *in vivo* have been published (Coogan et al., 2010; Sakadzic et al., 2010). The independence of these lifetime images of oxygenation states of myoglobin or haemoglobin or of cytochrome or flavin redox states present great advantages over intensity measurements. Newer luminophores include luminescent metal complexes (d^6 or d^8 transition metals), e.g., those based on Ru(1 1), Os(11), Re(1), Ir(1 1 1) and Pt(1 1) with highly conjugated ligands give intense charge transfer bands with long-lived excited states (10^{-7}–10^4 s). Toxicity can be minimized by embedding in polymer nanoparticles: electroporation facilitates homogeneous intracellular distribution, although endosomal uptake (for instance into cultured breast cancer cells has been observed) (Coogan et al., 2010). Lanthanide probes are usefully responsive to metal ions, oxyanions and pH of the intracellular environment, and an Europium (1 1 1) complex has been used as an efficient singlet O_2 probe. Spectrophotometry (Coogan et al., 2010), two photon confocal laser scanning microscopy with pinhole shifting and picosecond time-resolved epiphosphorescence (Shahidi, Wanek, Blair, Little, & Wu, 2010) have all been used for measuring intracellular O_2 using phosphorescence quenching, with good consistency between the latter two methodologies (Williams et al., 2012).

2.7 Oximetry

Electron paramagnetic resonance (EPR) (Khan et al., 2009; Lloyd et al., 2014) used in conjunction with compounds that stabilize or sensitize free radicals are used to detect tissue O_2. Thus in recent studies of the effects of radiation and chemotherapy on the oxygenation states of subcutaneous tumours in mice, crystals of lithium phthalocyanine were injected before placing anaesthetized animals into the cavity and obtaining spectra (Khan et al., 2009; Lloyd et al., 2014). Repeated assessment of localized changes in tissue pO_2 during treatment is thereby by a minimally invasive procedure. With cultured cells, simultaneous detection of other free radical species (e.g. $NO^{\cdot}$) is also possible.

2.8 Minimally Invasive Continuous Readout of Cellular Functions

Combinations of minimally invasive physical methods with continuous monitoring of controlled environmental conditions, provides powerful new approaches to functional analysis of metabolic performance. The scale of investigation can be chosen so as to be appropriate at the single cell level, in exposed tissues, or in highly concentrated populations of cultured cells or organisms in large-scale fermenters.

Continuous readout methods provide data that are much more amenable to analysis than are a series of discrete time measurements. As well as providing high quality information for kinetic analyses, especially valuable for the interpretation of periodic phenomena, automated data acquisition with frequent sampling of continuous output enables reliable assessments of long-term behaviour (days, weeks, months or years). Where multiple parameter monitoring is feasible, new insights into biological processes and activities *in vivo* become evident.

2.9 Open System Studies at Low O_2

Maintenance of low O_2 environments for organisms during experimental procedures over extended time periods, necessitate systems open for gas flow (Lloyd et al., 1981). Both the measuring method and the biomass consume O_2 and thus in a closed system O_2 levels fall continuously (Fig. 2A). O_2 control for extended periods and measurement in a steady state requires an open system (Fig. 2B): the slight positive pressure provided by the mobile gas flow helps conserve low O_2 conditions. Stainless steel reaction vessels designed for different multiparameter applications (fluorimetry, spectrophotometery) are fitted with quartz glass observation windows.

Flow lines are also of stainless steel (e.g. 3.2 mm o.d., 1 mm i.d.), as are all fittings. Black butyl rubber is highly gas impermeable, whereas silicone rubber tubing is highly permeable for air gases: therefore, its use is to be avoided. Traces of O_2, almost always present in cylinders of N_2 or Ar, must be removed by passage over heated Cu, through Fieser's solution (20 g KOH, 2 g anthraquinone-β-sulphonate, 100 ml water, 15 g Na dithionite (hydrosulphite, $Na_2S_2O_4$) absorbs 788 ml O_2 until it changes from clear to dull red or brown), through a thick suspension of baker's yeast or by using a commercially available O_2-scavenging column (e.g. those supplied by Alltech).

Measurement of dissolved O_2 in a suspension of organisms stirred at constant rate under a mobile gas phase kept at fixed O_2 tension in a thermostatically enclosed reaction vessel, enables long-term experiments to be carried out (Muppuraju et al., 2011). Stirring using a cross-shaped impellor produces a stable liquid vortex for gas–liquid exchange. The level of dissolved

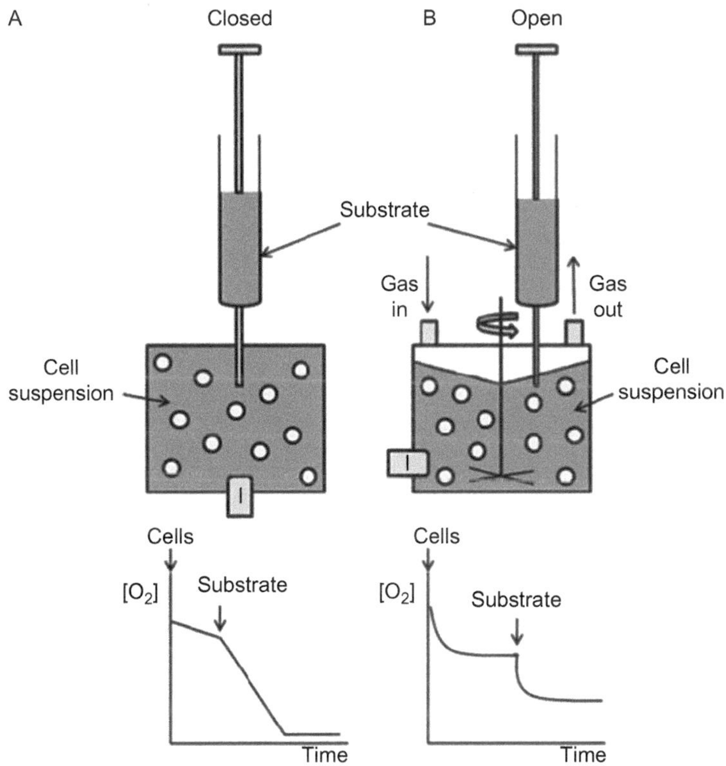

Figure 2 Systems for study of cells or organisms at controlled pO_2. (A) In a closed system, O_2 concentration decreases during its consumption, (B) whereas in a system open, to gas flow it is maintained at a steady state level by using an appropriate cell number and constant stirring under a mobile gas phase of defined composition. I indicates the position of O_2 electrode (or membrane inlet of mass spectrometer). *Adapted from Lloyd (2002) with permission from Elsevier Science (USA).*

O_2 attained represents the balance between supply and demand, so that by choosing appropriate gas phase and cell concentration, organisms can be maintained for extended periods at any required O_2 concentration. The accuracy of control attainable in such a system is extremely good, even for O_2 concentrations <1.0 μM. These systems can be constructed for liquid volumes at any scale (e.g. 5 ml 5 l). The practical limit for using small liquid volumes is reached when evaporative losses become considerable. Hence, a requirement for humidification of the mobile gas phase becomes critical (e.g. when a hanging-drop version of the system is used for imaging by microscopy). A working volume of 5 ml with a stirring rate of 200 rev min^{-1} gives a response time ($t_{1/2}$, the half time for equilibration in the liquid phase after a step change in gas phase O_2 in the absence of organisms) of about 1.5 min. Calculation of the respiration rate of organisms (V_r) in this system requires determination of the gas exchange coefficient (k):

$$k = \frac{\ln 2}{t_{1/2}} = \frac{0.693}{t_{1/2}}$$

$$V_r = k(T_G - T_L)$$

where T_G and T_L are the O_2 tensions in gas and liquid phases, respectively.

For these calculations, it is convenient to measure time in min and gas tensions as μM (for T_L) and as the equivalent of μM (for T_G), where 100% (i.e. gas saturation) is the gas solubility at the temperature of the experiment. This value is obtained from the published literature (Gnaiger & Förstner, 1983; Wilhelm et al., 1977). Dissolved O_2 is measured by an electrode, photobacterium or MIMS probe fitted through a port below the level of the stirred vortex, or by direct inclusion of photobacterium in the reaction mixture and measuring light emission at 495 nm through the window: fluorometric, spectrophotometric or photon counting observations are made through a second window, using appropriate optical filters. Alternatively, an EPR or nuclear magnetic resonance tube may serve as the reaction vessel for combined measurements.

3. GAS MIXING

A simple, reliable and versatile device for mixing gases (Degn, Lundsgaard, & Petersen, 1980; Lloyd, 2002; Lloyd et al., 1981) uses a combination of stainless steel capillary tubes acting as flow resistors to apportion the blending of a pair of commercially available gases from cylinders. A series of two-way switches can be used to change the composition of the gas stream; five switches enables generation of a mixture that can be altered from A to B in 5% steps. The two gases must be provided at equal pressures to their inlets; the most convenient method for this involves the use of water columns to provide pressure heads. The columns are adjusted in a ratio inversely proportional to the respective dynamic viscosities of the two gases. A further refinement that uses seven valves and automated switching has been described for the generation of gas gradients. Another automatic mixer that works on a similar principle, but involves time-shared energization of two valves, is invaluable for the fast changes required in computer-controlled gas composition supply to fermenters that operate under automated control. Mixing of three gases can be achieved by using two mixing boxes in series. This configuration is useful for generation of a gas mixture of fixed relative humidity that can be altered instantly from one value to another by switching gas streams.

For most purposes, adequate flow rates of gases (e.g. up to 100 ml min^{-1}) are obtained from the mixing devices described here by using short columns of water (e.g. <1 m) as pressure heads. Where higher gas flow rates are required (e.g. for fermentation vessels of 1 l liquid volume or more), longer water columns (2–3 m) are necessary.

4. CONCLUSIONS

In summary, defining physiological O_2 tensions is essential in order to ensure optimal performance of cells and tissues in culture. It has been demonstrated that traces of O_2 can be found in the most unlikely of environments, e.g. 3 mm Hg in the hind-colon of the rat (Degn & Lundsgaard, 1980), an environment commonly considered to be anoxic. Indeed, the cytochrome *bo′* of the gut bacterium *Escherichia coli* has an extremely high affinity for O_2 (K_m O_2 = 5 nM) (He et al., 1999), reflecting the very low O_2 present in some niches of the gut. Identifying and maintaining optimal O_2 tensions is likely to be an important factor when attempting to sustain the growth of currently "unculturable" bacteria *in vitro*. Thus, it is advised that preliminary investigations are first conducted to determine the optimal, physiological O_2 tension required by the cell type of interest (e.g. based on growth assays). Strict monitoring of O_2 tensions during routine culture and subsequent experiments should then follow in order to ensure that *in vitro* experiments are physiologically relevant.

REFERENCES

Acker, H., Lubbers, D. W., Purves, M. J., & Tan, E. D. (1980). Measurements of the partial pressures of oxygen in the in the carotid body of fetal sheep and newborn lambs. *Journal of Developmental Physiology*, *2*, 323–328.

Aon, M. A., Cortassa, S., & O'Rourke, B. (2010). Redox-optimized ROS balance: A unifying hypothesis. *Biochimica et Biophysica Acta*, *1797*, 865–877.

Banerjee, R., Becker, D., Dickman, M., Gladyshev, V., & Ragsdale, S. (2008). *Redox biochemistry*. Hoboken, New Jersey: John Wiley & Sons Inc.

Barr, R. E., & Roetman, E. L. (1974). Oxygen gradients in the anterior chamber of anaesthetised rabbits. *Investigative Ophthalmology & Visual Science*, *13*, 386–389.

Barzu, O. (1984). Measurement of oxygen consumption by the spectrophotometric oxyhemoglobin method. *Methods of Biochemical Analysis*, *30*, 227–267.

Bowen, J. C., & Garg, D. K. (1977). Effect of graded mechanical ischemia on oxygen tension and electrical potential in the canine gastric mucosa. A new experimental model. *Gastroenterology*, *73*, 84–88.

Brezis, M., Rosen, S., Silva, P., & Epstein, F. H. (1984). Renal ischemia: A new perspective. *Kidney International*, *26*, 375–383.

Cater, D. B., Garaltini, S., Marina, F., & Silver, I. (1961). Changes of oxygen tension in brain and somatic tissues induced by vasodilator and vasoconstrictor drugs. *Proceedings of the Royal Society of London. Series B. Biological Sciences Journal*, *155*, 136–157.

Chance, B., Noka, S., Warren, W., & Yurtsever, G. (2005). Mitochondrial NADH as the bell-wether of tissue O2 delivery. *Advances in Experimental Medicine and Biology*, *566*, 231–262.

Chance, B., & Quistorff, B. (1978). A study of tissue oxygen gradients by single and multiple indicators. *Advances in Experimental Medicine and Biology*, *94*, 331–338.

Coogan, M. P., Court, J. B., Gray, V. L., Hayes, A. J., Lloyd, S. H., Millet, C. O., et al. (2010). Probing intracellular oxygen by quenched phosphorescence lifetimes of nanoparticles containing polyacrylamide-embedded [Ru (dpp (SO3Na)2)3]Cl2. *Photochemical and Photobiological Sciences*, *9*, 103–109.

D'Mello, R., Hill, S., & Poole, R. K. (1995). The oxygen affinity of cytochrome bo' in Escherichia coli determined by the deoxygenation of oxyleghaemoglobin and oxymyoglobin: Km values for oxygen are in the submicromolar range. *Journal of Bacteriology*, *177*, 867–870.

Degn, H., Cox, R. P., & Toftlund, H. (Eds.), (1985). *Gas enzymology*. Dordrecht: Reidel. pp. 250+x.

Degn, H., & Lundsgaard, J. S. (1980). Dynamic gas mixing techniques. *Journal of Biochemical and Biophysical Methods*, *3*, 233–242.

Degn, H., Lundsgaard, J. S., & Petersen, L. C. (1980). Polarographic measurement of steady state kinetics of oxygen uptake by biochemical samples. *Methods of Biochemical Analysis*, *26*, 47–77.

Duling, B. R., & Berne, R. M. (1970). Longitudinal gradients in periarteriolar oxygen tension. *Circulation Research*, *27*, 669–678.

Edwards, S. W., Hallett, M. B., & Campbell, A. K. (1984). Oxygen-radical production during inflammation may be limited by oxygen concentration. *Biochemical Journal*, *217*, 851–854.

Eliasson, P., & Jonsson, J. I. (2010). The hematopoietic stem cell niche: Low in oxygen but a nice place to be. *Journal of Cellular Physiology*, *222*, 17–22.

Erecińska, M., & Silver, I. A. (2001). Tissue oxygen tension and brain sensitivity to hypoxia. *Respiratory Physiology*, *128*, 263–276.

Flamant, S., Lebastard, M., Pesher, P., Besmond, C., Milon, G., & Marchal, G. (2003). Enhanced cloning efficiency of mouse bone marrow macrophage progenitors correlates with increased content of CSF-1 receptor of their progeny at low oxygen tension. *Infection*, *5*, 1064–1069.

Gnaiger, E., & Förstner, H. (1983). *Polarographic oxygen sensors* (pp. 1–370). Berlin: Springer-Verlag.

He, G., Shankar, R. A., Chzhan, M., Samouilov, A., Kuppuasamy, P., & Zweier, J. L. (1999). Non-invasive measurement of anatomic structure and intraluminal oxygenation in the gastrointestinal tract of living mice with spatial and spectral imaging. *Proceedings of the National Academy of Sciences of United States of America*, *96*, 4586–4591.

Honig, C. R., & Gayeski, T. E. J. (1987). Comparison of intracellular pO2 and conditions for blood-tissue O2 transport in heart and working red skeletal muscle. *Advances in Experimental Medicine and Biology*, *215*, 309–327.

Kessler, M., Lang, H., Sinagowitz, E., Rink, R., & Höper, J. (1973). Homeostasis of oxygen supply in liver and kidney. *Advances in Experimental Medicine and Biology*, *37*, 351–360.

Khan, N., Mupparajui, S. P., Hou, H., Lariviere, J. P., Demidenko, E., Swartz, H. M., et al. (2009). Radiotherapy in conjunction with 7-hydroxystaurosporine: A multimodal approach with tumor pO2 as a potential marker of therapeutic response. *Radiation Research*, *172*, 592–597.

Lash, J. M., & Bohlen, H. G. (1987). Perivascular and tissue PO2 in contracting rat spinotrapezius muscle. *American Journal of Physiology. Heart and Circulatory Physiology*, *252*, H1192–H1202.

Lloyd, D. (2002). Noninvasive methods for the investigation of organisms at low oxygen levels. *Advances in Applied Microbiology*, *51*, 155–183.

Lloyd, D., Bohátka, S., & Szilágyi, J. (1985). Quadrupole mass spectrometry in the monitoring and control of fermentations. *Biosensors*, *1*, 179–212.

Lloyd, D., James, K., Williams, J., & Williams, T. N. (1981). A membrane-covered photobacterium probe for oxygen measurements in the nanomole range. *Analytical Biochemistry*, *116*, 17–21.

Lloyd, D., Mellor, H., & Williams, J. L. (1983). Oxygen affinity of the respiratory chain of Acanthamoeba castellanii. *Biochemical Journal*, *214*, 47–51.

Lloyd, D., Williams, C. F., Vijayalakshmi, K., Kombrabail, M., White, N., Hayes, A. J., et al. (2014). Intracellular oxygen: Similar results from two methods of measurement using phosphorescent nanoparticles. *Journal of Innovative Optical Health Sciences*, 7, 1350041.

Messmer, K. L., Sunder-Plassmann, F., Jesch, L., Görnandt, L., Sinagowitz, E., & Kessler, M. (1973). Oxygen supply to the tissues after limited normovolemic hemodilution. *Research in Experimental Medicine*, *159*, 152–156.

Mohyeldin, A., Garzón-Murdi, T., & Quiñones-Hinojosa, A. (2010). Oxygen in stem cell biology: A critical component of the stem cell Niche. *Cell Stem Cell*, 7, 150–161.

Muppuraju, S., Hou, H., Lariviere, J. P., Swartz, H., Jounaidi, Y., & Khan, N. (2011). Repeated tumor oximetry to identify therapeutic window during metronomic cyclophosphamide treatment of 9L gliomas. *Oncology Reports*, *26*, 281–286.

Nair, P., Whalen, W. J., & Buerk, D. (1975). PO2 of cat cerebral cortex: Response to breathing N2 and 100 per cent O2. *Microvascular Research*, *9*, 158–165.

Sakadzic, S., Roussakis, E., Yaseen, M. A., Mandeville, E. T., Srinivasan, V. J., Araik Ruvinskaya, S., et al. (2010). Two-photon high-resolution measurement of partial pressure of oxygen in cerebral vasculature and tissue. *Nature Methods*, 7, 755–759.

Sato, N., Kamada, T., Kawano, S., Hayashi, N., Kishida, Y., Meren, H., et al. (1983). Effect of acute and chronic ethanol consumption on hepatic tissue oxygen tension in rats. *Pharmacology Biochemistry and Behaviour*, *18*, 443–447.

Semenza, G. L. (1998). Hypoxia-inducible factor 1: Master regulator of O2 homeostasis. *Current Opinion in Genetics & Development*, *8*, 588–594.

Semenza, G. L. (2000). Oxygen-regulated transcription factors and their role in pulmonary disease. *Respiratory Research*, *1*, 159–162.

Shahidi, M., Wanek, J., Blair, N. P., Little, D. M., & Wu, T. (2010). Retinal tissue oxygen tension imaging in the rat. *Investigative Ophthalmology and Visual Science*, *51*, 4766–4770.

Smith, R. H., Guilbeau, E. J., & Reneau, D. D. (1977). The oxygen tension field within a discrete volume of cerebral cortex. *Microvascular Research*, *13*, 233–240.

Stucker, M., Struk, A., Altmeyer, P., Herde, M., Baumgärtl, H., & Lübbers, D. W. (2002). The cutaneous uptake of atmospheric oxygen contributes significantly to the oxygen supply of human dermis and epidermis. *Journal of Physiology*, *538*, 985–991.

Tamura, M., Hazeki, O., Nioka, S., & Chance, B. (1989). In vivo study of tissue oxygen metabolism using optical and nuclear magnetic resonance spectroscopies. *Annual Reviews of Physiology*, *51*, 813–834.

Tiede, L. M., Cook, E. A., Morsey, B., & Fox, H. S. (2011). Oxygen matters: Tissue culture oxygen level affect mitochondrial function and structure as well as responses to HIV viroproteins. *Cell Death & Disease*, *2*, e246.

Toussaint, O., Weemaels, G., Debacq-Chainiaux, F., Scharffetter-Kochanek, K., & Wlaschek, M. (2011). Artefactual effects of oxygen on cell culture models of cellular senescence and stem cell biology. *Journal of Cellular Physiology, 226,* 315–321.

Tsai, A. G., Johnson, P. C., & Intaglietta, M. (2003). Oxygen gradients in the microcirculation. *Physiological Reviews, 83,* 933–963.

Vanderkooi, J. M., Erecińska, M., & Silver, I. A. (1991). Oxygen in mammalian tissue: Methods of measurement and affinities of various reactions. *American Journal of Physiology, 260,* C1131–C1150.

Wang, Y., Yang, J., & Yi, J. (2011). Redox sensing by proteins: Oxidative modifications on cysteines and the consequent events. *Antioxidants & Redox Signaling, 6,* 649–657.

Weiss, I. K., Isenberg, S. J., McArthur, D. L., Del Signore, M., & McDonald, J. S. (2011). Monitoring the conjunctiva for carbon dioxide and oxygen tensions and pH during cardiopulmonary bypass. *Journal of Extracorporeal Technology, 43,* 13–18.

Weiss, H. R., & Winbury, M. W. (1972). Effects of intracoronary nitroglycerin, pentaerythritol trinitrate and dipyridamol on intramyocardial oxygen tension. *Microvascular Research, 4,* 273–284.

Wenger, R. H. (2002). Cellular adaptation to hypoxia: O2-sensing protein hydroxylases, hypoxia-inducible transcription factors, and O2-regulated gene expression. *Federation of American Societies for Experimental Biology Journal, 16,* 1151–1162.

Whalen, W. J., & Nair, P. (1970). Skeletal muscle PO2: Effect of inhaled and topically applied O2 and CO2. *American Journal of Physiology, 218,* 973–980.

Wilhelm, E., Battino, R., & Woodcock, R. J. (1977). Low pressure solubility of gases in liquid water. *Chemical Reviews,* 77, 219–262.

Williams, C. F., Kombrabail, M., Vijayalakshmi, K., White, N., Krishnamoorthy, G., & Lloyd, D. (2012). Technical design note: Evaluation of two novel methods for assessing intracellular O2. *Measurement Science and Technology, 23,* 084005.

Wittenberg, B. A., & Wittenberg, J. B. (1989). Transport of oxygen in muscle. *Annual Reviews of Physiology, 51,* 857–878.

Yu, D. Y., & Cringle, S. J. (2005). Retinal degeneration and local oxygen metabolism. *Experimental Eye Research, 80,* 745–751.

Zhou, S., Cui, Z., & Urban, J. P. G. (2004). Factors influencing the oxygen concentration gradient from the synovial surface of articular cartilage to the cartilage-bone interface: A modeling study. *Arthritis and Rheumatology, 50,* 3915–3924.

AUTHOR INDEX

Note: Page numbers followed by "*f*" indicate figures and "*t*" indicate tables.

A

B

C

D

E

H

I

J

K

N

O

P

Q

R

S

T

U

V

W

X

Y

Z

SUBJECT INDEX

Note: Page numbers followed by "*f*" indicate figures and "*t*" indicate tables.

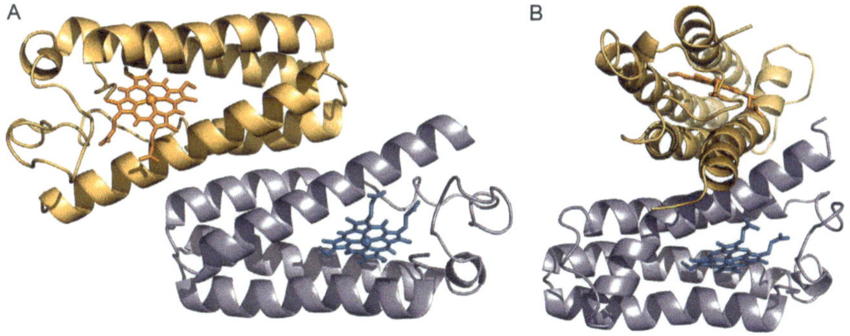

Michael A. Hough and Colin R. Andrew, Figure 3 Homodimer arrangements in (A) RCCP and (B) RMCP. Note the X-shaped dimeric arrangement in the type 1 RMCP and the near-parallel arrangement in the type 2 RCCP.

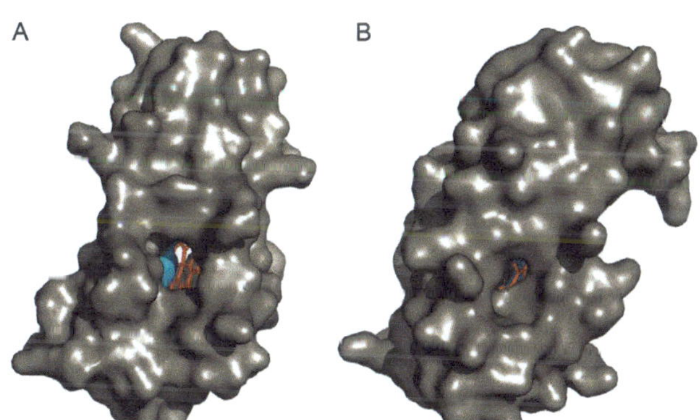

Michael A. Hough and Colin R. Andrew, Figure 4 Surface representations of monomers of (A) group 1 RCCP and (B) group 2 RMCP. Note the large access channel to the distal haem pocket in RCCP that is not present in the group 2 cytochromes *c*′.

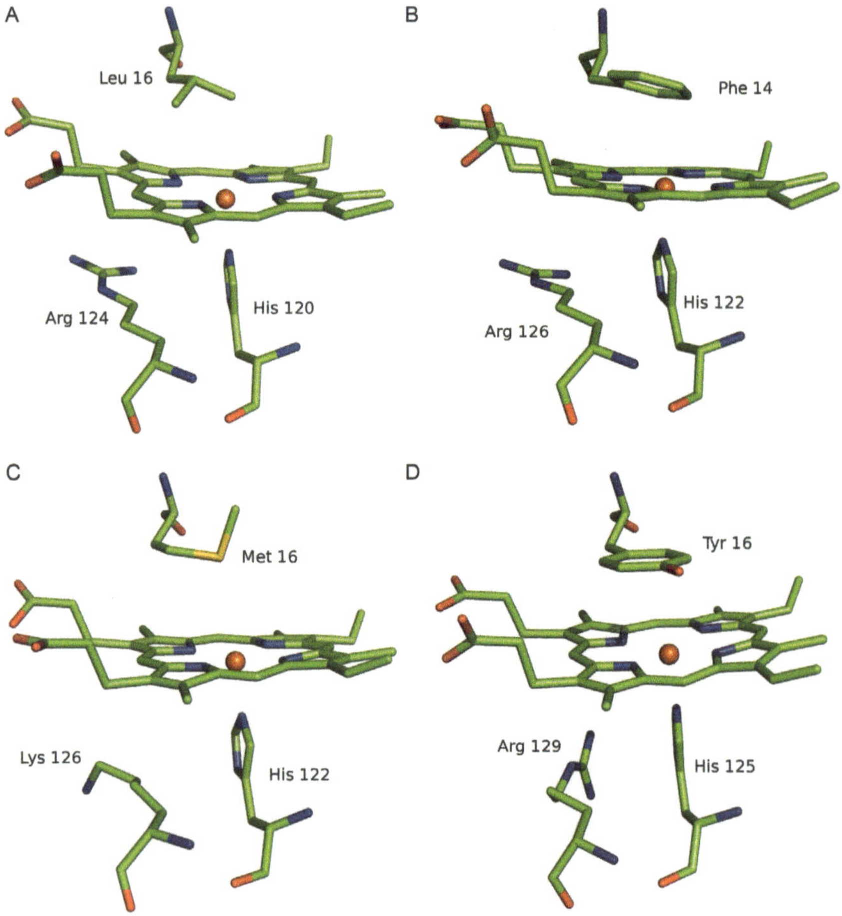

Michael A. Hough and Colin R. Andrew, Figure 5 Distal (top) and proximal (bottom) pocket residues in (A) AXCP, (B) RCCP, (C) RMCP and (D) CVCP with occluding residues Leu, Phe, Met and Tyr, respectively.

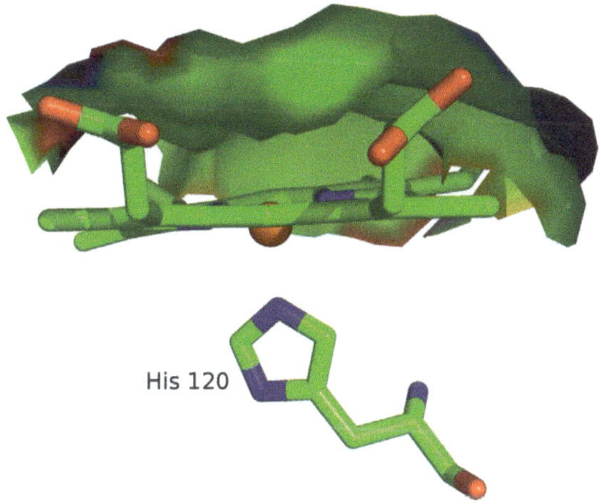

Michael A. Hough and Colin R. Andrew, Figure 6 The distal pocket of AXCP. Carbon atoms are shown in green, oxygen in red, nitrogen in blue and sulphur in yellow.

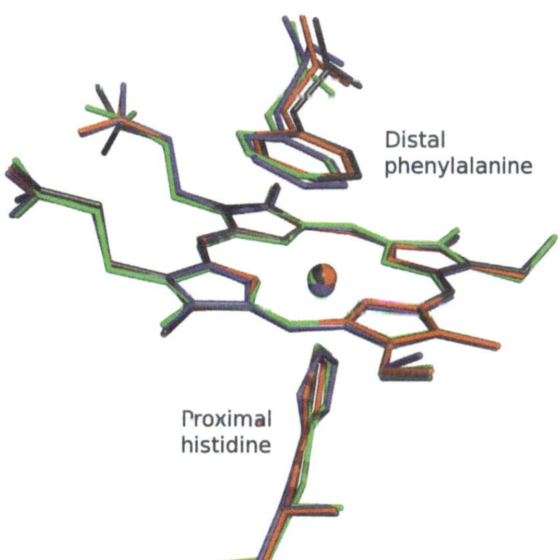

Michael A. Hough and Colin R. Andrew, Figure 7 Occlusion of the Fe distal ligand binding site by Phe residues in SFCP (red), RGCP (grey), RCCP (green) and RSCP (blue). Note that the Phe positions are clustered into two groups.

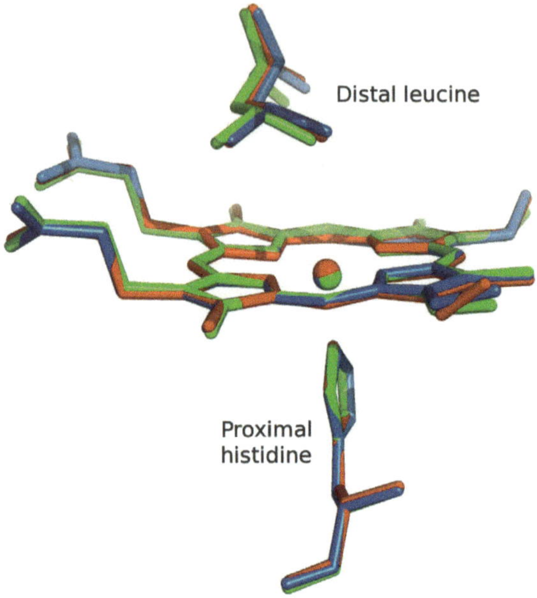

Michael A. Hough and Colin R. Andrew, Figure 8 Superposition of the haem environment in AXCP (red), ADCP (blue) and RPCP (green). Note that coordinates are not available for RRCP which also has a distal Leu residue.

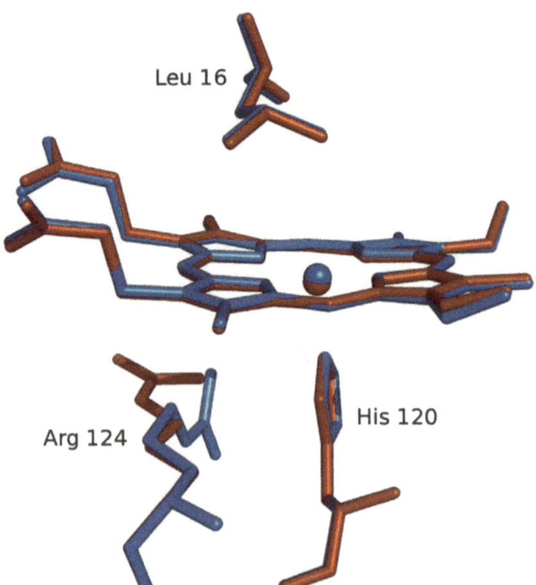

Michael A. Hough and Colin R. Andrew, Figure 9 The haem environment in ferric (red) and ferrous (blue) AXCP.

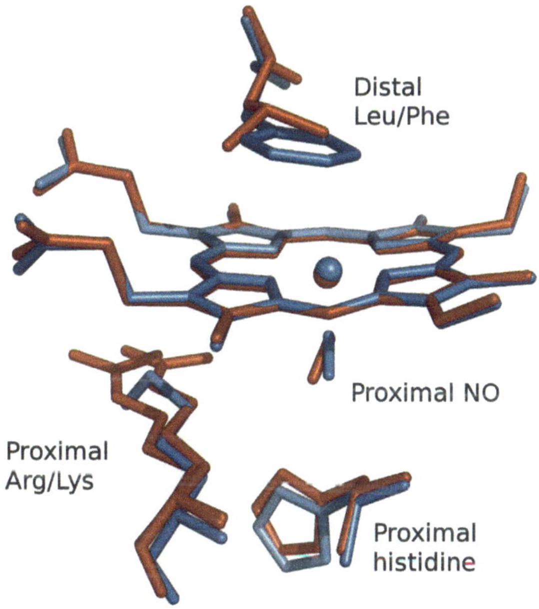

Michael A. Hough and Colin R. Andrew, Figure 11 Superposition of the 5cNO complexes in AXCP (red) and SFCP (blue).

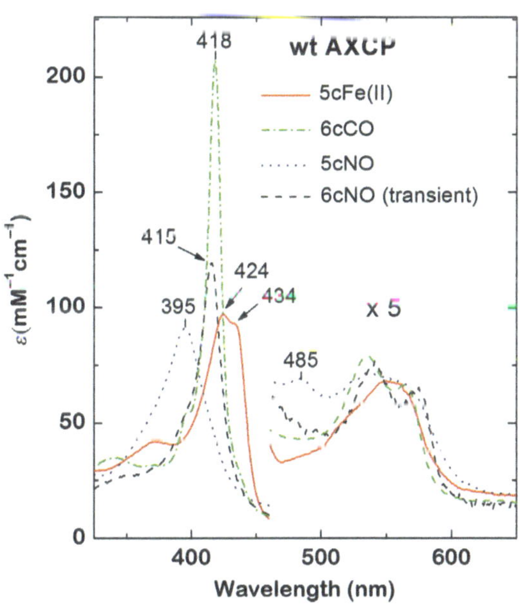

Michael A. Hough and Colin R. Andrew, Figure 12 Spectroscopic distinction of AXCP complexes by UV–vis absorbance.

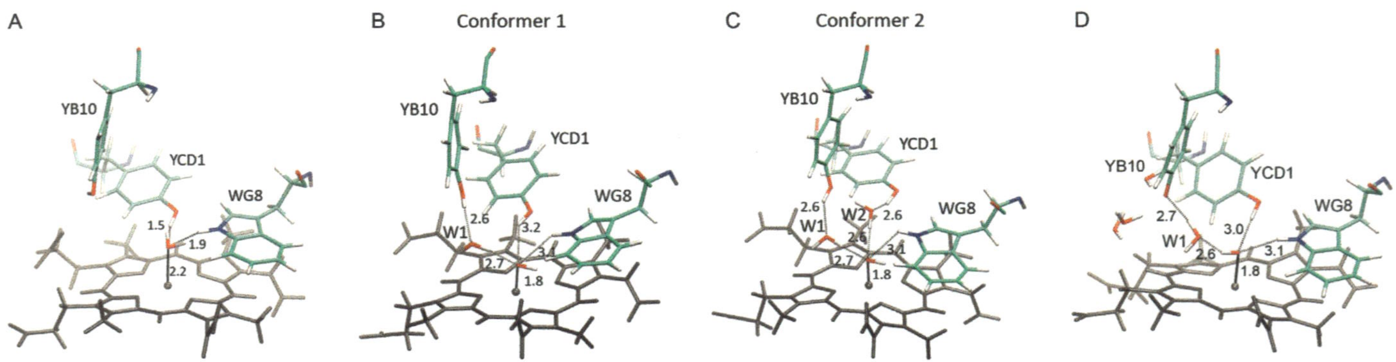

Barry D. Howes *et al.*, Figure 3 Schematic representation of the distal side of wild-type (WT) Tf-trHb at neutral (A) and alkaline pH (B–D). The dotted lines represent the hydrogen-bond network stabilizing the iron-bound H_2O (A) or the iron-bound OH^- in the two conformations of WT Tf-trHb at alkaline pH when both YB10 and YCD1 are protonated (B and C), or when only YB10 is deprotonated (D), on the basis of MD simulations. Distances are in angstroms.

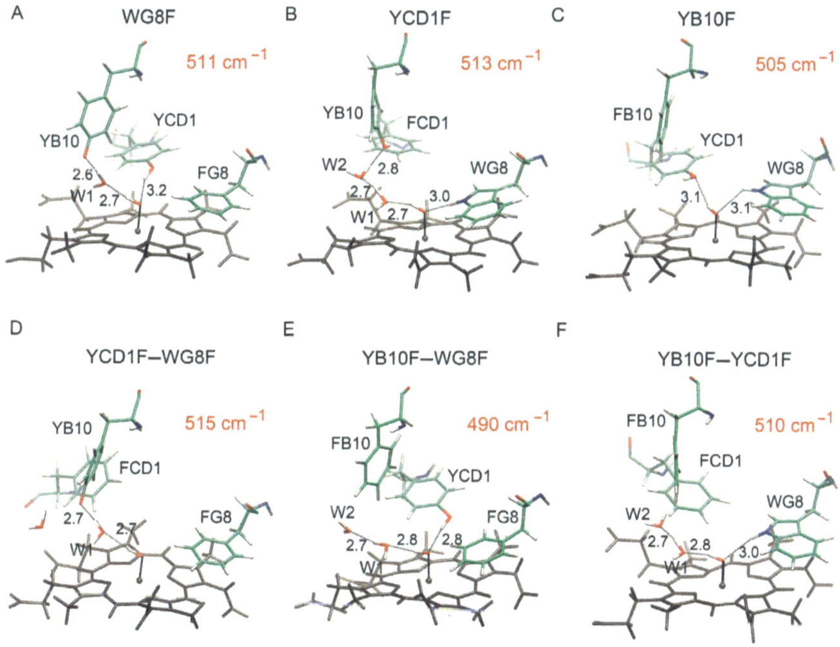

Barry D. Howes *et al.*, Figure 4 Schematic representation of the distal haem cavity of the mutated proteins at alkaline pH showing, on the basis of MD simulations, the hydrogen-bond network (dotted lines) stabilizing the iron-bound OH^- for single (A–C) and double (D–F) mutant proteins. As TyrB10 is deprotonated it acts as a H-bond acceptor, whereas WG8 and YCD1 are H-bond donors. Distances are in angstroms. The RR ν(Fe–OH) stretching mode frequencies for each mutant are shown in red. *Reprinted with permission from Nicoletti et al. (2014). Copyright (2014) American Chemical Society.*

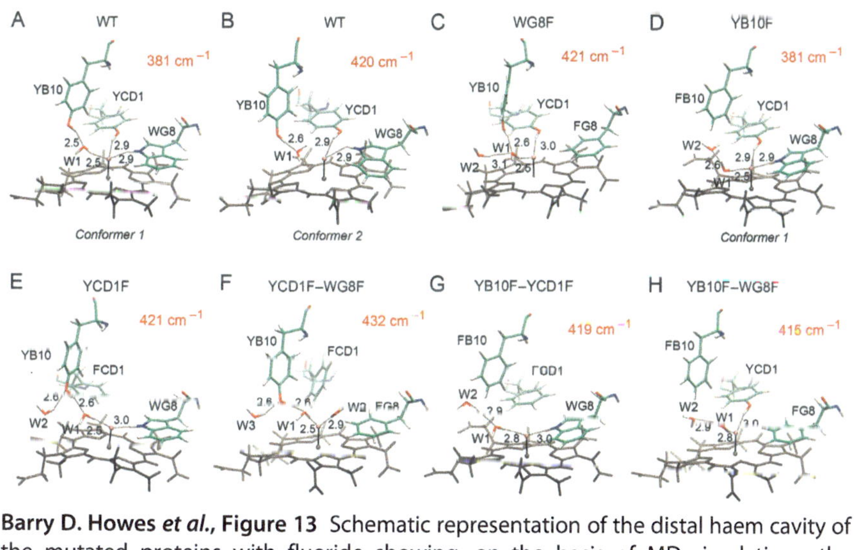

Barry D. Howes *et al.*, Figure 13 Schematic representation of the distal haem cavity of the mutated proteins with fluoride showing, on the basis of MD simulations, the hydrogen-bond network (dotted lines) stabilizing the fluoride for WT (A and B), single (C–E), and double (F–H) mutant proteins. TyrB10, WG8, and YCD1 are H-bond donors. Distances are in angstroms. The RR ν(Fe–F) stretching frequencies for each mutant are shown in red. For the YB10F mutant, only conformer 1 is shown.

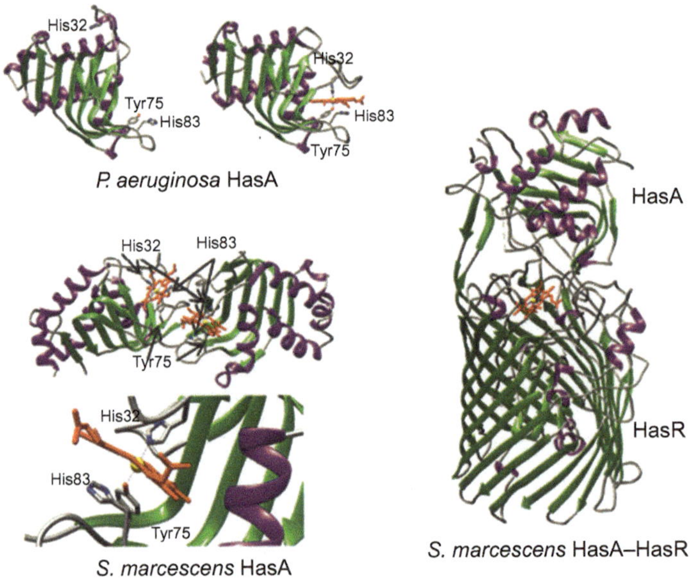

Paolo Ascenzi *et al.*, Figure 2 Crystal structures of *S. marcescens* and *P. aeruginosa* haem uptake systems. Ligand-free (left) and haem-bound (right) *P. aeruginosa* HasA (PDB ID: 3MOK and 3ELL, respectively). The haem axial ligands His32 and Tyr75, and the His83 residue stabilizing the tyrosinate form of Tyr75 are shown in stick representation. Swapped homo-dimeric haem-bound HasA from *S. marcescens* (PDB ID: 2CN4). The haem axial ligands His32 and Tyr75 residues of both monomers are shown in stick representation. Hetero-dimeric haem-bound HasA–HasR complex from *S. marcescens* (PDB ID: 3CSL). Helices are in purple, strands in forest green, coils in grey, haem in red, and the haem–Fe atom in gold. *Pictures drawn by UCSF-Chimera (Pettersen et al., 2004).*

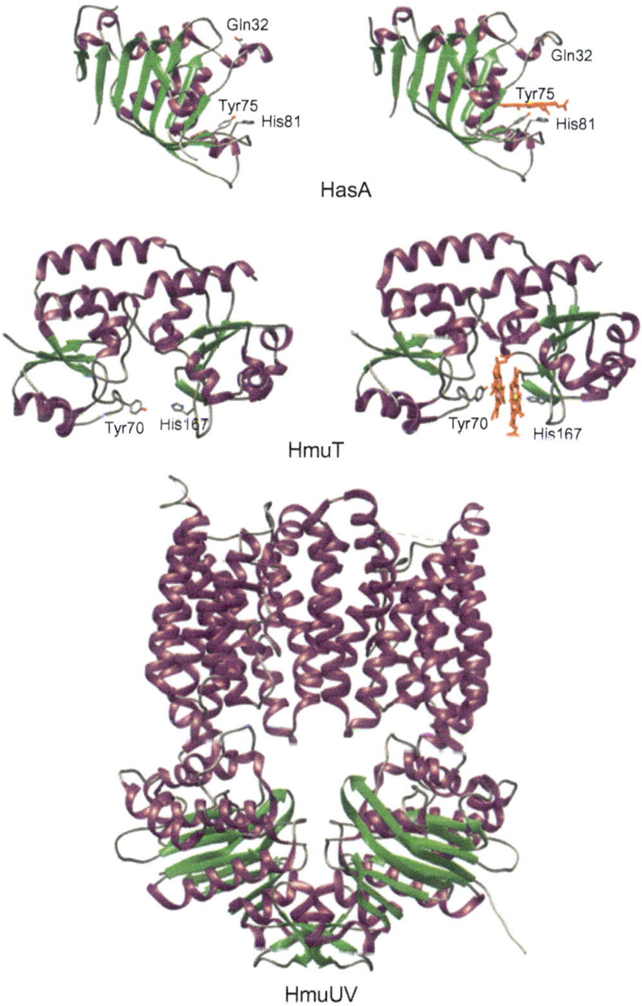

Paolo Ascenzi *et al.*, Figure 3 Crystal structures of *Y. pestis* haem uptake system. Ligand-free (left) and haem-bound (right) HasA (PDB ID: 4JER and 4JET, respectively). The haem axial ligand Tyr75 and the His81 residue stabilizing the tyrosinate form of Tyr75 are shown in stick representation. Ligand-free (left) and haem-bound (right) periplasmic haem-binding protein HmuT (PDB ID: 3MD9 and 3NU1, respectively). The haem axial ligands Tyr70 and His167 are shown in stick representation. Ligand-free haem transporter HmuUV (PDB ID: 4G1U). Helices are in purple, strands in forest green, coils in grey, haem in red, and the haem–Fe atom in gold. *Pictures drawn by UCSF-Chimera (Pettersen et al., 2004).*

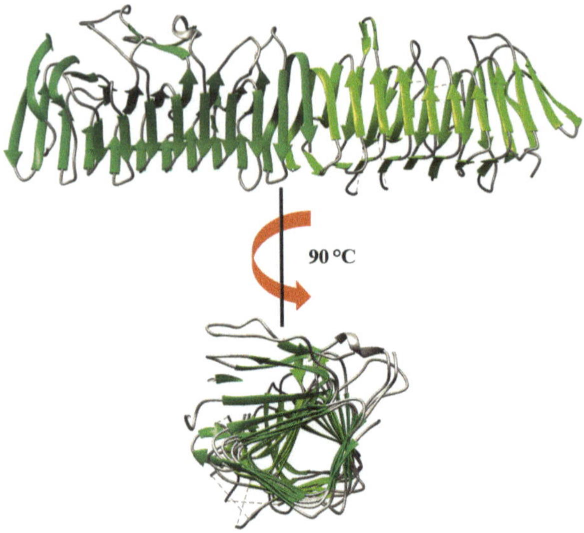

Paolo Ascenzi *et al.*, Figure 5 Crystal structure of *H. influenza* haem uptake system. Dimeric *H. influenzae* ligand-free HxuA (PDB ID: 4I84). The two pictures are rotated by 90°. For clarity, the two β-helix subunits are in forest green and chartreuse (coils in grey). *Pictures drawn by UCSF-Chimera (Pettersen et al., 2004).*

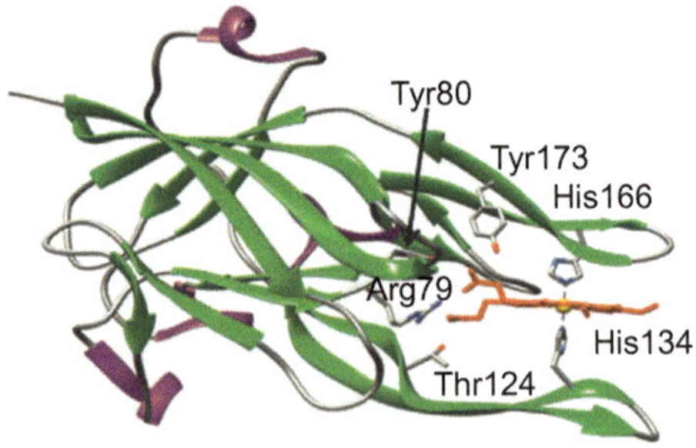

Paolo Ascenzi *et al.*, Figure 7 Crystal structure of *P. gingivalis* HmuY system. Haem-bound HmuY (PDB ID: 3H8T). The haem axial ligands His134 and His166, and other relevant residues, are shown in stick representation. Helices are in purple, strands in forest green, coils in grey, haem in red, and the haem–Fe atom in gold. *Picture drawn by UCSF-Chimera (Pettersen et al., 2004).*

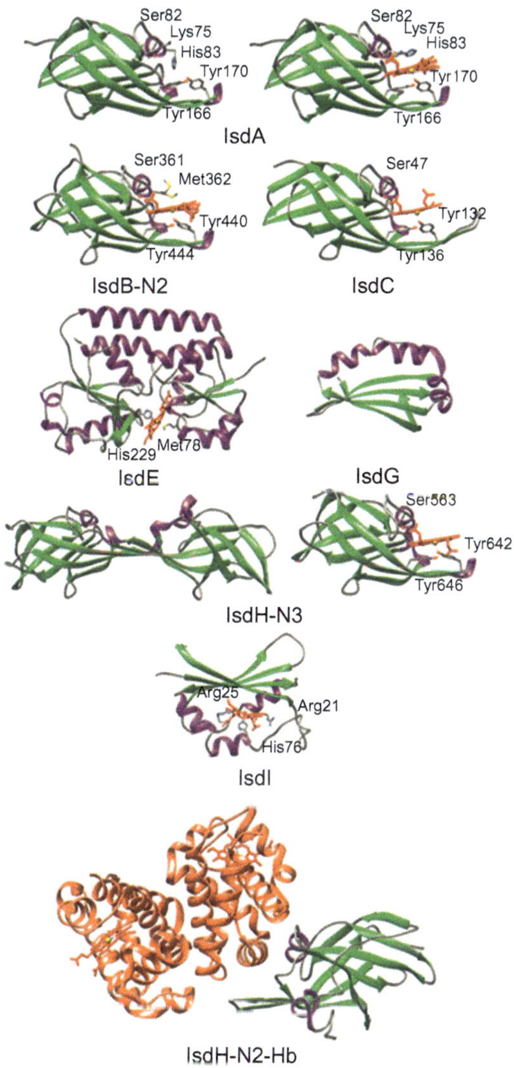

Paolo Ascenzi *et al.*, Figure 9 Crystal structures of *S. aureus* Isd system. Ligand-free (left) and haem-bound (right) IsdA (PDB ID: 2ITE and 2ITF, respectively), haem-bound IsdB-N2 (PDB ID: 3RTL), haem-bound IsdC (PDB ID: 2O6P), haem-bound IsdE (PDB ID: 2Q8Q), ligand-free IsdG (PDB ID: 1XBW), ligand-free (left) and haem-bound (right) IsdH-N3 (PDB ID: 2E7D and 2Z6F, respectively), and haem-bound IsdI (PDB ID: 3QGP). IsdH-N2 is complexed with αβ-Hb (PDB ID: 4FC3). Relevant residues are shown in stick representation. Helices are in purple, strands in forest green, coils in grey, haem in red, the haem–Fe atom in gold, and Hb in old rose. *Pictures drawn by UCSF-Chimera (Pettersen et al., 2004).*

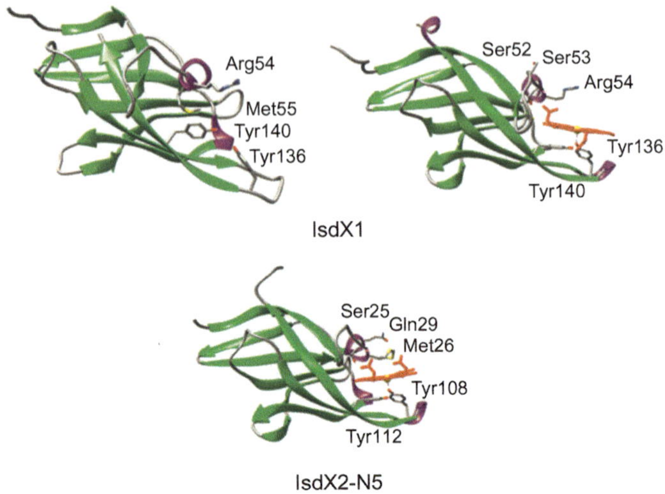

Paolo Ascenzi *et al.*, Figure 11 Crystal structures of *B. anthracis* Isd system. Ligand-free (left) and haem-bound (right) IsdX1 (PDB ID: 3SZ6 and 3SIK, respectively) and haem-bound IsdX2-N5 (PDB ID: 4H8P). The haem axial ligands of IsdX1 (Tyr136) and IsdX2-N5 (Tyr108) and other relevant residues are shown in stick representation. Helices are in purple, strands in forest green, coils in grey, haem in red, and the haem–Fe atom in gold. *Pictures drawn by UCSF-Chimera (Pettersen et al., 2004).*

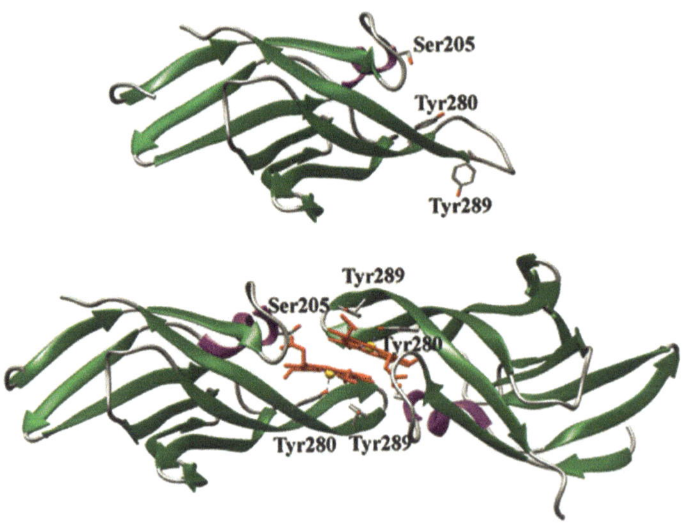

Paolo Ascenzi *et al.*, Figure 13 Crystal structures of *L. monocytogenes* Hbp2 system. Monomeric ligand-free (top) and homo-dimeric haem-bound (bottom) Hbp2 (PDB ID: 4NLA and 4MYP, respectively). The haem axial ligand Tyr280 and the Tyr289 residue stabilizing the tyrosinate form of Tyr280 are shown in stick representation. Helices are in purple, strands in forest green, coils in grey, haem in red, and the haem–Fe atom in gold. *Pictures drawn by UCSF-Chimera (Pettersen et al., 2004).*

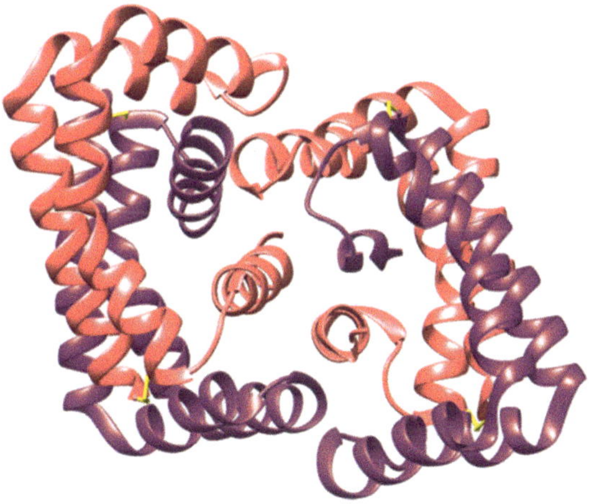

Paolo Ascenzi *et al.*, Figure 15 Crystal structure of *M. tuberculosis* haem carrier protein Rv0203. Hetero-tetrameric ligand-free Rv0203 (PDB ID: 3MAY). For clarity, two subunits are in purple and two subunits in hot pink. Disulphide bridges are in yellow. *Pictures drawn by UCSF-Chimera (Pettersen et al., 2004).*

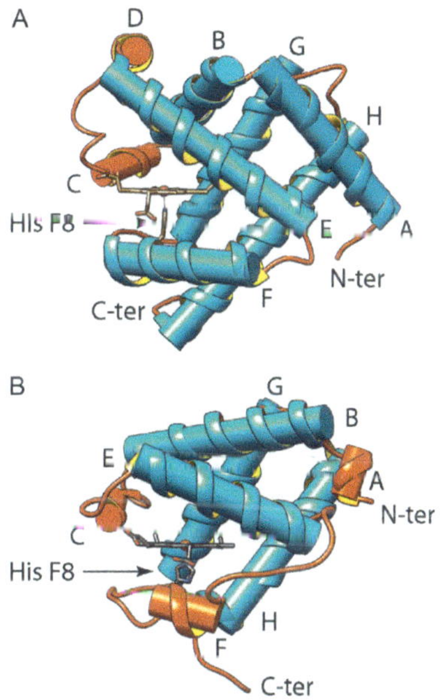

Eric A. Johnson and Juliette T.J. Lecomte, Figure 1 The two main topologies of the haemoglobin superfamily. (A) The 3-on-3 fold illustrated by sperm whale myoglobin (1JP6) (Urayama, Phillips, & Gruner, 2002). The α-helical sandwich is formed by the A, B, E, F, G, and H helices depicted in blue. (B) The 2-on-2 fold illustrated by the haeme domain of *Chlamydomonas eugametos* LI637 (1DLY) (Pesce et al., 2000). The α-helical sandwich is formed by the B, E, G, and H helices depicted in blue. Haemes and proximal histidines are represented with sticks. The proximal histidine is denoted as F8 (eighth position in the F helix) by alignment with the canonical 3-on-3 haemoglobins (Perutz, 1979).

```
                         B10                             E10
SYN6803   TLYEKLGGTTAVDLAVDKFYERVLQDDRIKHFFA-DVDMAKQRAHQKAFLTYAFGGTDKYDG-------
  LI637   SLFAKLGGREAVEAAVDKFYNKIVADPTVSTYFS-NTDMKVQRSKQFAFLAYALGGASEWKG-------
  LI410   SLFAKLGGREAVEAAVDKFYNKVVADPTVSVFFS-KTDMKVQRSKQFAFLAYALGGAAEWKG-------
             |A|  |-------B-------|     |-C-|     |-------E------|
   THB1   SLYSRMGGEAAVEKAVDVFYERIVADPQLAPFFA-NVDMKKQRRKQVAFMTYVFGGSGAYEG-------
   THB2   SIFDRLGGEAAVEKAVDVFYERIVADPQLAPFFE-GVDMRTQRRKQQAFMTYAFGGATGYTG-------
   THB3   CVYESMGGWETLSAAVESLYNRMRGDGRCASLFS-EGNEQQLKTHMLEFLTCALGGKARFAS-------
   THB4   TLHAKLGGAAAVAATVDVFYKKLMNDPDLEPFFR-GVDMVTLIAKQNRFLAYAFGATTHYHG-------
  THB5A   GIVERLGGADRVHDVIQAFYRKLFQNAMVRHYFE-NLSSDRMRSKQMKFMRYIMGGPGTYTG-------
  THB5B   PLVAKLGGPEPVQGIITAFYRRLFSSEELKYFFV-GLSTERLRTMQFRFMRYLLAGPGWYATT--------G
  THB5C   GLVRRVGGPEAAYELVRRLYSRLFADEVLKGFFA-GLSVERLKTKQLRFMRYLLGGPPAYAAA------TGG
  THB5D   SIYERLGGEEVLCDLVDMWAGKLGSDPALAHIYSGGVDMPRQRAHQLAFLRQAFGGSGGGGASRPGSA-VGA(ins1)
   THB6   HKRIKNSEEFSMKFLVDKFYGKVLKIKEIKPFFD-GVDMHKLKYHQEALMLLVFGGQELLDDELPGLS----
   THB7   TLLAKLGGNDILKQVAERWYAKLKRDERLGTFLA-DQDTIHLRARLSAYLANLFGQKHDPMAALLRKQ-GRC(ins2)
   THB8   TLLKRIGGEKGMKDVVEMFYRKIYANERLRHFLK-DKDVSLLRSKQAAFMTWLFGPSTATYT---------G
   THB9   PLLDRMGGADYVKRVAERFYHKLYADEQLAPFLG-EHDVVSLRAKQNAFLIWAFGPPNRPYP---------G
  THB10   PLLARVGGIDNVKRVVEGFYRRLYADERMDKYLS-GFDMPYLRAKQSAFMGWLFGPASVPYS---------G
  THB11   PLLQRVGGLDVVKKVVELFYRKLYADPQLIKYLH-DQDPMHLRAKQSMFVSWLFGPPNVPYT---------G
  THB12   PLLQRVGGSDVVKHVVELFYRHLYADPSLIKFLH-DQEMTHLRAKQSAFISWLFGPPNVPYT---------G

               F8
SYN6803   ---RYMREAHKELVENHGLNGEHFDAVAEDLLATLKEMGVPEDLIAE-VAAVAGAPA-----HKRDVLNQ--------
  LI637   ---KDMRTAHKDLVPH--LSDVHFQAVARHLSDTLTELGVPPEDITD-AMAVVASTR-----TEVLNMPQQ-------
  LI410   ---KDMRTAHKDLVPH--LTDVHFQAVVRHLSDTLAELGVTPGDIAD-AMAVVASTK-----TEVLNMPRQQGAESNR
             |-----F----|   |-------G-------|     |------------H------------|
   THB1   ---RDLGASHRRLIREQGMNHHHFDLVAAHLDSTLQELGVAQELKAE-AMAIVASAR-----P--LIFGT--------
   THB2   ---RDLAAAHRRLIRDKGLKEEHFDMVAGHLAGTLQLLGIGADLVDE-VIALVATTK-----P--VIFERV-------
   THB3   ---STLLTSQRDQLRQHGFGVGQFDILLQHMKAVLDEIGVQQETAAS-AIALLRCHK-----YLF-ERRQ--------
   THB4   ---KDIVMGHAHLIINRGLNLTHFDKVAGHFVDSLKEMGVGQELIDE-AAGVLIGVR-----PLFDP---E-------
  THB5A   ----NLRCIHAQMVKEEGLDKAKFQVVIGMLLDTFKELNVPQDVVSE-ILGHVESAK-----SAIFTPQP--------
  THB5B   --GGNMRCVHARMIAEGGLDLPKFQAVTGMLRAACDEHLVSPPLIRE-IMLNVKAAE-----QAIFTPGP--------
  THB5C   TGGANLRCPHLKLMVKQPAAAAAAGADD--------------DLHRS-STGGTEEGGGGASTAKAAGGGRDVLDGTK
  THB5D   VGAMRLEIAAAHLARDKALAPAQFERIADHLTACLKKEGASPELVAE-VRAMQGPGR-----CLLWPPSS--------
   THB6   ---ADLRLIHLHLLF-EGLNLSHWQIFADTFCETLDELPQIPADVKAHAKAYMRATK-----QH-FRPI---------
   THB7   GRCHHLRRVHLRLIQQMGFSPADFDAGLGYFRESLQELGAPEWLVDE-VMLKLVEVK-----DLVFDDPD--------
   THB8   --GRSVRCAHLRMIKQRGFSEDDFLLGMSCFEEAMRDYGAPQRLTNE-IMAKILPFK-----DIIFSP-
   THB9   --GRHLRTAHLRLIKQKGFTVEHFEMGLKHFEEALKEMDAPPSVVRD-ALMKLRPYK-----TVVFTPSSA-
  THB10   --GKHLRTVHLKMIKQQGFSPEDFEVGMDYFEESMKELGAPETIISE-VMSKIRPFK-----AVIFTPNA-
  THB11   --GKSVRIAHLRIIKQRGFSPEDFDLGMKYFEEAMTELGAPEVLRGE-VMRRMLPYK-----DAIFTP-
  THB12   --GKSVRIAHLRIIKQRGFSPEDFDLGMKYFAESMRELGAPEILINE-VLRRITPYK-----DAIFTP-

  THB5D ins1    A-C-----------------------------------------PALG---GTAGAAGGM
  THB7  ins2    ATSQALQHQRSSVLQHQPTSNGLALHSSQPSLVTSRAPSHAMSRLPPSYSLHHQATSSRVGA
```

Eric A. Johnson and Juliette T.J. Lecomte, Figure 3 The amino acid sequences of *C. reinhardtii* haemoglobins. For comparison, the top three sequences are for the cyanobacterium *Synechocystis* sp. PCC 6803 (UniProt identifier P73925), the globin domain of *C. eugametos* LI637 (UniProt identifier Q08753), and the globin domain of *C. eugametos* LI410 (UniProt identifier P52334). Only the globin domains are included. The secondary structure of THB1 (Rice et al., 2015) is indicated above its own sequence. THB5D and THB7 contain long insertions in the EF loop (ins1 and ins2, respectively). These inserts have been extracted from the domain sequences and are listed at the bottom of the figure. Residues B10, E10, and F8 are highlighted along with the location of the Gly-Gly motifs characteristic of TrHb1s.

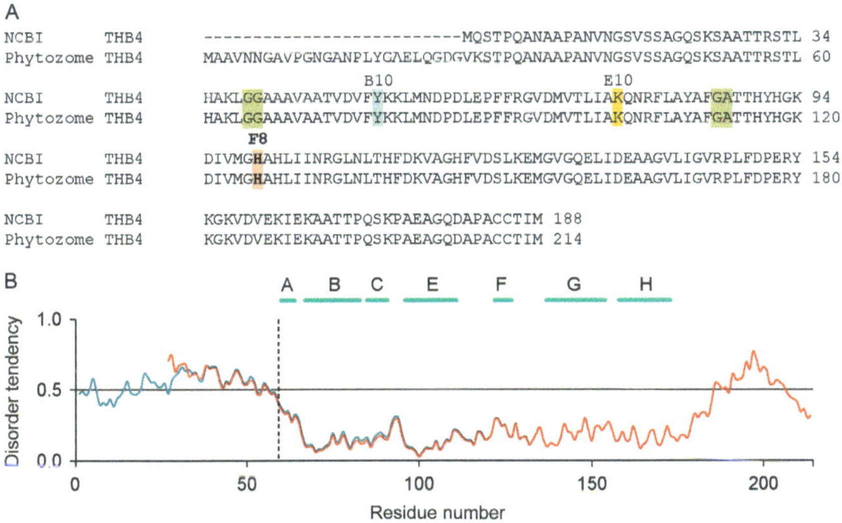

Eric A. Johnson and Juliette T.J. Lecomte, Figure 4 (A) Sequence alignment of two possible *THB4* gene products. (B) The IUPred profile (Dosztanyi, Csizmok, Tompa, & Simon, 2005) for THB4 (UniProt identifier A8ISP8) (long disorder calculation). The locations of helices in THB1 are marked.

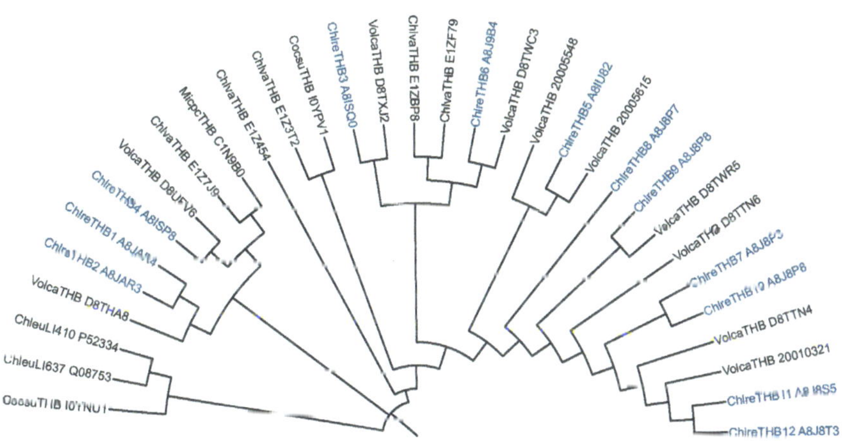

Eric A. Johnson and Juliette T.J. Lecomte, Figure 5 Phylogenetic tree for chlorophytic haemoglobins prepared with iTOL software (Letunic & Bork, 2007, 2011). Proteins are from *C. reinhardtii* (Chlre, blue font), *C. eugametos* (Chleu), *V. carteri* (Volca), *C. variabilis* (Chlva), *Micromonas pusilla* (Micpc), and *Coccomyxa subellipsoidea* (Cocsu).

```
                                                                                 B10
tr|A8ISP8|A8ISP8_CHLRE    MQSTPQANAAPANVNGSVSSAGQSKSAATTRSTLHAKLGGAAAVAATVDVFYKKLMNDPD
tr|D8UFV6|D8UFV6_VOLCA    ---------------------------MTARATLHAKLGGAAAVAATVDIFYKKLMGDPD
                                                      *:*:*****************:****** ***

                                           E10                     F8
tr|A8ISP8|A8ISP8_CHLRE    LEPFFRGVDMVTLIAKQNRFLAYAFGATTHYHGKDIVMGHAHLIINRGLNLTHFDKVAGH
tr|D8UFV6|D8UFV6_VOLCA    LAPFFHGVDMVTLIAKQNRFLAYAFGATSHYHGKDIVMGHAHLIINRGLNLSHFDKVAGH
                          * ***:**********************:**********************:********

tr|A8ISP8|A8ISP8_CHLRE    FVDSLKEMGVGQELIDEAAGVLIGVRPLFDPERYKGKVDVEKIEKAATTPQSKPAEAGQD
tr|D8UFV6|D8UFV6_VOLCA    FVDSLKEMGVTQELIDDAAGVLIGVRPLFDPERYKGKIDVEKIEKAAVVPQTKDTEGVKE
                          ********** *****:********************:*********..**:* :*. ::

tr|A8ISP8|A8ISP8_CHLRE    APACCTIM
tr|D8UFV6|D8UFV6_VOLCA    TPACCTVM
                          :*****:*
```

```
tr|A8ISQ0|A8ISQ0_CHLRE    ------------------------------MPKVDGTPSALDRTAAISAATESLIASLED
tr|D8TXJ2|D8TXJ2_VOLCA    MAQDDVNTIHQAVTDWHEKLLSLPAMHGVLLHGPRQAPPPEERGAAVALAAQHLATCLED
                                                        :     :*   :* **:: *:: * :.***

tr|A8ISQ0|A8ISQ0_CHLRE    KQLPDGLIDDILSVVDVFTTLCGLSGPSPSDSFGAGGS---------TRSGAPLNPAKLK
tr|D8TXJ2|D8TXJ2_VOLCA    KNLPDGLINDLLSVCETFTALCGVSLPNNS--LGALSSPAHGALWVGWGSNRPLKAAKVK
                          *:******:*:*** :.**:***:* *. *  :** .*           *  **: **:*
                                                                      B10
tr|A8ISQ0|A8ISQ0_CHLRE    LLKAATGLPVHTTSIGRRASASYCVYESMGGWETLSAAVESLYNRMRGDGRCASLFSEGN
tr|D8TXJ2|D8TXJ2_VOLCA    LLKAATGLPVHTTSIGRRASASYCVYESMGGWETLSAAVESLYNRMRNDQRCEVMFADAN
                          *********************************************** * **  :*::.*
                                  E10                      F8
tr|A8ISQ0|A8ISQ0_CHLRE    EQQLKTHMLEFLTCALGGKARFASSTLLTSQRDQLRQHGFGVGQFDILLQHMKAVLDEIG
tr|D8TXJ2|D8TXJ2_VOLCA    EQNLKTHTLEFLTCALGGKARFASSTLLTSQRDQLRNHGFGAGQFDALLQHMKAVMDELG
                          **:**** ****************************:****.**** ********:**:*

tr|A8ISQ0|A8ISQ0_CHLRE    VQQETAASAIALLRCHK-----------------------------------
tr|D8TXJ2|D8TXJ2_VOLCA    IHPEVAASAIALLRCHKYMFEKRPSDDGEEEAEEELSPTPTSASSPTAEASG
```

Eric A. Johnson and Juliette T.J. Lecomte, Figure 6 The sequences of *V. carteri* proteins homologous to THB4 (top) and THB3 (bottom).

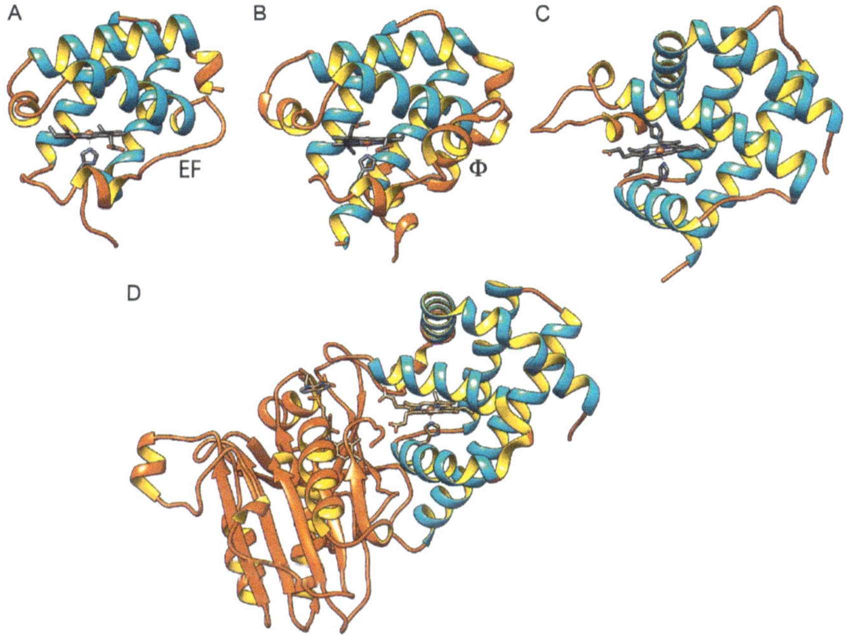

Eric A. Johnson and Juliette T.J. Lecomte, Figure 7 Ribbon structures of representative haemoglobins. (A) TrHb1: *C. eugametos* LI637 haemoglobin domain (1DLY) (Pesce et al., 2000). Distinctive features are the short A and F helices and the long EF loop. (B) TrHb2: *Arabidopsis thaliana* GLB3 (4C0N) (Reeder & Hough, 2014). The TrHb2 architecture includes an extended EF loop folded into a helix denoted by Φ. (C) SDgb: *Arabidopsis thaliana* AHb1 (3ZHW) (Mukhi et al., 2013). This protein is an example of *bis*-histidine coordination of the haeme iron. (D) FHb: *Escherichia coli* Hmp (1GVH) (Ilari et al., 2002). The long H-helix (33 residues) allows the C-terminal flavin domain to make close contact with the globin domain, which guarantees efficient electron transfer from the FAD molecule to the haeme group. The helices forming the 2-on-2 and 3-on-3 topologies are highlighted in blue. Haemes and proximal histidines are represented with sticks.

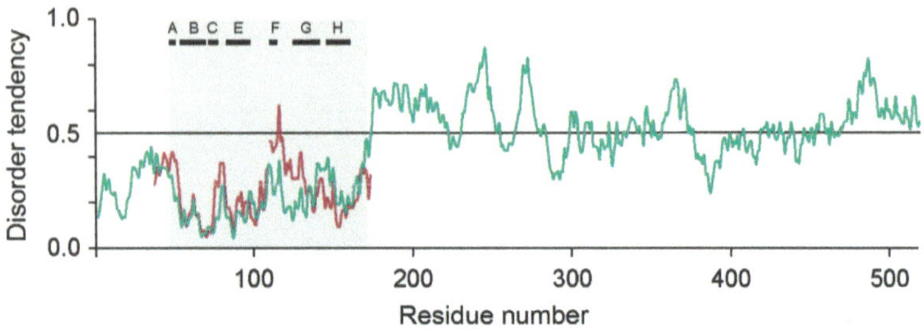

Eric A. Johnson and Juliette T.J. Lecomte, Figure 8 The IUPred profile (Dosztanyi et al., 2005) of THB1 (UniProt identifier A8JAR4, magenta) and THB8 (UniProt identifier A8J8P7, blue) (long disorder calculation). The haemoglobin domain is shaded in light blue. The folded region of THB8 is readily recognised.

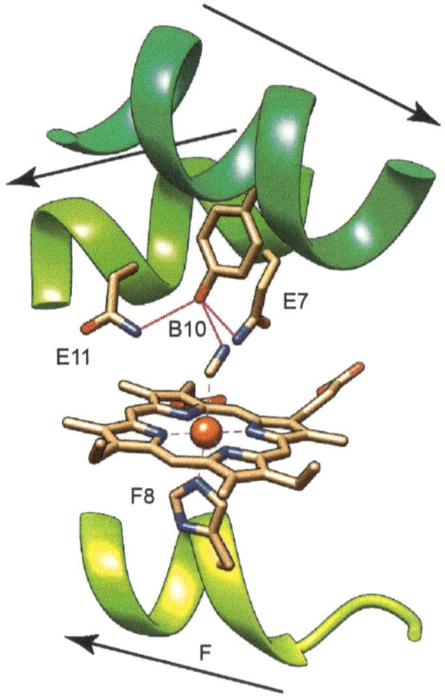

Eric A. Johnson and Juliette T.J. Lecomte, Figure 9 The haeme cavity of CtrHb (1DLY) (Pesce et al., 2000), showing portions of helices B, E, and F and the hydrogen bonds stabilising the bound cyanide. Tyr B10, Gln E7, Gln E11, and His F8 are depicted with sticks. The direction of the helices is marked with an arrow.

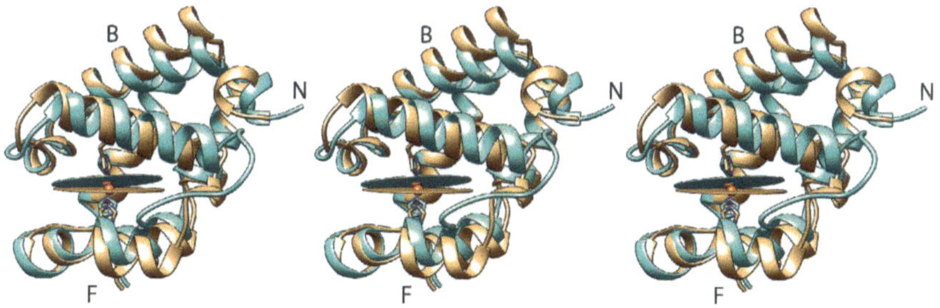

Eric A. Johnson and Juliette T.J. Lecomte, Figure 10 Superimposition of *Synechococcus* sp. PCC 7002 ferric GlbN in the *bis*-histidine state (4MAX, beige) and cyanomet state (4L2M, cyan) (Wenke et al., 2014). The left pair of the stereo triptych is for wall-eyed viewing. The haeme groups are represented by a disc with iron atom as a ball. The proximal histidines are below the haeme planes.

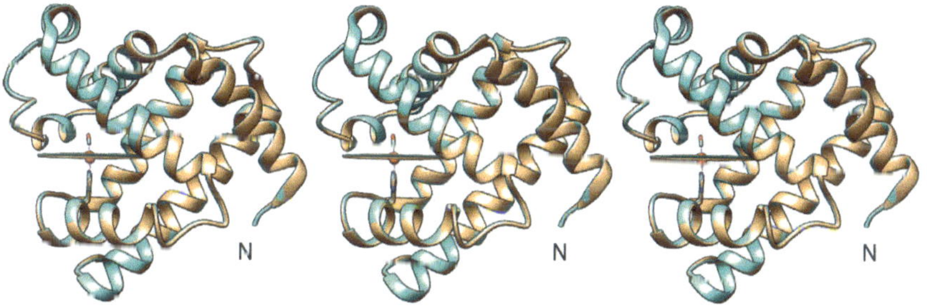

Eric A. Johnson and Juliette T.J. Lecomte, Figure 11 Superimposition of sperm whale myoglobin in the deoxy state (1BZP, beige) and the carbonmonoxy state (1BZR, cyan) (Kachalova, Popov, & Bartunik, 1999). The left pair of the stereo triptych is for wall-eyed viewing. The haeme groups are represented by a disc with iron atom as a ball. The proximal histidines are below the haeme planes.

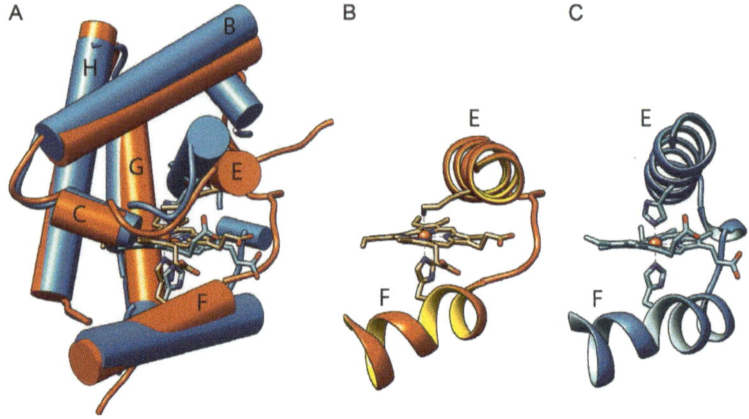

Eric A. Johnson and Juliette T.J. Lecomte, Figure 12 Superimposition of the structures of THB1 (orange) (4XDI) (Rice et al., 2015) and *Synechococcus* sp. PCC 7002 GlbN (blue, 4MAX) (Wenke et al., 2014) in the absence of exogenous ligand. (A) Cylinder diagrams emphasising the reorientation of helices. (B) The E and F helices of THB1 showing the histidine-lysine coordination. (C) The E and F helices of GlbN, showing the *bis*-histidine coordination. (B) and (C) are in the same orientation as in (A).

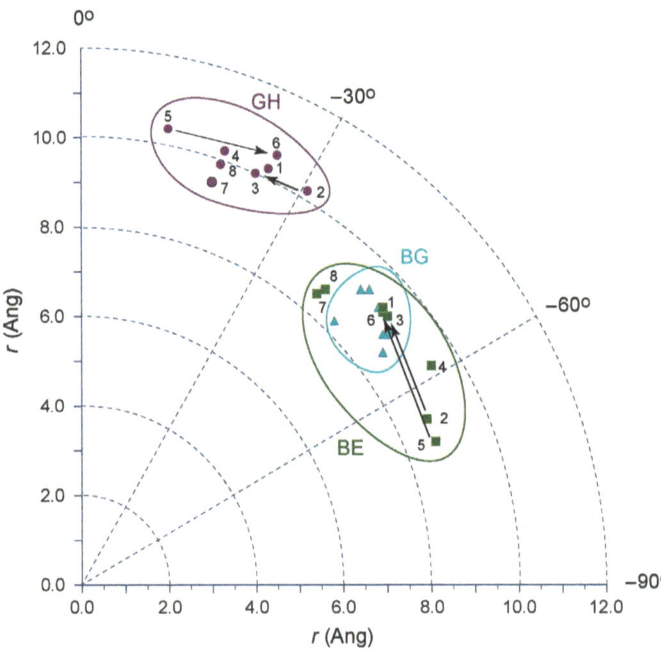

Eric A. Johnson and Juliette T.J. Lecomte, Figure 13 Helix crossing parameters for selected TrHbs. Helical axes were calculated with UCSF Chimera (Pettersen et al., 2004) for helices defined by DSSP (Kabsch & Sander, 1983). The distance *r* is obtained by finding the vector perpendicular to a pair of axes. Crossing angles were obtained by projection into a plane perpendicular to the *r* vector. A negative sign is given to pairs for which the near helix is rotated clockwise. The data points are for CtrHb-CN (1DLY, 1), *Synechocystis bis*-histidine GlbN (1RTX, 2), *Synechocystis* GlbN-CN (1S69, 3), *C. reinhardtii* histidine-lysine THB1 (4XDI chain A, 4), *Synechococcus bis*-histidine GlbN (4MAX chain C, 5), *Synechococcus* GlbN-CN (4L2M chain A, 6), *Arabidopsis thaliana* GLB3-OH (4C0N, 7), and *Thermobifida fusca* TrHb-acetate (2BMM, 8).

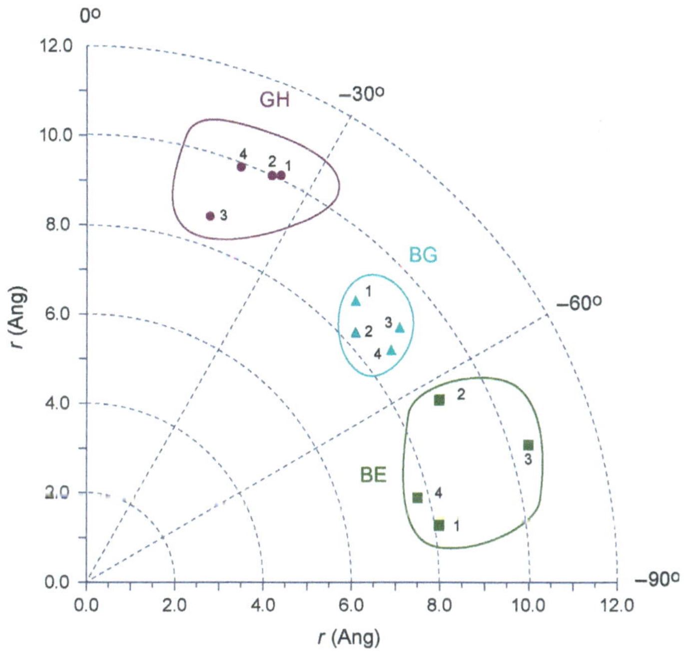

Eric A. Johnson and Juliette T.J. Lecomte, Figure 14 Helix crossing parameters for selected 3-on-3 haemoglobins (see Fig. 13 for details). The data points are for *Arabidopsis thaliana bis*-histidine AHb1 (3ZHW, 1), *Methylacidiphilum infernorum* Hb1-O_2 (3S1I, 2), *Escherichia coli* flavohaemoglobin (1GVH, 3), and *Physeter catodon* Mb-O_2 (1MBO, 4).

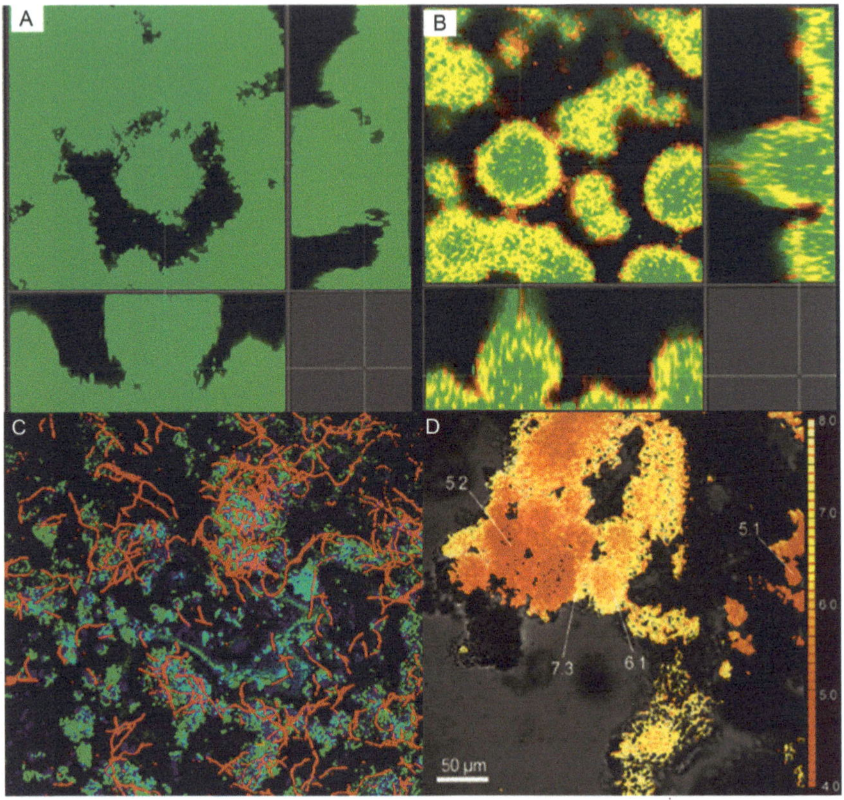

Joke Donné and Sylvia Dewilde, Figure 6 Biofilm physiology and the effect of the microenvironment. The physical, chemical and biological heterogeneity that can develop in a biofilm population is shown. (A) Wild-type *P. aeruginosa* biofilms, grown in *in vitro* flow cells, visualised by confocal laser scanning microscopy (Bjarnsholt et al., 2013). (B) The same wild-type *P. aeruginosa* biofilm as (A) treated with 100 mg/ml tobramycin for 24 h. Green bacteria are alive and yellow/red bacteria are dead. The AB cannot diffuse to the interior of the biofilm population because of mechanisms of tolerance (Bjarnsholt et al., 2013). (C) Micrograph of 18-h-old multispecies biofilm grown in flow cells. *Fusobacterium nucleatum* (red), *Aggregatibacter actinomycetemcomitans* (green), *Veillonella* sp. (dark blue) and *Streptococcus oralis* (light blue) (Häussler & Parsek, 2010). (D) *E. coli* PHL628 biofilm confocal and tomographic pH imaging, with the PH scale shown at the right. The biofilm population is composed of microenvironments of different pHs (Hidalgo et al., 2009).

FSC
www.fsc.org
MIX
Papier | Fördert gute Waldnutzung
FSC® C083411

Kolibri 360 GmbH
Ferdinand-Jühlke-Straße 7
99095 Erfurt, Deutschland
produktsicherheit@kolibri360.de